...e: 1595. La quelle fut tres manifique

MUSIC, DISCIPLINE, AND ARMS IN EARLY MODERN FRANCE

MUSIC, DISCIPLINE, AND ARMS IN EARLY MODERN FRANCE

KATE VAN ORDEN

The University of Chicago Press
CHICAGO AND LONDON

Kate van Orden is associate professor of music at the University of California, Berkeley.

The University of Chicago Press, Chicago 60637
The University of Chicago Press, Ltd., London

Printed in the United States of America
14 13 12 11 10 09 08 07 06 05 5 4 3 2 1
ISBN (cloth): 0-226-84976-7

A Weiss/Brown Publication Subvention Award from the Newberry Library is gratefully acknowledged, as is a Lloyd Hibberd Publication Endowment from the American Musicological Society.

Library of Congress Cataloging-in-Publication Data

Van Orden, Kate.
Music, discipline, and arms in early modern France / Kate van Orden.
p. cm.
Includes bibliographical references (p.) and index.
ISBN 0-226-84976-7 (cloth : alk. paper)
1. France—Civilization—1328–1600. 2. France—Civilization—17th century. 3. Music—France—History. 4. Music and war—France—History. 5. France—Social life and customs—16th century. 6. France—Social life and customs—17th century. 7. Nobility—France—History—16th century. 8. Nobility—France—History—17th century. 9. Military education—France—History—16th century. 10. Military education—France—History—17th century. I. Title.
DC33.3.V36 2005
944′.03—dc22

2004020976

This book is printed on acid-free paper.

ENDPAPERS: Entry of Henry IV into Lyons (1595), Bibliothèque nationale de France, Cabinet des estampes, Qb1-1595

For Mom,
her love of hard science, assays, music, and slide rules,
and her ability to marvel at those of us who
wrest other truths from books

CONTENTS

ILLUSTRATIONS

FIGURES

EXAMPLES

PLATES (FOLLOWING PAGE 194)

ACKNOWLEDGMENTS

IT IS WITH special pleasure that I thank the friends, colleagues, and readers for the press who offered such invaluable advice on my manuscript: Geoffrey Burgess, Donald Chae, Marie-Alexis Colin, Richard Crocker, Jonathan Dewald, Georgie Durosoir, Veit Erlmann, Martha Feldman, Mark Franko, Barbara Haggh, Michel Huglo, Steve Lehning, Susan McClary, Davitt Moroney, Nicolas Paige, Lisa Perella, Martha Pollack, Orest Ranum, Patricia Ranum, Downing Thomas, and Philippe Vendrix. The usual disclaimer applies here in full force: much of what is right is owed to them, and the mistakes are entirely my own. I was also greatly aided by suggestions in the conversations I had along the way with Jane Bernstein, Frédéric de Buzon, Tim Carter, Frank Dobbins, James Helgeson, Isabelle His, Donna Jackson, John McClelland, Anthony Newcomb, Patrick O'Brien, Timothy Reiss, and Treva Tucker, as well as the comments of the members of the University of California Performance Studies Research Group and participants at several conferences: "Musique et Mathématique," held at the Centre d'Études Supérieures de la Renaissance, Tours, France, in February 2000; "Culture and Authority in the Baroque: Communion, Community, Concealment," held at the Clark Library, UCLA Center for Seventeenth- and Eighteenth-Century Studies, in June 2001; and "L'équitation à la Renaissance," held at the École Nationale d'Équitation in Saumur, France, in October 2002.

This book could not have been researched without the resources of numerous libraries and archives. For the welcome I received at the CESR in Tours, the Centre de Musique Baroque de Versailles, and the Warburg Institute, I offer my special thanks, and words of gratitude to Thomas Le Conte at the CMBV, Elizabeth McGrath, the curator of the photographic collection at the Warburg, and Carla Zecher at the Newberry Library, Chicago. For the marvelous collections of the British Library and the Bibliothèque nationale de France, not to mention those of our own Hargrove Music Library and Bancroft Library at the University of California, Berkeley, I am especially grateful.

Serious work on this project began during a postdoctoral year at the Columbia Society of Fellows in the Humanities and a formative term at the Warburg Institute on a Frances A. Yates Fellowship, two institutions to which my debt is profound. Further research has been generously supported by the University of California in the form of research grants, a substantial award from the Hellman Faculty Fund, a President's Research Fellowship in the Humanities, and a fellowship year with a wonderfully stimulating group at the Townsend Center for the Humanities. The American Association of University Women came through with a Post-Doctoral American Fellowship at a crucial moment early on in my research. The final stages of work were completed at the CESR with the support of the Studium Fellowship from the Centre National de la Recherche Scientifique. Je vous remercie, tous, de tout mon coeur.

This book has received the Weiss/Brown Publication Subvention Award from the Newberry Library, which supports scholarship concerning European civilization before 1700 in the areas of music, theater, French and Italian literature, and cultural studies. It is made to commemorate the career of Howard Mayer Brown, who was my dissertation advisor at the University of Chicago. Let this book also commemorate his career. I am further indebted to the Lloyd Hibberd Publication Endowment Fund of the American Musicological Society for its support.

Finally, I must thank Robert Cole, the director of CalPerformances at the University of California, Berkeley, for his courage and unwavering commitment to producing my reconstruction of the horse ballet that inspired this book, as well as to the other mad visionaries who collaborated on *Le carrousel du roi* in 2000 and again in 2002: Creeky Routson and Teresa Trull, our choreographers, Richard Cheetham, our music director, Laura Abrams, Thierry Bosquet, Cass Carpenter, and all the musicians, riders, and horses, particularly Chelsey Sibley and the magnificent Vosmaer. It was the 1612 ballet that attracted me to this subject in the first place, and the beauty of the performance that compelled me to hold on as I was dragged by my heels through the rest of the book project. They helped realize this as well.

MUSIC, DISCIPLINE, AND ARMS IN EARLY MODERN FRANCE

1
MUSIC IN A TIME OF WAR

Il ne faut aussi que votre Magesté s'esmerveille si ce livre de meslanges . . . est composé des plus vieilles chansons qui se puissent trouver aujourdhuy, pource qu'on a tousjours estimé la Musique des anciens estre la plus divine, d'autant qu'elle a esté composée en un siecle plus heureux, et moins entaché des vices qui regnent en ce dernier age de fer.

[Your Majesty should not be surprised if this collection . . . is composed of the oldest songs one can find today, because people have always judged the Music of the ancients to be the most divine, all the more since it was composed in a century happier and less tainted by the vices that reign in this last age of iron.]

In his preface to the *Livre de meslanges* (1560), Pierre de Ronsard drew a gritty comparison between ancient music and the songs of his own "Iron Age."[1] His allusion to harsh times recalled the heavy armaments of sixteenth-century warfare and the smoky gunpowder that powered them, the great blasts from cast iron cannons and muskets, and the thick armor worn by French knights. But in 1560 Ronsard seems also to have sensed that an unprecedented and more vicious kind of warfare was on the horizon, one distinguished by the wholesale abandonment of moral codes. This was civil war. He had no doubt witnessed the Protestants' attempt to kidnap the young Francis II from Amboise in March of that year and seen their bodies hanging from the walls of the chateau after the plot was found out. And this was only the beginning. By the end of the century, religious strife had eradicated the flower of the French nobility, destroyed the economy, ravaged the countryside, and brought death to tens of thousands from massacre, siege, and starvation. Already in 1560, the Muses seemed to be retiring from France.

Ronsard was right that his unhappy age was inhospitable to music. The Wars of Religion that began in 1562 devastated French production of polyphonic music. In the decade of the 1580s, the Royal Printers of Music, Le Roy et Ballard, produced only thirty-some new books of music, half the number they had put out in the

1. Pierre de Ronsard, "Préface," in *Oeuvres complètes,* ed. Jean Céard, Daniel Ménager, and Michel Simonin, 2 vols. (Paris: Gallimard, 1993–1994), 2:1173.

1570s. Even more telling, of the new music published in the 1580s, well over half was composed by foreigners.[2] Native composers were struggling. Claude Le Jeune, we know, led an extremely peripatetic life as he moved from one Huguenot patron to the next. He not only served and traveled with some of the foremost military commanders of the age (François de La Noue, the Duke of Anjou, and Henry of Navarre), but he may have been trapped in Antwerp when Anjou was ambushed there in 1583, and he certainly almost met his end during the siege of Paris in 1590. As the story goes, he was rescued from the prison of the League by Jacques Mauduit, who also saved his *Dodecacorde* (still in manuscript) from the guards' bonfire. By that time, Claude Goudimel, the other bright star of the Protestant firmament, had long since perished in Lyons during the St. Bartholomew's Day Massacre of 1572. And Catholics had it no easier—Anthoine de Bertrand was martyred in Toulouse sometime around 1580 by those who detested his hymns and *chansons spirituelles.*

In 1589 French music publishing collapsed entirely. The assassination of Henry III left Le Roy et Ballard without a king to serve, and they printed no music at all in 1589, 1590, and 1592. A few reprints of earlier editions came out in 1591 and 1593. Only in 1594, with the crowning of Henry IV and his seizure of Paris, did new music appear from the royal printers—Le Jeune's *Airs mis en musique* and a book of *chansonnettes rimées.* But the press barely limped along. In association with Ballard's widow, Le Roy managed to produce only seven new books of music between 1590 and 1598, the year Henry issued the Edict of Nantes and signed the peace treaty with Spain. Presses in Lyons, Paris's rival, produced but one.

These facts suggest the period of this study as a particularly bad one for French music. Yet it *was* music that attracted me to the years Ronsard condemned as vile. The decades of the religious wars—1562–1629—saw the birth of *ballet de cour, air de cour,* and *musique mesurée,* three genres trading in an exquisite rhythmic style that would come to define French music for the next century. Indeed, *ballet de cour* and *musique mesurée* owed their very invention to the political needs of a court confronting civil war, for both explicitly aimed to restore concord among a nobility rent by infighting. The political nature of these genres was observed long ago by Frances A. Yates in her magisterial study of the French academies, where she argued that Catherine de' Medici used musical fetes to soften hearts and bring together those of opposite opinions—to harmonize unlike wills and stabilize relationships among courtiers with conflicting desires. Yates elucidated the role of musical reform in the moral and civic reform hoped for by Catherine, whose con-

2. Indeed, two were books of madrigals, one was a book of *villanelle* and *moresche,* and still others included significant numbers of Italian and even Spanish songs. See François Lesure and Geneviève Thibault, *Bibliographie des éditions d'Adrian Le Roy et Robert Ballard, 1551–1598* (Paris: Société Française de Musicologie, 1955).

ciliatory policies fit well with the goals of the academicians she patronized. The statutes of the Academy of Poetry and Music, founded in 1570, state that "where music is disordered, there morals are also depraved, and where it is well ordered, there men are well disciplined morally."[3] From political necessity emerged new musical forms that would sustain an absolute state which had itself been forged in the crucible of that Iron Age.

The political ambitions of the academy always intrigued me, for they seem to force the history of music into direct confrontation with the history of power. It is not just that music operated politically, but that music was taken to be so powerful in an age dominated by the most brutal forms of political struggle. During the Wars of Religion, politics were pursued via regicide, assassination, slaughter, and armed revolt, and court spectacles were not taken to be above the fray. More than one of Catherine's contemporaries saw in the first *ballets de cour* the violence by which power was attained. In 1572 the narrative of Catherine's *Paradis d'amour* incited Huguenots to accuse her of having ordered the Saint Bartholomew's Day Massacre, and even ballets performed at less incendiary moments prompted observers to liken the queen mother's lovely ladies-in-waiting to soldiers. Given that the dancers made entrées to "airs de guerre," marched, and fell into various formations, it would seem that the creators of *ballets de cour* fully intended them as some sort of military display.

Despite the genesis of *ballet de cour* and *musique mesurée* in an age of war, scholars have been adverse to considering how music was allied with a politics of armed aggression. The idea that music contributed to the pursuit of war seems, at first blush, wholly at odds with the rhetoric of Neoplatonism. After all, Ronsard juxtaposes peace and war, the music of the ancients and music of his day, divinity and vice. The separation seems complete, and scholars who have studied the performing arts and the exercise of power in Renaissance France have accepted the oppositions drawn by humanists such as Ronsard. Yates, whose first lectures on the French academies were given in the unheated rooms of the Warburg Institute in London during the winter of 1940, saw her own time as analogous to that of the academicians and found in their Neoplatonic philosophy "a steadying subject for contemplation in a disintegrating world."[4] Violence occupies a very restricted position in her study, as it does in Roy Strong's studies of court festivities and Margaret M. McGowan's studies of *ballet de cour.* War functions as a negative property, a wrong to be righted by the festive rule of music, song, dance, and stagecraft. Yet this pacific vision of music as healing—"music hath charms to soothe the savage breast"—

3. Cited in Frances A. Yates, *The French Academies of the Sixteenth Century,* 2nd ed. (London: Routledge, 1988), 23, 319–322.

4. Ibid., vii.

captures neither the full range of powers music was understood to possess nor its full function in court spectacle and civil society. Unlike the pastoral mode, for instance, which did tend to be conceived as war's inverse, music traversed the boundaries between war and peace. During that hostile time, music was sent to war: drums kept troops together in battle and trumpets quickened the ire of knights and their steeds. As we shall see, Neoplatonists were not shy about extending their speculations to these vicious domains with analyses of choleric musical modes. Certainly Neoplatonic philosophy taught that music was a force of universal order, but it was not in itself *peaceful.*

This book examines a series of practices in which military action and music coincide in order to elaborate a history of fete, divertissement, and ceremonial that intersects with the battles they so often celebrated. Dueling met court balls in the *ballet de cour* (chapter 3); armed marches by Catholic extremists intersected the *sacre et couronnement* in the Te Deum ceremonial (chapter 4); infantry maneuvers crossed ballet in pyrrhic dance, a form of ensemble swordplay choreographed to music (chapter 5); and in the carrousel and equestrian ballet, cavalry formations met public spectacle (chapter 6). This history of physical practices charts the terrain upon which music met war and peace.

Portions of this book are set amid the inky smoke of early modern battlefields, but not many. Its main concern is with the activities of an aristocracy on other campaigns that were hardly less hostile. Kings, *grands,* and aristocrats continued their power struggles off the field of battle, in winter quarters and during times of peace. Indeed, during the sixteenth and seventeenth centuries, aristocrats increasingly adopted nonviolent means of settling disputes and expanding their dominions. The century preceding the reign of Louis XIV saw a transformation of the relationship between the French aristocracy and the crown as aristocrats relinquished their traditions of waging politics through private violence. Dueling, armed rebellion, and even political dissidence ultimately gave way to an ideal of the French state that privileged order above these historic rights. Force was exerted in new ways. The reign of Louis XIV is beyond where this book will end, but it is useful to place it on our horizon from the start, for if the absolutist government associated with his reign was achieved at all, it was through compromises between the state and the aristocracy that were always in the process of being negotiated. Its hard-won and achingly slow rapprochements began during the Wars of Religion.[5]

How did music contribute to the reordering of society that took place as civil war was overcome by the more cooperative interactions characteristic of absolutism?

5. See Fanny Cosandey and Robert Descimon, *L'absolutisme en France* (Paris: Seuil, 2002), and David Parker, *The Making of French Absolutism* (London: Edward Arnold, 1983).

These political transformations went hand in hand with a cultural transformation of the aristocracy, one that found its most dramatic expression at Versailles. Here we can see politics operating via cultural practices. Power struggles expanded to new territories. The relationship between violence and culture was first explored by Norbert Elias in *The Civilizing Process,* where he showed how the emphasis on manners in early modern Europe helped to control violence and contain outbursts of armed aggression in civil society. Manners promoted self-censorship in the form of a ceaseless attention to style, to politeness, and to the minutiae of self-fashioning. And by encouraging gentler, considered behavior, the practice of *civilité* essentially policed society from the inside out, bringing with it the benefit of keeping subjects in check. For as violence lost its cultural credibility, the monarchy could more easily monopolize control of it, marshaling its force with police and state army. Scholars since Elias—notably Philippe Ariès and Roger Chartier—have pursued this study of how culture shaped social relations, and Michel Foucault's now classic study of the interaction of self-discipline and military training in *Discipline and Punish* launched a related field of inquiry into the relationship between physical coercion and the social body.

My object is to situate music in the broad process of behavioral and cultural disciplining that the French aristocracy underwent beginning in the late sixteenth century. I have chosen, for several reasons, to concentrate on the nobility and its culture of arms. First, even if they did not pursue military careers, most nobles tended to identify with the traditional calling of their estate, and many finished their education at riding schools and military academies. As I explain below, arms were central to noble identity. Second, those men who actually did take up military commands wielded considerable political power. In the most obvious sense, studying music in the culture of arms brings us closer to understanding how it civilized the very nobles who had the greatest tendency to express their political independence by violent means. Happily, pyrrhic dance, equestrian ballet, and the lute instruction students received at military academies present clear cases of music's place in their lives. But this history of music also sheds light on the modes of the civilizing process more generally. Music had a long-standing place in noble culture before civility promoted it as a grace expressive of emotional and social refinement. The lute and dance were accepted as fundamental parts of a noble education long before the *studia humanitatis* began to promote letters, foreign languages, and the classics as educational standards. So before literary games, wordplay, and eloquence became the accoutrements of the courtier, men-at-arms were plucking out little gaillardes on their lutes and dancing in balls at court. Musical modes suited those with a military disposition and, ultimately, the practices of civility, which meant not only that music was a constant denominator for nobles during a period of great cultural transformation, but that this transformation can be sighted early on in musical activities, as though civility first took hold through musical means.

THE CULTURE OF ARMS

Machiavelli observed in *The Prince* that France was different from other countries. Invaders could occasionally make incursions with the aid of this or that belligerent duke whose interests crossed those of the king, but one always ran up against problems with the French nobility. "It is simply not sufficient to kill the ruler and his close relatives, for the rest of the nobility will survive to provide leadership for new insurrections. You cannot win their loyalty or wipe them out."[6] Theoretically, the French system of noble privilege and allegiance gave the kingdom a particular resilience against external military aggression. Indeed, military service and noble culture were so inextricably tied together in France that its feudalistic nobility was called the *noblesse d'épée,* or nobility of the sword.

In traditions received from the Middle Ages, the sword symbolized a nobleman's obligation to levy soldiers from his seigneurial estate, to equip himself and a band of supporting horsemen, and to fight for the crown with the courage, leadership, and heroism appropriate to the elevated moral standards of his class. Nobles had a clear function in society, as expressed in the contemporary order of the three estates. The first estate (the clergy) prayed, the second estate (the nobility) fought, and the third estate (the commoners) labored. Machiavelli admired the French *gendarmerie,* for these knights constituted a highly useful standing army of horsemen, but he also recognized the perils of a militarized nobility in a country as vast as France, for the same counts, dukes, and barons who served the king might turn their armies against him. We will be reminded of this ever-present danger many times in the chapters that follow.

The military orientation of noble culture in France fostered an ideal of noble mien that centered on the body. Facility with the thrusting sword and lance on horseback, strength to wield the two-handed sword in hand-to-hand combat, a sure seat in the saddle, and a good eye for terrain, fortification, and ballistics armed the nobleman for a successful career. Morality, too, was physical, for the expression of noble virtue lay in deeds. The nobleman deemed virtuous was ambitious, courageous, swift to discern the best course of action, decisive, disciplined, and perseverant—all qualities of action and motion, qualities expressed in the face, voice, gait, carriage, and gaze. Noble life as a whole was a *vita activa* in France. Those born into noble families were obliged to act virtuously—that is, to fight—and, conversely,

6. Niccolò Machiavelli, *Selected Political Writings,* trans. David Wootton (Indianapolis: Hackett, 1994), 16.

those who fought valiantly in battle were by definition noble.[7] An ordinary captain who distinguished himself in war might thus be ennobled by the king. Although commoners rarely received such rewards for military heroism, and although the already ennobled held fast to the privileges of their class even when disdaining arms, the equation between military virtue and nobility in France was so complete that Baltassar Castiglione perhaps rightfully sneered in *Il cortegiano* that the French recognized only the nobility of arms and thought nothing of letters.[8]

Castiglione's critique of the *noblesse d'épée* had its sympathizers in France among a new class of nobles that arose during the course of the sixteenth century and called for a less militaristic definition of virtue. Humanists in outlook and education, these new nobles formed the upper crust of what was essentially a class of bureaucrats newly minted by French kings to serve the state. Many were the sons of wealthy merchants or bourgeois who could afford to attend *collèges* such as the Collège de Navarre and the Collège du Plessis in Paris.[9] They studied letters and law, afterward entering the exploding paper chase of government as jurists, lawyers, administrators, and parliamentarians who sported the robes of the legal profession. They rubbed elbows with the *nobles d'épée* at court and were even quite a bit wealthier on average, for while nobles had to refrain from retail commercial activity, among commissioners, notaries, bailiffs, doctors, professors, scribes, and clerks, gold coins fell "thicker than rain."[10] Worse yet for established nobles, the jurists that made up this expanding stratum of society began to pass laws that permitted them to gain noble titles.

Versed in letters but unskilled in arms, these men had money, clothes, and land to rival those of the *nobles d'épée,* and they came to constitute a new nobility, the *noblesse de la robe* (so called for their legal attire). With their abundance of cash,

---

7. On the military profession of noblemen, see Ellery Schalk, *From Valor to Pedigree: Ideals of Nobility in France in the Sixteenth and Seventeenth Centuries* (Princeton: Princeton University Press, 1986). On advancement based on bravery in battle, see Blaise de Monluc, *Commentaires, 1521–1576,* ed. Paul Courteault (Paris: Gallimard, 1964), 22–23.

8. Baltassar Castiglione, *The Book of the Courtier,* trans. George Bull (London: Penguin Classics, 1976), 88, 90, 93.

9. On the college system and the new association between education, moral improvement, and upstanding citizenship that fostered its development, see George Huppert, *Public Schools in Renaissance France* (Urbana: University of Illinois Press, 1984); on the new noble class that followed, see idem, *Les Bourgeois Gentilshommes* (Chicago: University of Chicago Press, 1977). For arguments that the expansion of the nobility, its changing military function, and new ideals of behavior reshaped noble culture, see Davis Bitton, *The French Nobility in Crisis, 1560–1640* (Stanford: Stanford University Press, 1969), and Schalk, *From Valor to Pedigree.* Though the causal arguments are often overstated, Schalk and Bitton draw attention to issues that have been actively pursued in subsequent scholarship.

10. Nicolas de Cholières, "Des lettrez et guerriers," in *Oeuvres,* ed. Edouard Tricotel (Paris: Librairie des Bibliophiles, 1879), 288–289.

they purchased noble titles that an economically embattled monarchy was only too happy to sell. Henry III, for example, attempted to vend a thousand *lettres d'anoblissement* during the financial crisis precipitated by the fifth religious war in 1576. The titled nobility suppressed this plan to "sell nobility," and in 1583 an ordinance was passed to limit the number of ennoblements a ruler might vend. But Henry IV proved more lax in this regard than his predecessor.[11] Even the marks of nobility suffered devaluation: by 1580 so many men had acquired the collar of the Order of Saint Michael that courtiers referred to it as the *collier à toutes bestes,* leading Henry III to establish the more exclusive Order of the Holy Spirit and limit its membership to one hundred.[12]

The tension between the new and old nobles crystallized into a fight between competing ideologies of nobility in which the *nobles d'épée* faced off against the *nobles de la robe.* "The pen versus the sword" and "arms versus letters" were the clichés of a pamphlet war over the definition of true nobility. Jurists maintained a calculated confusion as to whether a bourgeois might be made into a *gentilhomme* through official *anoblissement* by the crown and focused instead on beautiful speech, ability in languages, knowledge of classical literature and culture, sensitivity to music and poetry, and eloquent self-expression as the marks of noble character. The military men, meanwhile, looked upon these newcomers as mere "parchment gentlemen," no matter how persuasive and glib they were.[13]

Much—too much, many historians would argue—has been made of the arms-versus-letters debate. The studies of James Collins, Jonathan Dewald, J. H. M. Salmon, and Donna Bohanan, among others, have shown how permeable the categories of robe and sword actually were in early modern France.[14] The political

11. Jacqueline Boucher, *La cour de Henri III* (Rennes: Ouest France, 1986), 148.

12. Ibid., 194.

13. Huppert, *Bourgeois Gentilshommes,* 9.

14. The arms-versus-letters literature has traditionally been taken to signify the decline of the aristocracy: the internal strife created between competing classes of sword and robe, it was argued, allowed royal authority to expand unopposed, leading to absolutism during the reign of Francis I. But recent studies have reevaluated theories of a nobility in decline and of severe differences between sword and robe. On the latter, see especially James B. Collins, *Classes, Estates, and Order in Early Modern Brittany* (Cambridge: Cambridge University Press, 1994); J. H. M. Salmon, *Society in Crisis: France in the Sixteenth Century* (London: Benn, 1975); James B. Wood, *The Nobility of the Election of Bayeux, 1463–1666: Continuity through Change* (Princeton: Princeton University Press, 1980); Jonathan Dewald, *Aristocratic Experience and the Origins of Modern Culture: France, 1570–1715* (Berkeley and Los Angeles: University of California Press, 1993); Robert Descimon, "The Birth of the Nobility of the Robe: Dignity versus Privilege in the Parlement of Paris, 1500–1700," in *Changing Identities in Early Modern France,* ed. Michael Wolfe (Durham, N.C.: Duke University Press, 1997), 95–123; and Donna Bohanan, *Old and New Nobility in Aix-en-Provence, 1600–1695: Portrait of an Urban Elite* (Baton Rouge: Louisiana State University Press, 1992). Also see Mack P. Holt, ed., *Society and Institutions in Early Modern France* (Athens: University of Georgia Press, 1991). This collection, which is dedicated to

interests of robe and sword often coincided, and the members of each class often came from similar backgrounds or took similar educational and career paths. Collins shows, for example, that in early modern Brittany, title, seigneurial land, and military office formed a set of interrelated attributes by which nobles dominated society.[15] Royal officers and lawyers, though not noble, often bought substantial amounts of land, which brought them into a social order just below that of the nobility. They lacked seigneurial rights, but they were still landlords, which made these men of letters more like the nobility, from whose ranks came most military officers and cavalrymen in the *compagnies d'ordonnance.*

ALONGSIDE these social similarities, the cultural priorities of robe and sword often coincided as well. Setting aside the rhetoric of opposition generated by the arms-versus-letters debate allows us to see exactly how extensive the cultural hegemony of the sword really was. The growing emphasis on civility and, eventually, noble birth was not at all at odds with a military career, and arms remained the profession of choice for gentlemen. Castiglione's Count Lodovico announced that the first and true profession of the courtier must be that of arms, Michel de Montaigne concurred, and even Nicolas Faret, a professional courtier whose *L'honneste homme ou l'art de plaire à la cour* (1630) defined nobility for a later generation, echoed Montaigne when he asserted that "there is no profession more honest, nor more essential to a Gentleman than that of arms."[16] Even if one did not enter military service, the emphasis on action, the body, and the physical manifestation of virtue that derived from the culture of arms remained strong.

Some of the writers and philosophers most admired today as humanists identified sooner with the sword than the pen, and the recently titled did not always champion letters, no matter how they had come by the money to purchase their titles and seigneurial estates. An excellent example of just such a self-styled *noble d'épée* is Montaigne himself.[17] It is tempting to think of Montaigne as a great humanist author, happiest at the desk in his tower library. Yet Montaigne descended

J. Russell Major, presents a rich array of evidence to support the theory that in the sixteenth century both cities and aristocrats enjoyed significant powers that precluded absolute authority for the king.

15. Collins, *Classes, Estates, and Order,* 2–13.

16. Michel de Montaigne, *Essais,* ed. Pierre Villey and V.-L. Saulnier, 3 vols. (Paris: Presses Universitaires de France, 1988), 2:384. Nicolas Faret, *L'honneste homme ou l'art de plaire à la court,* ed. Maurice Magendie (Paris: Presses Universitaires de France, 1925), 12. "Or, comme il n'y a point d'hommes qui ne choisissent une profession pour s'employer, il me semble qu'il n'y en a point de plus honeste, ny de plus essentielle à un Gentil-homme que celle des armes."

17. Montaigne presents an interesting case of the uncertainty surrounding recent ennoblement, as his status as a *noble d'épée* was compromised not just by the specter of bourgeois roots, but by his work as a jurist. The *Essais* illuminate a tension between two conflicting notions of virtue—one based on valor, the other on classical erudition. Montaigne is never overly deterministic about whether virtue

from a family of merchants whose painstaking acquisition of noble status began with the purchase of an estate in 1477 and continued with the service of Michel's father on Italian military campaigns.[18] Montaigne clearly thought of himself as a chevalier fulfilling a traditional calling by pursuing the active life of the nobleman. "I do not willingly dismount when I am on a horse," he said, "for it is the seat in which I feel the best, whether healthy or ill." [19] So, too, did René Descartes, anxious to obtain the military finish of a *noblehomme,* quickly join the army of Maurice of Nassau after completing his law degree in Poitiers. We should also note his reputation among his friends in Paris as an excellent fencer and that he wrote an *Art of Fencing* (now lost) at some point between 1622 and 1629.[20] Even Ronsard portrayed himself as a conquering Caesar in Roman armor and a laurel wreath in the frontispiece to his first book of sonnets. Virtue meant military virtue, even for lyric poets. It is fair to conclude that while college education and civic standing produced a humanistically inclined class of *nouveaux nobles,* the outward manifestation of nobility remained tied to the sword.

Finally, several facts suggest that the cultural significance of the sword actually increased from the sixteenth to the seventeenth century. To begin with, dueling as an assertion of noble honor escalated in the first decades of the seventeenth century. Duels had the effect of prolonging the civil warfare from which the country was trying to recover, and Henry IV attempted to stem their tide by issuing bans and requiring duelists to seek royal sanction ahead of time.[21] It is no coincidence that

was to be achieved with the pen or the sword, but he clearly subscribed to a natural idea of nobility. In exploring how the self is invested with nature's intrinsic nobility, Montaigne searches for means of developing virtue both through humanistic ideals and military ones. His extremely complex ideas about nobility have occupied a number of scholars, see especially James J. Supple, *Arms versus Letters: The Military and Literary Ideals in the* Essais *of Montaigne* (Oxford: Clarendon Press, 1984). My work is indebted both to Supple's enlighening study and to its bibliography of primary sources for the arms-versus-letters debate.

18. See George Hoffmann, *Montaigne's Career* (Oxford: Clarendon Press; New York: Oxford University Press, 1998), and Philippe Desan, *Les commerces de Montaigne: le discours économique des* Essais (Paris: Nizet, 1992), 47–81.

19. Montaigne, *Essais,* 1:289. "Je ne démonte pas volontiers quand je suis à cheval, car c'est l'assiette en laquelle je me trouve le mieux, et sain et malade."

20. See René Descartes, *Oeuvres de Descartes,* ed. Charles Adam and Paul Tannery, augmented new ed., 11 vols. (Paris: Vrin, 1974–1989), 10:535–538. On humanists and the martial arts, see Sydney Anglo, *The Martial Arts of Renaissance Europe* (New Haven: Yale University Press, 2000), 27–30.

21. On the increase in dueling after 1600 and its association with nobility (and the subversion of monarchic authority), see François Billacois, *Le duel dans la société française des XVIe–XVIIe siècles: essai de psychosociologie historique* (Paris: École des Hautes Études en Sciences Sociales, 1986); and Pascal Brioist, Hervé Drévillon, and Pierre Serna, *Croiser le fer: violence et culture de l'épée dans la France moderne, XVIe–XVIIIe siècle* (Seyssel: Champ Vallon, 2002).

dueling proved a cultural denominator for nobles in the same years during which France first edged toward a strong and lasting peace, for the duel allowed valor, action, and honorable deeds to make the transition from war to peacetime. The monarchy's regulation of dueling and prosecution of abuses as *crimes de lèse-majesté* certainly illustrate the fundamental incompatibility of the duel's jurisdiction with that of the king, but at a time when many dukes and barons were still quite capable of raising small armies for private campaigns, dueling was hardly the greatest threat to royal authority. Indeed, commentators such as Guillaume de Chevalier believed that a stronger monarchy might itself be responsible for the surge in dueling: because good government resulted in fewer wars and battles, nobles ended up expending their extra energy fighting with each other.[22] Some thought that duels helped prevent civil war, others that they kept nobles fit for battle.

As the seventeenth century wore on, the establishment of the state-commission army in 1635 and its tremendous expansion in the following decades created more and more opportunities for brilliant military careers. By the time of the Nine Years' War at century's end, the theoretical wartime strength of the French army was 420,000, a five- to eightfold increase over the peak strengths of sixteenth-century armies.[23] So intoxicating was the desire for the glory arms could bring that despite the perils, nobles risked their fortunes to acquire commissions as a colonel or captain. Add to this the personal costs to these officers of maintaining the regiment, the chances that it might be disbanded and the investment lost, and the diminishing autonomy of commanders as the crown demanded increasing loyalty—to say nothing of the dangers of combat—and we gain some sense of how great the honor of a military commission must have been to make it worth such sacrifice.[24]

MUSIC AND MILITARY VIRTUE

Music was a formative part of education for a *noble d'épée.* Indeed, in the copious sixteenth-century literature on noble comportment that was generated as part of the arms-versus-letters debate, rarely does one find criticisms of music as effeminizing or otherwise opposed to military ideals.[25] On this point, the French agreed with

22. See Guillaume de Chevalier, *Discours des querelles et de l'honneur* (Paris: Matthieu Guillemot, 1598), 29–32.

23. John A. Lynn, *Giant of the* Grand Siècle: *The French Army, 1610–1715* (Cambridge: Cambridge University Press, 1997), 56–57.

24. Ibid., 228–229.

25. For an analysis of the sources, see Kate van Orden, "Vernacular Culture and the Chanson in Paris, 1570–1580" (Ph.D. diss., University of Chicago, 1996), chap. 6. The research I presented there was one starting point for this study and has already born fruit in the work of other scholars. See, for example, Jeanice Brooks, *Courtly Song in Late Sixteenth-Century France* (Chicago: University of Chicago Press, 2000), chap. 3, "The Conjunction of Arms versus Letters." Her reading—unlike my

Castiglione's Count Lodovico, who rushed to defend music when Signor Gaspare claimed that it made men soft, fearful, and unfit for battle.[26] The Lacedaemonians played flutes as they marched into battle, Count Lodovico observed, launching into a long list of musically inclined warriors that included Achilles, Epaminondas, and the Cretans. And there was also the famous story of Alexander the Great, whose legendary sensitivity to music caused him to seize his arms when his musician, Timotheos, played in the Phrygian mode. Timotheos could calm the great king only by changing his tune to the Hypophrygian. Alexander's soul was so finely attuned to musical harmonies that they caused violent reactions in the passions animating his body. In this story, music overcomes reason at first, but Castiglione's contemporaries saw Alexander's fury as a *marque de noblesse* signaling both the sensitivity of the virtuous man and the decisive action of the hero.

The legend was conspicuously updated in the tale of a musical rehearsal for the wedding festivities of the Duke of Joyeuse in 1581. The musicians were singing an *air* composed by Claude Le Jeune when a gentleman present in the hall suddenly laid hand to his arms, swearing all the while that he could not refrain from fighting someone; only when the singers began a song in the Hypophrygian did he grow tranquil, as before.[27] The infuriating effect attributed to the Phrygian mode caused such stories to be told time and time again across the sixteenth century and into the next, verifying that music could imitate the most masculine passions.

At the French court, Achilles' education under the wise centaur Chiron beautifully illustrated the place of music within the military lifestyle for generations of French princes who passed through the Gallery of Francis I at Fontainebleau (see plate 1). In one fresco Chiron teaches Achilles to swim, hunt, fence, and play the lute, the last an emblem of self-knowledge, social accord, and the ability of epic to immortalize heroic deeds.[28] In contrast to the trumpet calls that stirred knights' courage as they charged into battle, the lute and lyre represented "songs" such as Homer's *Iliad* and private music-making of a sort that fortified the warrior's soul before battle. Ronsard described Charles de Guise, lute at his belly, singing of his forefathers' acts and the new battles won by his brothers in this way, "like Achilles did to soothe himself a little, even when Hector threw fire into the army

---

own—concentrates on literary presentations of military virtue rather than a study of the noble lifestyle itself.

26. Castiglione, *Book of the Courtier,* 94–95.

27. For the passage from Artus Thomas describing the incident see Frances A. Yates, *Astraea: The Imperial Theme in the Sixteenth Century* (London and Boston: Routledge & Kegan Paul, 1975), 154.

28. Also see Ronsard's exploration of the education of Achilles, "Institution pour l'adolescence du Roy Tres-Chrestien Charles IXe de ce nom," in *Oeuvres complètes,* 2:1007, and the preface to the *Livre de melanges* in ibid., 2:1173. Castiglione also refers to the myth of Chiron and Achilles (*Book of the Courtier,* 95). For an account of the many sixteenth-century French references to the musical education of Achilles, see Brooks, *Courtly Song,* 133–150.

of the Greeks, and the horrible noise intoned by the trumpet cried against that of the lyre being played."[29] Here the poet portrayed Achilles on the verge of a battle that would end with his merciless slaughter of Hector and massacre of the Trojans, his gentle plucking of the lyre seemingly at odds with his capacity for unbounded savagery. Other poets, too, described Achilles after the battle, playing his lyre with bloody fingers. These extremes—intimate musical reflection and a bloodlust consistent with the most terrifying ends of the heroic code—mark out the full range of a warrior's inflamed passions, and it is wrong to see them as incompatible. Chiron taught Achilles the cithara as strength training for the passions, and in its ability both to kindle animalistic desires and to tame them, one finds the same useful combination of brutality and humanity embodied by Chiron himself.

MUSIC WAS not just for princes and kings, though we do know more about their musical activities than those of minor aristocrats. At the seventeenth-century military academies I will discuss in chapter 2, music fit easily into a curriculum featuring riding, fencing, and dance, and before 1600 those who espoused the military ideal likewise recommended that nobles study music, especially the lute. One of the staunchest military men of the time, François de La Noue—an illustrious general known as "Bras de Fer"—recommended music-making for young *nobles d'épée* in his *Discours politiques et militaires* (1587). The training he suggests for the French gentleman includes "exercises of the mind" such as reading classical texts in translation and studying mathematics, geography, and fortification, along with whatever vernacular languages might eventually be useful. And because life is made up of work and rest, the student should have some honest occupations to hold and please the spirit during leisure time so that he does not succumb to bad thoughts.[30] La Noue reminds his reader that Aristotle ordered musical instruction for young people, including lessons on instruments, in order to guard against *otium* or laziness, the principal danger of leisure.[31] Aristotle considered music not just an honest pastime, but a means of ethical education, and readers of Plato would have known that he took music seriously enough to ban all but the Dorian and Phrygian *harmoniai* from his republic because they were the only modes that imitated the

29. Ronsard, "L'Hymne de Charles Cardinal de Lorraine," in *Oeuvres complètes,* 2:503. "Comme Achille faisoit pour s'alleger un peu / Bien qu'en l'ost des Gregeois Hector ruast le feu / Et que l'horrible effroy de la trompe entonnée / Criast contre le bruit de la lyre sonnée."

30. I paraphrase François de La Noue, *Discours politiques et militaires,* ed. F. E. Sutcliffe (Geneva: Droz, 1967), 154.

31. On music and *otium,* see Kate van Orden, "Sexual Discourse in the Parisian Chanson: A Libidinous Aviary," *Journal of the American Musicological Society* 48 (1995): 1–42, and Richard Wistreich, "Giulio Cesare Brancaccio and Solo Bass Singing in Sixteenth-Century Italy" (Ph.D. diss, Royal Holloway College, University of London, 2002), chap. 4.

"utterances and the accents" of a man engaged in warfare or works of peace and of bravery, temperance, and modesty (*Republic* 3:399a–c).[32] Through music, students could take on the attributes of that which was being imitated, learning in this way to discern virtue from vice. Following these classical paradigms, La Noue advised parents to engage music masters for their sons' education, and Isabelle His has suggested that La Noue's son Odet was tutored by none other than Claude Le Jeune, whose long-standing relationship with the family is attested in the dedications of his *Dix pseaumes* (1564) to François and of his *Livre de melanges* (1585) to Odet, who played lute and keyboard, composed his own melodies for psalms, and followed in his illustrious father's footsteps with a career in arms.[33] In this same vein, seventeenth-century military academies employed lutenists who gave lessons after lunch, the students having exhausted themselves riding in the morning.

Further evidence of such advice well taken is to be found in contemporary praise for the musical skills of notable military men. François Le Poulchre de La Motte-Messemé urged nobles to follow the examples of François, the Duke of Anjou, who could sing "tresbien une basse contre," and the young Duke of Mayenne, who, at the siege of Poitiers, sang a delicate superius and later became a good bass.[34] Mayenne also excelled at the lute.[35] The young Henry II enjoyed singing Marot's translations of the psalms so much that he made up his own tune for Psalm 128.[36] And Henry's nemesis, Charles V, reportedly beat time for his chapel singers and sang along with them in part music.[37] Charles IX liked to sing tenor or *dessus* with

32. Plato, *The Republic,* trans. Paul Shorey, 2 vols., Loeb Classical Library (London: William Heinemann; Cambridge, Mass.: Harvard University Press, 1982–1987), 1:247–249.

33. See Isabelle His, *Claude Le Jeune, v. 1530–1600: un compositeur entre Renaissance et baroque* (Arles: Actes Sud, 2000), 30–36. On Odet's career, see Gédéon Tallemant, sieur des Réaux, *Les historiettes de Tallemant des Réaux,* 3rd ed., ed. Louis Jean Nicolas Monmerqué and Alexis Paulin Paris, 9 vols. (Paris: Techener, 1854–1860), 4:199–200.

34. François Le Poulchre de La Motte-Messemé, *Le passe-temps,* 2nd ed. (Paris: Jean le Blanc, 1597), book 2, fol. 34r. "Nous prenons plaisir à la Musique, nous faisons si peu de cas des chantres que Philippus demandoit à Alexandre son fils, s'il n'avoit point de honte de chanter si bien, c'est un honneste exercice toutefois, & ay veu François fils & frere de Roy dernier Duc d'Anjou chanter tresbien une basse contre en sa chambre: j'ay veu au siege de Poitiers Monsieur de Mayenne fort jeune chanter delicatement un superius, il chante aujourd'huy une basse tres-bien."

35. Jacqueline Boucher, "Mayenne," in *Histoire et dictionnaire des guerres de religion,* ed. Arlette Jouanna et al. (Paris: Robert Laffont, 1998), 1089.

36. Louis Régnier de La Planche, *La Legende de Charles, Cardinal de Lorraine* (Reims: Jacques Martin, 1576), fol. 32v.

37. Charles Burney, *A General History of Music from the Earliest Ages to the Present Period,* (London, 1789), ed. Frank Mercer, 2 vols. (New York: Dover, 1957), 1:800–801. See Prudencio de Sandoval, *Historia de la vida y hechos del Emperador Carlos V,* 2 vols. (Pamplona: Bartholome Paris, 1634), 2:828.

his chapel, and Henry III also reputedly sang well.[38] Knights in Henry III's Confraternity of the Holy Spirit were required to sing *ad alternatim* with the choir during services and were probably expected to own the Book of Hours for the order published at its inception by Le Roy et Ballard, even if they could not read the plainchant therein. To take one last example, Louis XIII began his lute studies with Robert Ballard at age ten and later whiled away the empty hours at the siege of La Rochelle penning motets and directing his clerics in the chanting of vespers for Pentecost.[39] For the most part, singing was considered a princely activity in the tradition of King David, but not all reports of kings' singing are favorable. One evening in 1597 Gabrielle d'Estrées shushed Henry IV when he joined everyone in singing Psalm 78 at her sickbed, where Vaumesnil was playing lute to console her. Perhaps the company was too public for a musical performance that included the king, or perhaps he sang off key. Either way, the clerics present were angry that he was prevented from singing God's praises.[40]

These anecdotes indicate the kinds of music made by noble amateurs: they sang hymns, psalms, and chansons whose tunes they might have memorized and more often than not played the lute, though some could read part-songs and sacred polyphony. It may be that a good many amateurs could read tablature for the lute, cittern, or guitar, but not mensural notation, and even that tablature was the very thing recommending the lute. Because tablature showed fingerings instead of absolute pitches, it enabled students to avoid all study of clefs, accidentals, and *musica ficta,* not to mention the troubles beginners had coordinating the parts of polyphonic songs and motets. It also meant that the element of mensural notation amateur lutenists *could* read was rhythmic values, which is a point of some interest when we consider how well they must have known dance music.

Dance lessons, balls, and ballets familiarized aristocrats with a large repertoire of dance tunes, and it would seem that publishers catered to this taste for dances when pulling together collections of music intabulated for lute and guitar, which includes a wealth of dances, dances based on songs, and—as Daniel Heartz has shown—texted dances masquerading as songs.[41] Le Roy et Ballard got their music publishing business up and running with tablatures for lute and guitar, dividing the repertoire more or less evenly between psalms, chansons, and dances, with a few fantasies

38. Pierre Bourdeille de Brantôme, *Oeuvres complètes,* ed. Ludovic Lalanne, 11 vols. (Paris: Mme Ve Jules Renouard, 1864–1882), 5:284–285.

39. Jean Héroard, *Journal,* ed. Madeleine Foisil, 2 vols. (Paris: Fayard, 1989), 2:2049.

40. Pierre de L'Estoile, *Mémoires-journaux de Pierre de L'Estoile,* ed. Gustave Brunet et al., 12 vols. (Paris: Alphonse Lemerre, 1888–1896), 7:82–83.

41. See Daniel Heartz, "*Voix de ville:* Between Humanist Ideals and Musical Realities," in *Words and Music: The Scholar's View; A Medley of Problems and Solutions Compiled in Honor of A. Tillman Merritt by Sundry Hands,* ed. Laurence Berman (Cambridge, Mass.: Harvard University Press, 1972).

and motets as well. Intabulations thus verify the anecdotal evidence about what kinds of music nobles enjoyed even as they remind us of the centrality of dance at all levels of society. Anyone who danced knew a number of dance melodies by heart, and when people asked the band to play a particular dance, they either called it out by name or hummed a few bars. For many, the melodies and measured quality of dance must have framed their entire musical understanding; for others, the musical instruction they received during dance lessons may have doubled as their only formalized musical training.

WITH WARRIORS strumming the lute at sieges and victory parades greeting their return from battle, bellicose music abounded. Yet looking over the repertoire published by Le Roy et Ballard and the subsequent Ballard dynasty, most of what came off the presses in Paris between 1562 and 1629 was sacred music and love songs.[42] It is fine to note that the Phrygian mode provokes anger, but the pieces in E or A that might have represented it for contemporaries are fairly scarce, and those that do exist treat the tribulations of love sooner than the trials of Mars.[43] Where is the music associated with arms?

42. For the sixteenth century, see Lesure and Thibault, *Bibliographie des éditions d'Adrian Le Roy et Robert Ballard.* For the seventeenth century, see Laurent Guillo, *Pierre I Ballard et Robert III Ballard: imprimeurs du roy pour la musique, 1599–1673,* 2 vols. (Liège: Mardaga, 2003).

43. I use the analytical method established by Harold Powers, "Tonal Types and Modal Categories in Renaissance Polyphony," *Journal of the American Musicological Society* 34 (1981): 428–470, by which one can sometimes posit what mode contemporaries might have assigned to a piece of music based on an analysis of three features (signature [♭ or ♮], cleffing [high = g2 or low = c1, c2], and final [the bass or "root" of the final chord]). While French writers on music from Pontus de Tyard and Ronsard to Marin Mersenne tended to agree on the choleric affects of the Phrygian, the musical materials comprising the ancient Greek modes were never codified in a single system. Tyard, who was a member of the Pléiade, active in the second half of the century, adopted the theories of Heinrich Glareanus, whose *Dodecachordon* (Basel, 1547) prominently linked the ancient modes and their effects to church modes. Like Glareanus, Tyard defined the Phrygian as mode 3, having a range from e to e′ and a final of E (Pontus de Tyard, *Solitaire second* [Lyons, 1555], ed. Cathy M. Yandell [Geneva: Droz, 1980], 208). The reality was, of course, more complex, particularly given the muddled distinctions between mode 3 (Phrygian) and mode 4 (Hypophrygian) in practice. In the analytic of Powers, the tonal types g2-♮-E, g2-♮-A, c1-♮-E, c1-♮-A, and occasionally c2-♮-E might all have represented modes 3 or 4. Moreover, the practice of sixteenth-century composers shows that they thought of the "Phrygian mode" as a one with a lamenting, rather than choleric, affect, rather like Tyard's definition of the Hypophrygian as "propre aux paroles lamentables et appaisantes l'ire" (ibid.). Josquin des Prez's "Milles regretz" is a good example of the inconsistencies between the theoretical attempts to explain modal ethos in terms that took the church modes to be equivalent of Greek ones and actual practice. See Jean-Pierre Ouvrard, "Modality and Text Expression in Sixteenth-Century French Chansons: Remarks Concerning the E Mode," *Basler Jahrbuch für historische Musikpraxis* 16 (1992): 89–116.

First, one must look beyond the polyphonic repertoire. Monophonic chansons written to the tunes of well-known songs often treat the subjects excluded from *fin amors,* such as death, crime, and war. They fill the pages of modest *chansonniers* whose titles advertise their unseemly subject matter. To cite but one of many, in 1580 the Lyonnais publisher Benoist Rigaud put out *La Fleur des chansons nouvelles, Traittans partie de l'amour, partie de la guerre, selon les occurrences du temps present* (The Flower of New Songs, Treating Partly Love and Partly War, According to Current Occurrences). A good share of Rigaud's business came from printing war songs.[44]

Second, we should recognize that psalms and hymns were battle cries of the highest order during wars fought over sectarian devotion. Huguenots in Sancerre sang Psalm 144 to mark their initial resistance during the siege of 1572, finding its insurrectionary text perfect for a holy war: "Blessed be the Lord my strength which teacheth my hands to war and my fingers to fight."[45] As we shall see, the hymn *Te Deum laudamus* was likewise taken to the streets by Catholics desperate to purify the country of heresy, while masses and motets blessed the work of chivalric orders that vowed to crusade against the Huguenots. In ultra-Catholic Burgundy, confraternities of the Holy Ghost began to crop up in the 1560s, their primary missions being communal prayer and forcing conversions from heretics. It is not clear whether the *L'homme armé* masses traditionally heard by the Order of the Golden Fleece were employed by these newer orders as well, but at the Sainte-Chapelle in Dijon, clerics celebrated daily masses for that order until as late as 1626, and had other confraternities wanted to hear a *Missa L'homme armé,* there were over thirty masses on the rollicking fight song to choose from, including two new ones by Palestrina.[46]

---

44. See Kate van Orden, "Cheap Print and Street Song Following the Saint Bartholomew's Massacres of 1572," in *Music and the Cultures of Print,* ed. Kate van Orden, with an afterword by Roger Chartier (New York: Garland, 2000), 271–323.

45. Barbara B. Diefendorf, *Beneath the Cross: Catholics and Huguenots in Sixteenth-Century Paris* (New York: Oxford University Press, 1991), 136–144.

46. See especially Craig Wright, *The Maze and the Warrior: Symbols in Architecture, Theology, and Music* (Cambridge, Mass.: Harvard University Press, 2001), chaps. 6 and 7; William F. Prizer, "Music and Ceremonial in the Low Countries," *Early Music History* 5 (1985): 113–153; Barbara Haggh, "The Archives of the Order of the Golden Fleece and Music," *Journal of the Royal Musical Association* 120 (1995): 1–43; and Richard Taruskin, "Antoine Busnoys and the *L'homme armé* Tradition," *Journal of the American Musicological Society* 39 (1986): 255–293. Taruskin argues that the origins of the *L'homme armé* tradition rested with Busnoys, who worked for Charles the Bold. The meetings of the order featured a cycle of religious services that lasted five days and included three masses, a Requiem Mass, the Office of the Dead, and three vespers services, thus affording prime occasions for the celebration of

Finally, we must take full account of unwritten repertoires, especially of the dance music, marches, trumpet calls, and drum patterns played by professional instrumentalists, many of them in military service. Minstrels controlled their repertory and employment through a guild system based on orality and secrecy, and little of it was written down. Masters taught chansons and dance tunes to their apprentices by rote, circumventing musical literacy and thereby preserving trade secrets. Apart from the dances intabulated for amateurs to play on the lute, guitar, or cittern, little dance music comes down to us. Even the beautiful and rightfully famous prints of *ballets de cour* such as the *Balet comique de la Royne* and *La délivrance de Renault* include only vocal music and virtually none of the instrumental numbers for the dances. Yet we know balls to have been a hugely important forum for socializing in high society and ballets to have been a favorite musical genre at court. In the end, this book must often take descriptions of music-making as its material basis and weave into these accounts the fragments of written music that we do have.

But before we turn primarily to dance music—as we will—let us begin squarely in the repertoire of polyphonic vocal music, for of all the war songs, *chants triomphals,* odes, and victory hymns of every sort, none could rival the stunning popularity of Clément Janequin's songs on famous battles, polyphonic works that marked out the place of arms in the catalogs of music printers.[47] The most beloved one was "Escoutez tous gentils gallois" (or *La guerre*), a magisterial work on the scale of a motet celebrating the victory of Francis I at Marignano in 1515. Its extraordinary presence in the repertory makes it a necessary point of orientation for any study of music and arms, for it defined the musical style of war for sixteenth-century audi-

masses on this tune before the "armed men" of the order. The *L'homme armé* tune was worked into over thirty polyphonic masses by composers who included Busnoys, Dufay, Ockeghem, Basiron, Tinctoris, Vaqueras, Brumel, Obrecht, Compère, La Rue (two), Josquin (two), Pipelare, Mouton, Peñalosa, Morales (two), Guerrero, and even Palestrina (two), who dedicated the publication containing his five-voice mass on the *L'homme armé* tune to Philip II in 1570. In 1584 the Burgundian Mayenne and the Guises formed an alliance with Philip II. Also see Alejandro Enrique Planchart, "Guillaume Du Fay's Benefices and His Relationship to the Court of Burgundy," *Early Music History* 8 (1988): 117–171, and Owen Rees, "Guerrero's *L'homme armé* Masses and Their Models," *Early Music History* 12 (1993): 19–54.

47. For Janequin's chansons on the siege of Metz, the surrender of Bologna, the battle of Renty, and so forth, see his *Verger de musique* (Paris: Le Roy & Ballard, 1559). Other composers celebrated specific victories with polyphonic chansons. See, for example, Nicolas de La Grotte, "Tel qu'un petit aigle sort," on a poem Ronsard wrote to celebrate the victory of Henry de Valois at Jarnac in 1569 (La Grotte, *Chansons de P. de Ronsard, Ph. Desportes, et autres* [Paris: Le Roy & Ballard, 1569]). Also see the monophonic songs, particularly on the siege of Metz, in the Chansonnier Maurepas, F. Fr. 12616–12659, Bibliothèque nationale de France; and the Chansonnier Clairambault, F. Fr. 12676–12743, Bibliothèque nationale de France.

ences. Pierre Attaingnant published Janequin's four-voice setting in 1528;[48] the following year he published an intabulation in his "very brief and easy" lute tutor, and his widow issued a four-part *Pavanne de la guerre* with three gaillardes in 1556. By then Nicolas du Chemin had already published a new version of the polyphonic chanson with a fifth voice that had been added by Verdelot (1555), intabulations abounded, and it had become a hit as a dance tune. Just as quickly, Janequin worked the song into a *Missa La bataille,* which was published by Jacques Moderne in Lyons in 1532. Although Marignano brought no military gains, the song it inspired became a monument to the king's strength of arms. Ultimately, it developed into a universal emblem of victory in war, and battle masses became so popular that they attracted the attention of Tridentine reformers, who worried that such music "refresh[ed] the ear more than the mind and seem[ed] to incite lasciviousness rather than religion."[49]

La guerre's wonderfully noisy evocation of clanking armor, drum tattoos, rimshots, trumpet fanfares, calls to arms, harquebus fire, and marching troops was strikingly original, particularly given the general resistance of the so-called Parisian chanson to any sort of text illustration in music. The song dramatizes the actual progress of the battle, beginning with the assault of the advance guard—Francis I rode at its center—and, in the second part, narrating the attack of the main army or *bataille,* with its artillery and infantry battalions that turned the tide at Marignano and brought the victory that returns the song to the more lyric strains typical of the Parisian chanson at its end. For the battle, Janequin matched the text's bullet commands, short lines, and quickly reiterated rhymes ("courage prenez après suyvez, frapez, ruez") with much actual military music. Fusae set long strings of nonsense syllables ("fan frere le le fan fan fan feyne") in rhythms that could have been lifted directly from the tremendous side drums that issued infantry commands in the field, so closely do they resemble the military drumming recorded by Thoinot Arbeau in *Orchésographie* (1589) in both rhythm and the vocal syllables with

48. Clément Janequin, "Escoutez tous gentils gallois," in *Chansons de maistre Clement Janequin* (Paris: Pierre Attaingnant, [1528]); modern edition in idem., *Chansons polyphoniques,* ed. A. Tillman Merritt and François Lesure, 6 vols. (Monaco: Éditions de l'Oiseau-Lyre, 1965–1971), 1:23–53. Here and where possible—exx. 1.4, 1.6, 1.7—I use the standard modern editions. All other transcriptions are my own. For the many instrumental arrangements of *La guerre,* see Howard Mayer Brown, *Instrumental Music Printed before 1600: A Bibliography* (Cambridge, Mass: Harvard University Press, 1965).

49. See Craig Monson, "The Council of Trent Revisited," *Journal of the American Musicological Society* 55 (2002): 1–37, at 8. Other composers wrote parody masses on *La guerre,* a model particularly popular in Spain and the New World. See the masses by Francisco Guerrero, Joan Cererols, Sebastian Durón, Mateo Romero, Juan Bautista Comes, and Francisco Lopez Capillas, as well as Tomás Luis de Victoria's exquisite *Missa Pro Victoria.* Also see the introduction to Clément Janequin, *Missa La bataille,* ed. Frank Dobbins (Espoo, Finland: Fazer Music, 1995).

Example 1.1. Clément Janequin, *La guerre,* part 2, mm. 19–24. In *Chansons de maistre Clement Janequin* (1528)

which Arbeau underlays the patterns (exx. 1.1 and 1.2). So, too, the pitches Janequin chose match the conjunct noodling of the fifes that accompanied infantrymen on the march (ex. 1.3). When the cavalry commands to saddle up ("boute-selle"), mount ("a cheval"), and form up around the company's standard ("a l'estendart") appear in the text, Janequin sets them to figures characteristic of the natural trumpet (see fig. 1.1 and exx. 1.4 and 1.5). Although the French trumpet calls published by Marin Mersenne in his *Harmonie universelle* (1636) date from the seventeenth century, they give a good idea of the musical commands Janequin would have known.[50] The ascending fifth Janequin uses to set "boute-selle" is not far from the

50. Marin Mersenne, *Harmonie universelle* (Paris, 1636), facsimile ed., ed. François Lesure, 3 vols. (Paris: Centre National de la Recherche Scientifique, 1965), 3:264.

Example 1.1. (*continued*)

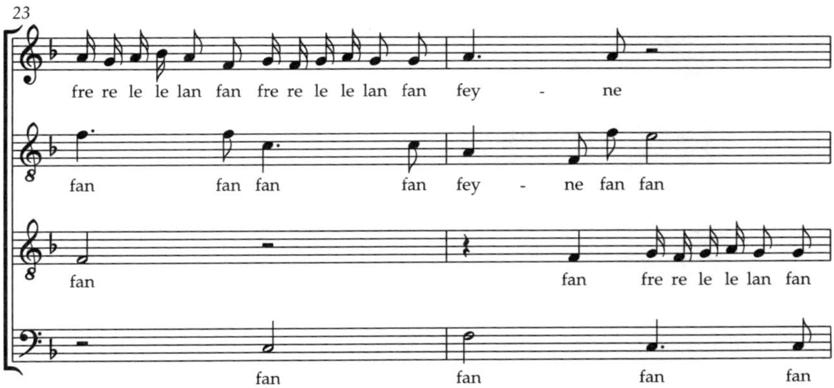

Example 1.2. Rhythms for military marches. Thoinot Arbeau, *Orchésographie* (1589)

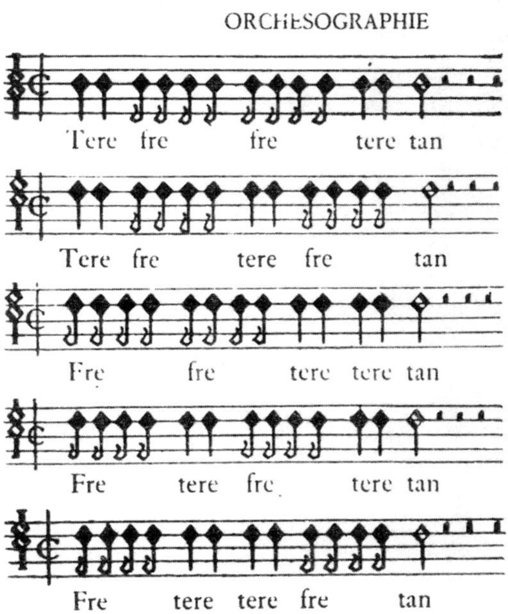

Example 1.3. Fife melodies. Thoinot Arbeau, *Orchésographie* (1589)

first "boute-selle" given by Mersenne. For listeners, the associations between the trumpet's brassy tones, mounted knights, and the king were unequivocal. In some places the use of trumpets was even restricted by sumptuary laws. In French cities trumpets announced royal proclamations and the arrival of the king, while in the military each company of horsemen had one or two trumpeters who lodged with the captain and worked as scouts, spies, envoys, and guards.[51] Finally, Janequin makes notable use of the rhythm ♩𝅗𝅥♩𝅗𝅥 in the first part of the song (mm. 43–59). The iambs seem calculated to evoke the canter of heavy steeds charging into battle while the text exhorts: "nobles jump in the saddle, buckle your arms, lively minions, lance at the ready, bold and quick."[52]

51. This sort of organization was formalized along with the formation of the first true cavalries, which replaced the sixteenth-century bands of mounted knights and light horsemen. On the qualities of a good trumpeter see Lodovico Melzo, *Regole militari sopra il governo e servitio particolare della cavalleria* (Antwerp: G. Trognoesius, 1611); I cite the French-German translation: Melzo, *Reigles militaires du chevalier frere Louys Melzo/Kriegs Regeln dess Ritters L. Meltzo* (Frankfurt: Caspar Rodtel, 1625), 21–23.

52. See Monika Woitas, "*Combattimento* & Co.: Choreographie und *imitatio* in den Torneo-Tänzen um 1600 und in Monteverdis *Combattimento*," in *Claudio Monteverdi und die Folgen: Bericht über das Internationale Symposium, Detmold 1993*, ed. Silke Leopold and Joachim Steinheuer (Kassel: Bärenreiter, 1998), 167–190. Woitas traces Monteverdi's rhythmic resources to similar idioms em-

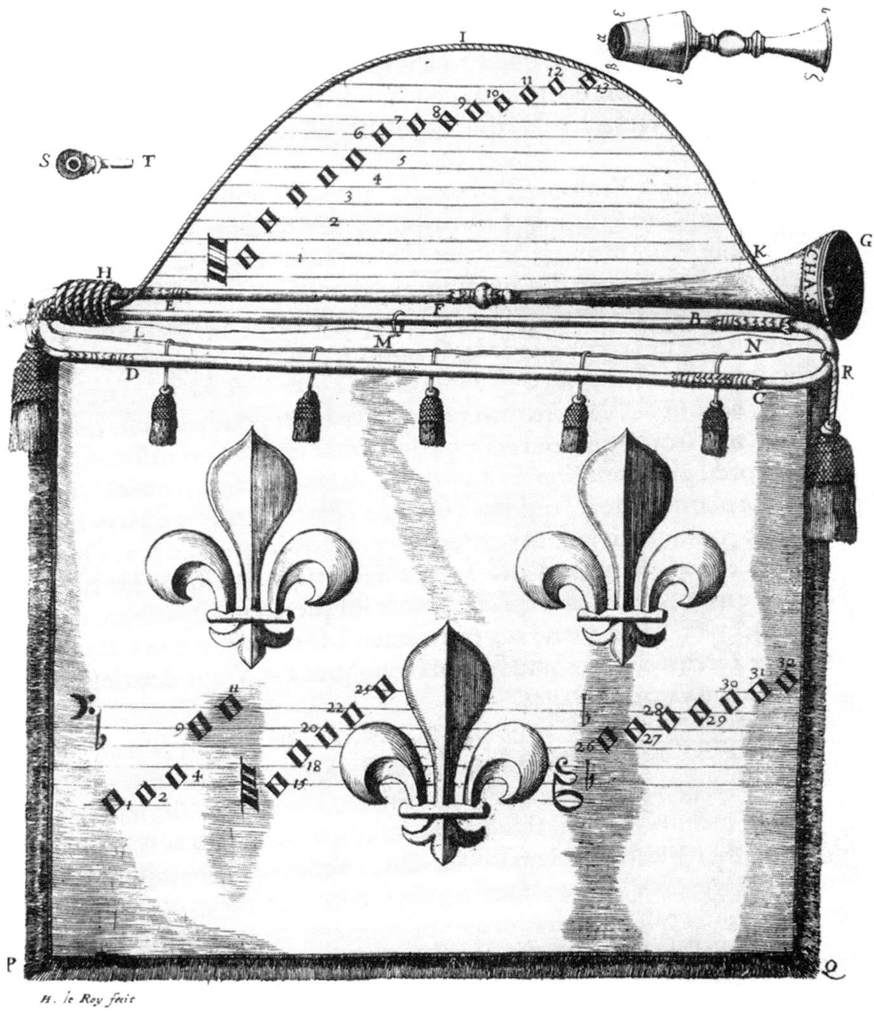

FIGURE 1.1. Harmonic series of the natural trumpet. Marin Mersenne, *Harmonie universelle* (1636)

So accurately did Janequin render the sounds of war that *La guerre* was said to quicken fury in male listeners. In an account obviously indebted to the well-worn story of Timotheos and Alexander introduced above, Noël Du Fail related that

ployed in the combats choreographed by Fabritio Caroso (*Il ballarino* and *Nobilità di dame*) and Cesare Negri (*Le gratie d'amore*). She suggests that the iambic meter Monteverdi uses for the "Trotto del Cavallo" (mm. 18–30) was a common emblem of battle.

Example 1.4. Clément Janequin, *La guerre,* part 2, mm. 9–12.
In *Chansons de maistre Clement Janequin* (1528)

"when, before the great Francis, one sang the war song Janequin wrote for the victory that Francis had had over the Swiss, there was not a man who did not look to see if his sword was safe in its scabbard and who did not stand on his toes to look braver and taller."[53] *La guerre* created a *stile concitato* for composers of art music a

53. Noël Du Fail, *Oeuvres facétieuses,* ed. Jules Assézat, 2 vols. (Paris: Paul Daffis, 1874), 2:124–126. "Comme par exemple, quand l'on chantoit la chanson de la guerre faicte par Jannequin devant ce grand François, pour la victoire qu'il avoit euë sur les Suisses, il n'y avoit celuy qui ne regardast si son espee tenoit au fourreau, et qui ne se haussast sur les orteils pour se rendre plus bragard et de la riche taille." Brantôme also relates a story about *La guerre,* albeit a very different one. Mlle de Limeuil, one of Catherine de' Medici's ladies-in-waiting, asked to have it sung to her as she lay on her deathbed. "Julien, prenez vostre violon et sonnez-moi tousjours jusques à ce que me voyez morte (car je m'y en vois) la *Defaitte des Suisses,* et le mieux que vous pourrez; et quand vous serez sur le mot *Tout est perdu,*

EXAMPLE 1.5. Cavalry signals. Marin Mersenne, *Harmonie universelle* (1636)

Continued on next page

century before Claudio Monteverdi claimed to have invented it, and just as Monteverdi would claim for his *Combattimento di Tancredi et Clorinda,* Du Fail attributed to Janequin's chanson the ability to imitate not just the action of battle, but the passions of those who fought.[54] His account witnesses a fundamental precept of

sonnez-le par quatre ou cinq fois, le plus piteusement que vous pourrez." When he reached those words, she sang along, then quitted this world. "Voilà une mort joyeuse et plaisante," Brantôme remarks. See Pierre Bourdeille de Brantôme, *Recueil des dames: poésies et tombeaux,* ed. Etienne Vaucheret (Paris: Gallimard, 1991), 719.

54. See the preface to Monteverdi, *Madrigali guerrieri, et amorosi con alcuni opuscoli in genere rappresentativo . . . Libro ottavo* (Venice: Alessandro Vicenti, 1638). On the *stile concitato* and Monteverdi's *Combattimento di Tancredi et Clorinda* in particular, see Gary Tomlinson, *Monteverdi and the End of the Renaissance* (Berkeley and Los Angeles: University of California Press, 1987), 201–209; Woitas, "*Combattimento* & Co."; and Stefano La Via, "Le *combat* retrouvé: Les 'passions contraires' du 'divin Tasse' dans la représentation musicale de Monteverdi," in *La Jérusalem délivrée du Tasse: poésie, peinture,*

Example 1.5. (*continued*)

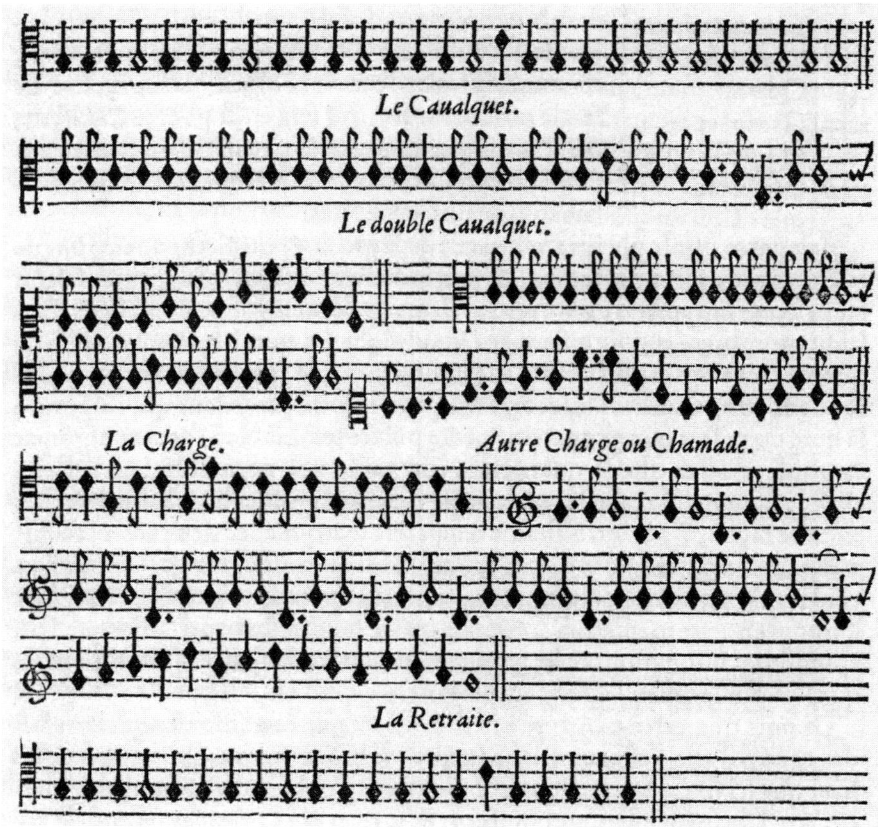

musical affect, one especially familiar to sixteenth-century humanists from Aristotle's *Politics* and Marsilio Ficino's *De triplici vita* (1489), in which Ficino says:

> remember that song is a most powerful imitator of all things. It imitates the intentions and passions of the soul as well as words; it represents also people's physical gestures, motions, and actions, as well as their characters and imitates all these and acts them out so forcibly that it immediately provokes both the singer and the audience to imitate and act out the same things.[55]

musique, ballet, ed. Giovanni Careri (Paris: Klincksieck, 1999), 109–158, who analyzes the progression of emotions in the piece and the large-scale tonal and rhythmic structures articulating them.

55. Marsilio Ficino, *De vita libri tres; De triplici vita; Apologia; Quod Necessaria Sit ad Vitam Securitas* (Florence, 1489), ed. and trans. Carol V. Kaske and John R. Clark as *Three Books on Life* (Binghamton, N.Y.: Medieval and Renaissance Texts and Studies, in conjunction with the Renaissance Society of America, 1989), 358–359. On Ficino and the imitative properties of music, see Gary Tom-

This philosophical linking of music and action (via the passions) underpins Du Fail's account of how the king's men-at-arms, struck by sounds of battle, puffed themselves up with courage or at least pretended to be brave as they anticipated the fracas that might ensue. The narrative, moreover, turns upon a series of cultural constructions with social implications, for the music that prompted the warriors' desire for glory before the king sanctioned the military force by which Francis secured his reign. In this way *La guerre* imitated the violence of the king's victory in a form that was not just socially condoned but even encouraged. Du Fail tells us that *every* man looked to his sword, that *every* man stood on his toes, his account insisting not just on the effect of the music, but on the passions and reactions proper to a nobleman.

It is extremely significant that so many stories describe not just the calming effect of music, but its ability to incite violence. This masculine virtue certainly justified the usefulness of pyrrhic dance, even among critics who would otherwise do away with dance altogether. Likewise, the famous story mentioned earlier about Le Jeune's *air* is one in which music causes a nobleman to act aggressively before undoing its own effect.[56] But how was music understood to operate?

Frances A. Yates first identified the specific work described in the anecdote about rehearsals for the Joyeuse *magnificences* as Le Jeune's *La guerre,* and although she did not read it against the background of Janequin's *La guerre* and its similarly fabled effects, we can see that the two pieces are related.[57] It was certainly with Janequin's chanson in mind that Le Jeune began his colossal suite of fourteen *chansons mesurées à l'antique,* which opens with the same descending figure Janequin had used to call listeners to attention at the beginning of his song. Le Jeune also employs the galloping rhythm to which Janequin set the text "Avanturiers, bon compagnons, ensemble croisez vos bastons" (m. 43) for the similarly turned phrase "Bataille, compagnons, bataille allons camper" (see ex. 1.6).

Yates was understandably keen to pinpoint the spot in the music where the modal shift from Phrygian to Hypophrygian worked its magic calming effect, but only Isabelle His has finally been able to connect the story with the music convincingly.[58] She argues that in the middle of the second *air* the affect turns choleric and at the end relaxes again (see exx. 1.6 and 1.7).

---

linson, *Music in Renaissance Magic: Toward a Historiography of Others* (Chicago: University of Chicago Press, 1993).

56. For the many sources of the story's retelling, from Mersenne to Jean-Jacques Rousseau, see His, *Claude Le Jeune,* 281–285.

57. See Yates, *Astraea,* 149–172. See also the excellent introduction by D. P. Walker and François Lesure to Claude Le Jeune, *Airs,* ed. D. P. Walker, 4 vols. (Rome: American Institute of Musicology, 1951–1959). The music can be found at 1:90–117.

58. His, *Claude Le Jeune,* 280–285, 291–295.

EXAMPLE 1.6. "Bataille compagnons, bataille allons camper." Part 2 of *La guerre,* second *air* ("Tost, tost, tost, bouteselle bride"). In Claude Le Jeune, *Airs* (1608)

Example 1.7. "Si de flâme briller cette bande se void," lines 1–3. Part 3 of *La guerre,* second *air* ("Tost, tost, tost, bouteselle bride"). In Claude Le Jeune, *Airs* (1608)

In "Bataille, compagnons," Le Jeune suddenly removes the B-flat that normally softens the tritone that would otherwise occur in F. His suggests that this harsh "natural Lydian" mode may have represented for Le Jeune "the Phrygian of his time"—a mode with a bellicose affect. The correction of the signature with the flat at the beginning of "Si de flâme briller" brings the piece back to a normative modality, an affect that is made all the more pronounced by the slower tempo implied by the move from ¢ to C, the heavier scoring for five voices, hesitant dominant tonalities at the ends of lines two and three, and the abandonment of iambic feet for a regular succession of anapests.

What is striking about this comparison of two clearly related songs is the very different methods by which the composers achieved reportedly similar affects. Janequin employs large infusions of *musique concrète,* extreme rhythmic values, shifting meters, and textures ranging from imitative polyphony and polyphonic cacophony to homophony. Le Jeune works within a musical language unfailingly committed to homophony and a vocabulary of minims and semiminims keyed to the stressed ("long") and unstressed ("short") syllables of his text. The text itself was written with an ear attuned to rhythm, and the musical setting, in following it, is marked by metrical variety (compare ex. 1.6 to ex. 1.7). Word-painting on "flâme," "feus," and "ardeur" in the tenor of example 1.7 produces some faster note values but never disrupts the pace of the declamation or the even deployment of these "code words" in each line. In sum, whereas Janequin's *La guerre* stirs the passions with extremes of texture, speed, and musical style incorporating everything from the standard lyric idioms of the Parisian chanson to the flat and insistent commands of infantry drums, Le Jeune's *La guerre* provokes its listeners in a more restrained manner. The story of stirring violent passions remained the same, but Le Jeune's style concentrates starkly on rhythm, meter, and tempo. It evinces a different relationship between words and music, in which the intention is not so much to depict the meanings of individual words using fiery melismas and other such devices, but to characterize the passions associated with whole swaths of poetry by the use of meters that matched the relative pace and agitation of these larger sections. There is no Phrygian here, just seven B-naturals in one section of a massive piece otherwise entirely in F, with its standard B-flat. The story of its affect attempts to explain Le Jeune's orphic powers with the well-accepted conceit of modal ethos, even though Le Jeune's own efforts concentrated on rhythm.

The style of Le Jeune's *La guerre* is, of course, exemplary of *musique mesurée à l'antique.* The result of experiments carried out by the Academy of Poetry and Music, *musique mesurée* aimed to retrieve the moral effects of music described by the Greeks by recreating ancient music itself.[59] These songs "in ancient measures" were

59. In addition to the studies of Le Jeune already cited, to which should be added Pierre Bonniffet, *Un ballet démasqué: l'union de la musique au verbe dans* Le printans *de Jean-Antoine de Baïf et de*

written with a newly invented orthography that made it possible to relate the stressed and unstressed syllables of the French language to the long and short syllables of Latin and hence to write in classical meters. Provided with quantitative verse, the musician's task was to match the long and short syllables of the words with notes of long and short values, translating poetic meter into musical rhythm. *Musique mesurée*'s claims to authenticity rested on the employment of classical models such as Sapphic or Anacreontic meter. Measuring word and music to Greek meters of beautiful proportions, the academicians believed that they had recaptured ancient music's ethical affects.[60] Thus ancient lyric provided forms that were transformed into rhythmic orders through which morals might be regulated.

The preface to *Le printans* (1603), Le Jeune's famous collection of *chansonettes mesurées,* makes it clear that rhythmic ethos, and not modal ethos, motivated their design:

> Harmony was so little known to the ancients that they used only the consonances of the octave, the fifth, and the fourth, with which they composed a certain chord on the Lyre, to the sound of which they sang their verse. Rhythm, on the contrary, was brought by them to such perfection that they produced marvelous effects with it, moving the souls of men with it to whatever passions they wished, which they wanted to show to us with the fables of Orpheus and Amphion, who softened the treacherous courage of the most savage beasts, and animated the woods and stones to the point of moving them and placing them where they pleased.[61]

Claude Le Jeune (Paris: Champion-Slatkine, 1988), see the classic scholarship on *musique mesurée* in Yates, *French Academies,* and of D. P. Walker, whose essays are collected in *Music, Spirit, and Language in the Renaissance,* ed. Penelope Gouk (London: Variorum Reprints, 1985), as well as the contributions of Brooks, *Courtly Song,* 293–315. For the clearest expression of musical Neoplatonism in the circle of the Pléiade—of which Baïf was a member—see Pontus de Tyard, *Solitaire premier,* ed. Silvo F. Baridon (Geneva: Droz, 1950), and idem, *Solitaire second.* The reception of Platonic teachings was tremendously influenced by the interpretations of Marsilio Ficino, on whom see Tomlinson, *Music in Renaissance Magic.*

60. For a reading of the Pléiade's conception of poetic measure and the relationship between their attitudes and contemporary mathematics see John McClelland, "Measuring Poetry, Measuring Music: From the Rhétoriqueurs to the Pléiade," in *Poetry and Music in the French Renaissance,* ed. Jeanice Brooks, Philip Ford, and Gillian Jondorf (Cambridge: Cambridge French Colloquia, 2001), 17–32.

61. Claude Le Jeune, *Le printemps,* ed. Henry Expert, Maîtres Musiciens de la Renaissance Française 12–14 (Paris: Leduc, 1900–1901). The dedicatory poems and preface are also reprinted in His, *Claude Le Jeune,* 456–460. "L'Harmonique a esté si peu cogneuë d'eux [les anciens], qu'ils ne se sont servis d'autres consonances que de l'octave, la quinte, & la quarte: dont ils composoyent un certain accord sur la Lyre, au son duquel ils chantoient leurs vers. La Rythmique au contraire a esté mise par eux en telle perfection, qu'ils en ont fait des effects merveilleux: esmouvans par icelle les ames des hommes a telles passions qu'ils vouloient: ce qu'ils nous ont voulu representer sous les fables d'Orphée, & d'Amphion, qui adoucissoyent le courage felon des bestes plus sauvages, & animoyent les bois & les pierres, jusques à les faire mouvoir, & placer ou bon leur sembloit."

In this analysis, rhythm animates the soul, it *moves* the soul, and in the right hands (those of Orpheus or Amphion), it moves objects as well. The preface goes on to credit Le Jeune with recovering these ancient effects in his *chansons en vers mesurés à l'antique.* The link between music and the passions was thus made not primarily through words, but through rhythmic number and measure, a connection that was consistent with music's place in the quadrivium of mathematical sciences and the frequent assignment of poetry to the musical portion of the quadrivium based on its science of quantity.[62]

Faithful to the meters devised by Jean-Antoine de Baïf, who wrote the poetry, Le Jeune's *Le printans* contains only measured settings. But what is quite surprising, given the claims of classical recovery made for *musique mesurée,* is that the collection is permeated with dance rhythms—so much so that scholars have posited that the songs were *chansons à danser,* meant to be danced.[63] Indeed, the whole project of creating a genre of ancient music relied substantially on the idioms of dance as the source of affect, despite its overt fixation on classical poetic meters and the quantification of language. And why not? If the academicians hoped to seize the souls of unruly nobles and harmonize them, it makes sense for them to have adopted the strong meters of the gaillardes, pavanes, and branles that their audience regularly danced to at balls.

Much of this book evolved from a desire to explain the primacy of rhythm, meter, and dance in French music. In the first place, it suggests that Neoplatonists and perhaps others believed that dance music had the same ravishing effects as songs. Because Plato rejected instrumental music as a form of ethical instruction and because most of the French sources explaining the effects of music were written by poets, lyric tends to dominate our histories of French musical humanism. Writing in the tradition of Marsilio Ficino, both Pontus de Tyard and Pierre de Ronsard theorized that the ravishing effect of music began with the poet, whose verse was written in the throes of a divine furor sent by the Muses.[64] Singing such poems released the *fureur poétique* inscribed in them, a furor that then seized the souls of listeners and carried them off in a spiritual transport. Scholars have tended to take poets at

62. Consider the remarks of Pierre de La Primaudaye, *Academie françoise,* 3rd ed. (Paris: Guillaume Chaudière, 1580), fol. 37r. "La Musique est aussi mise pour un des membres des Mathematiques, comme une Science tiree des nombres: pour ce que par iceux la proportion harmonique a esté trouvee. Puis suit la Poësie qui est contribuee à la Musique." For vernacular poets, the tradition began with Eustache Deschamps, *L'art de dictier* (1392), was continued by the *grands rhétoriqueurs,* and culminated in the obsession of sixteenth-century poets with number. On this last subject, see McClelland, "Measuring Poetry, Measuring Music," 17–32.

63. Bonniffet, *Ballet démasqué.* For a similar argument made about mid-century songs with humanistic pretensions see Heartz, "*Voix de ville.*"

64. The classic French statement of this thesis is found in Tyard, *Solitaire premier.*

their word, framing musical analysis with the concept of "oratory" and analyzing lyric settings as "readings" of their texts. This approach works well for the Italian madrigal, monody, and *favola in musica,* but explains far less about French music, which so persistently rejects Italianate rhetoric and word-painting in favor of the blanker forms we find in Le Jeune. Indeed, in this respect, *musique mesurée* is not so different from French songs more generally, for many of the *voix de villes,* odes, and *airs de cour* produced by composers at or near the court tended toward a homophonic style that developed the embodied meters of dance music rather than pursuing language-oriented forms. Dance was definitive of French lyric, and even when elaborated in the form of *chansonnettes mesurées, airs,* and, later, the recitatives of *tragédie en musique,* the style was keyed at some point of origin to the rhythms of moving bodies. In sum, from the music itself a Neoplatonics of dance emerges, strains of which we likewise hear in discussions of dance and its moral effects. When Thoinot Arbeau spoke of "rhythmic virtue" and Heinrich Cornelius Agrippa warned against "dishonest rhythms" in dance songs, they emphasized the moral qualities of various rhythms, even in the absence of words.[65]

But Neoplatonic philosophies are only one point of departure for understanding the cultural practices music sustained in France, which brings me to a second point. Certainly music was central to the concerns of Renaissance humanists, but investigation into this limited cultural milieu has unfairly dominated our approach to late sixteenth-century music, to the point of overshadowing the cultural currency of music in other domains. After all, many of Castiglione's interlocutors espoused both music and arms (rather than letters), and in France, *robins* and other humanists often adopted the cultural practices of the *nobles d'épée,* whose morning riding instruction at French military academies was followed by afternoon lute lessons. This book works outward from humanism to consider music in military culture and the culture of civility that was part of its dialectic.

Finally, the theories of rhythmic ethos that I explore in the following chapters suggest a significant coincidence of music with political aspirations for social order. Taken together with the culture of arms, they delineate a theoretical framework that, for the purposes of this book, tends to privilege dance over lyric. We know that kings, queens, and nobles pursued politics through *ballets de cour* and carrousels, and it may even be that opera failed to take root in France for so long precisely because it put too much store in words. By contrast, dance music visibly set bodies in

65. Thoinot Arbeau, *Orchésographie* (Langres, 1589), facsimile of Paris 1888 reprint, ed. Laure Fonta (Bologna: Arnaldo Forni, 1969), fol. 5r. "Car sans la vertu rithmique, la dance seroit obscure & confuse"; Heinrich Cornelius Agrippa, *Declamation sur l'incertitude, vanité, et abus des sciences* (N.p.: Jean Durand, 1582), 91. "Là on saute d'une façon enragee avec grand trepignement de pieds au son mol & lascif d'un instrument, au chant de sales chansons & rithmes deshonnestes."

motion with meters that squared both with political ideals of unity and with civility's growing emphasis on "measure" as a social science comprehending everything from comportment to politeness and speech. Le Jeune's obsession with the rhythmic dimension, his concern with order and with limiting the expressive range of his music, was not unique; rather, it occurred in the context of a whole series of investigations into the percussive effect of sound on the passions, the animating force of beats and measures, and how music could move troops around. New stories rivaling those about the effect of the Phrygian mode were devised about a rhythmic mode called pyrrhic, after the battle dance of the Lacedaemonians.

Focusing on French nobles through the lens of the military ideals they themselves announced invites concentration on activities such as pyrrhic dance and equestrian ballet, to be sure, but the larger point to be made about all of these balletic practices—*ballet de cour* and pitched battles included—is that they projected a musical order across the social body. The groups of people who gathered for coronations, entries, and Te Deums likewise participated in ceremonies emphasizing social order with music. Through the action of music, social bodies cohered.

In some sense, then, the story of this book is a retelling of the stories about *La guerre,* one that proceeds from changing assumptions about music's effect upon bodies individual, social, and politic. If this history embraces violence, it is because music was the tool of a culture that was deeply ambivalent about violent passions. On the one hand, nobility was defined in terms of virtues evinced in armed aggression. On the other hand, in order for civil war to end, violence had to be contained, centralized in the form of a royal army, and redirected in order to secure the king's sovereignty at home and reap the spoils of war abroad. Music did not exact obedience any more than the state did; rather, it created conditions for a pleasurable cooperation that, far from quenching violence, often made it more effective.

2

JUSTE PROPORTION

Music as the Measure of All Things

MUSICAL EDUCATION AS PHYSICAL EDUCATION

THE SIXTEENTH century saw a revolution in public schooling.[1] *Gens de bien* in provincial cities—municipal officials, merchants, local tycoons, royal magistrates in larger towns, aldermen, councilors, judges, lawyers and so forth—took education in hand, setting up local schools for the young. Colleges in Paris and major centers such as Bordeaux and Toulouse offered full programs in the *studia humanitatis,* though many fathers preferred not to send their sons off to a distant metropolis. In the latter part of the century, the Jesuits established colleges in Paris, Lyons, Tournon, Dijon, Toulouse, and a handful of other places.

The proliferation of colleges attests to the great emphasis on formation during the Renaissance. Nonetheless, institutions catering specifically to nobles arose only in the seventeenth century, before which time upper-class parents faced difficult choices when deciding how to school their sons.[2] The lords of great households often hired private tutors in order to educate their children at home. Other nobles found boarding schools, such as the Jesuit Collège de Clermont in Paris, particularly appealing because they could send their sons to school with servants and a tutor. This entourage lodged with the boy and supplemented his Latin studies with

1. See George Huppert, *Public Schools in Renaissance France,* and Marie-Madeleine Compère and Dominique Julia, *Les collèges français, 16e–18e siècles,* 2 vols. (Paris: Centre National de la Recherche Scientifique, 1984–1988).

2. On noble education see J. R. Hale, "The Military Education of the Officer Class in Early Modern Europe," in *Cultural Aspects of the Italian Renaissance,* ed. Cecil H. Clough (New York: A. F. Zambelli, 1976), 440–461; Schalk, *From Valor to Pedigree,* chap. 8; Roger Chartier, Dominique Julia, and Marie-Madeleine Compère, *L'éducation en France du XVIe au XVIIIe siècle* (Paris: Société d'Édition d'Enseignement Supérieur, 1976); and Mark Motley, *Becoming a French Aristocrat: The Education of the Court Nobility, 1580–1715* (Princeton: Princeton University Press, 1990). On masters of arms, see Brioist, Drévillon, and Serna, *Croiser le fer,* chap. 2; and Anglo, *Martial Arts of Renaissance Europe,* esp. 7–18.

lessons in history, modern languages, and the social graces appropriate to his class. In this way, fears that mixing with commoners would destroy the education in manners and the vernacular he had received at home were alleviated.[3] Still, the emphasis on Latin grammar, rhetoric, humanities, and philosophy at the colleges hardly suited young men who aspired to careers in the military and at court, and while younger sons who were headed for the clergy often stayed the whole course, many aristocrats left college after just three years. Clearly some education in Latin was desirable, and familiarity with classical history, too, was seen as a noble attribute, but nobles complained that colleges were simply too bookish. In "De l'institution des enfans," Montaigne reflects on his own education from the age of six to thirteen at the Collège de Guienne in Bordeaux and concludes that he left without a single fruit that he could employ in his present life (1:175). Thanks to an exceptional professor, he escaped with his love of literature intact: had he not been so lucky, he imagines, he would have carried away from college only the hatred of books, as did most nobles. Like others, he believed that the long hours of study destroyed children's "natural" nobility, rendering them incapable of civil conversation and turning them away from the best occupations. Meanwhile, the punishment used to teach them politeness only twisted their spirits and left them more brutish when they graduated than when they arrived. "Harden the child to sweat and the cold, to the wind, sun, and hazards that he needs to scorn," not to chastisement, Montaigne admonished educators (1:165).[4]

The need to toughen boys into men led Montaigne to prescribe a program of intellectual studies paired with "honest exercise." He explained that exercise is important because man is made up of both body and soul: "as Plato says, one should not address the one without the other but guide them equally, like a pair of horses harnessed to the same shaft" (1:165). If anything, Montaigne favored the corporeal over the contemplative in the belief that exercises of the body benefit the mind, but those of the mind do not benefit the body. Thus his ideal education would have youths running, wrestling, and hunting, and learning music, dance, and the handling of horses and arms.

The grouping of music with sport and arms may seem unusual given the disciplinary boundaries that placed music in the scholastic quadrivium of mathematical

3. Motley, *Becoming a French Aristocrat,* chaps. 1 and 2.

4. Montaigne, *Essais,* 1:165. "Endurcissez le à la sueur et au froid, au vent, au soleil et aux hazards qu'il luy faut mespriser; ostez-luy toute mollesse et delicatesse au vestir et coucher, au manger et au boire; accoustumez le à tout. Que ce ne soit pas un beau garçon et dameret, mais un garçon vert et vigoureux." The essay was written for Diane de Foix, Countess of Gurson, during her pregnancy. The Foix were among France's most distinguished courtiers, clergymen, and warriors, thus the essay delineates the proper education of the archetypal *noble d'épée.*

sciences. But the formulation came from the Greeks and enjoyed significant currency in the sixteenth century. In both the *Laws* (II) and the *Republic* (III), Plato treated gymnastics and music side by side, maintaining that the best training in gymnastics would resemble training in music. Like other physical activities proper to a gentleman, music developed adroitness and agility.[5] Pontus de Tyard put it well in his music treatise, *Solitaire second,* when he said that just as wrestling, running, jumping, and fencing were essential for physical well-being, music exercised the soul.[6] Additionally, music shared with sport a nonlaborious quality that was socially distinctive. Since, by definition, noblemen avoided all manual labor with a commercial end, music, dancing, games, and sport could fill leisure hours with active recreation that would not compromise social standing. But they were more than pastimes. Rather, honest exercise was the foundation of an education that began in the body.

Montaigne's "De l'institution des enfans" delineated an education that could not easily be had in France. Short of hiring a battery of private instructors, the best alternative remained riding schools, most of which were in Italy. François de La Noue, whose *Discours politiques et militaires* was explicitly aimed at nobles, concurred that Italian schools offered training in many *exercices honnestes,* though he warned parents to wait until their sons had achieved some fundament of piety and love of honesty before sending them to Italy, where the lures of a thousand voluptuous pleasures awaited them.[7] In no case should young men stay longer than two years. Even a righteous Protestant such as La Noue had to admit that Italian training could not be beat, whatever the moral dangers. With this advice, he acknowledged a longstanding practice among French nobles, who saw in an Italian education the opportunity to pick up a useful vernacular, socialize with other elites, and learn the martial arts required to fulfill the *raison d'être* of their class. Frenchmen with means hied to riding schools catering to foreign clientele, the most famous of which were located in Naples.[8]

What France needed was military academies of her own. La Noue proposed es-

---

5. See Montaigne, *Essais,* 2:642.

6. Tyard, *Solitaire second,* 75. "Aussi sçavez-vous qu'ainsi que la luite, la course, le saut, l'escrime et tels autres exercices estoient receuz pour necessaire entretien de la santé corporelle, la Musique servoit d'exercice pour reduire l'ame en une parfaite temperie de bonnes, loüables, vertueuses moeurs."

7. François de La Noue, *Discours politiques et militaires,* 147–148.

8. On the continuing trend see, for example, Alexandre de Pontaymery, *L'academie ou Institution de la Noblesse Françoise,* in *Les oeuvres* (Paris: Jean Richer, 1599), fols. 3v–6r; Salomon de La Broue, *Le cavalerice françois,* 3 vols. (Paris: Abel l'Angelier, 1602), 1:2; Thomas Pelletier, *La nourriture de la noblesse* (Paris: veufve Mamert Patisson, 1604), fol. 96r–v; and Antoine de Pluvinel, *L'instruction du roy en l'exercice de monter à cheval,* ed. René de Menou (Paris: Michel Nivelle, 1625), 195, all of whom dislike the practice.

tablishing academies in all the provincial capitals, beginning with Paris, Lyons, Bordeaux, and Angers, as well as setting up academies in four royal chateaux (Fontainebleau, Moulins, Plessis de Tours, and Cognac).[9] Henry IV apparently took the advice of his general to heart, for immediately upon his conquest of Paris, he established France's first academy for nobles in the capital under the direction of his most accomplished *escuyer,* Antoine de Pluvinel. That the academy opened during the civil wars and only months after Henry IV retook the capital is no coincidence, for by overseeing the education of the nobility more directly, the academy promised to foster cooperation from a class whose traditions promoted political independence, private violence, and aristocratic rebellion. We should see Pluvinel's academy as more than a riding school: it was the flagship of a whole series of educational initiatives begun by Henry, which subsequently included the restructuring of the University of Paris, the establishment of the College of Nobles at La Flèche, and the opening of provincial colleges overseen by the Jesuits, who answered directly to the king. Indeed, the conditions that precipitated the establishment of the academy and shaped its program bring into focus the political relevance of the educational enterprise more generally. After all, the dressage taught by Pluvinel made the academy attractive to nobles, for whom horsemanship was virtually synonymous with the military calling of their estate. To the academy came the sons of a small number of tremendously powerful French aristocrats—precisely the young nobles who might pose the greatest threat to the monarchy in the future; there they received an education that not only furthered their own aims, but also helped align their behavior with that favored at court.

Pluvinel's academy opened in 1594.[10] It accepted only nobles as students and aimed, as Pluvinel said, "to render them capable of serving their Prince well, whether in peacetime or in war."[11] Richelieu and the Duke of Vendôme both studied there, and its program mirrored the princely education Louis XIII received at

9. François de La Noue, *Discours politiques et militaires,* 153. La Noue's advice was seconded with a detailed plan published by Samson de Saint-Germain, sieur de Juvigny, *Advis de l'établissement de quatre académies en la France* (Rouen: Raphaël du Petit Val, 1596).

10. On Pluvinel's academy and subsequent ones modeled on it see Albert Folly, "Les académies d'armes, XVIe et XVIIe siècles," *Bulletin de la Société du VIe arrondissement du Paris* 2 (1899): 163–171; Maurice Dumolin, "Les académies parisiennes d'équitation," *Bulletin de la Société Archéologique, Historique et Artistique* 26 (1925): 417–428; Commandant de La Roche, *Les académies militaires sous l'Ancien Régime* (Paris: Auguste Picard, 1929); Schalk, *From Valor to Pedigree,* chap. 8; and Motley, *Becoming a French Aristocrat,* chap. 3.

11. Pluvinel, *Instruction du roy,* 200. Pluvinel outlines the program of study, concluding that it aims, "bref, tascher par ce moyen de les rendre capables de bien servir leur Prince, soit en paix, soit en guerre."

the hands of many of the same masters.[12] Unsurprisingly, given its royal subventions, Pluvinel's school stood in the shadow of the Louvre, with its main building in the rue Saint-Honoré and its dressage court extending all the way to the Château des Tuileries.[13] A second academy, run by a sieur de Benjamin, quickly opened near the site of the Palais Royal, and in 1620 Pluvinel called for the establishment of *académies d'armes* in Paris, Lyons, Bordeaux, and Tours or Poitiers.[14] By the end of the reign of Louis XIII, Paris boasted five more military academies.[15] With peace and the new expenditures on education, Frenchmen no longer had to travel to Italian schools, and the French academies even began to attract foreigners to Paris.[16] It is worth noting that they sprang up on the same social and physical terrain as other institutions of civility surrounding the court: for example, Catherine de Vivonne held her salon in the Hôtel de Rambouillet, just a stone's throw from Pluvinel's academy. Growing cheek by jowl, military academies and the "schools" of politeness and belles lettres conducted in the salons shared overlapping constituencies, and one might fairly say that both aimed to banish the aggression of the war years.

The academy undertook to teach more than just horsemanship, yet, in keeping with older traditions, riding remained the foundation of aristocratic education. Here riding was a *profession,* at once basic training for a military career and the expression of a social order that had for centuries maintained horsemanship as a *marque de noblesse.* After a stay usually lasting one or two years and beginning no earlier than age fourteen, students sallied forth into society, the army, or both, where they commanded respect first and foremost through their noble bearing. One need only consider how regularly poise in the saddle was taken as a measure of the man to see that a good reputation demanded real training in equitation.[17]

12. On Louis XIII see Héroard, *Journal,* esp. 1:130–188. An excellent contemporary example is given in Roy Strong, *Henry, Prince of Wales, and England's Lost Renaissance* (London: Thames & Hudson, 1986).

13. Folly, "Académies d'armes," 164.

14. Pluvinel, *Instruction du roy,* 196–199. Others also called for the establishment of military academies, for which see Pontaymery, *Academie,* and David de Fleurance-Rivault, *Le dessein d'une academie,et de l'introduction d'icelle en la Cour* (Paris: Pierre le Court en l'Imprimerie de Robert Estienne, 1612). Pelletier tells us that the riding academy of sieur de Benjamin was organized according to Pluvinel's model (*Nourriture de la noblesse,* fol. 96r–v) and that others followed in the capital.

15. Folly, "Académies d'armes," 164.

16. René de Menou, seigneur de Charnizay, *La practique du cavalier,* 4th ed. (Paris: Jean Corrozet, 1629), 9–10.

17. Consider the following—typical—remarks on Henry de Montmorency: "Il estoit brave, riche, galant, liberal, dansoit bien, estoit bien à cheval, et avoit tousjours des gens d'esprit à ses gages" (2:307); on the Comte de Cramail, "Il a tousjours esté galant: il estoit propre, dansoit bien, et estoit bien à cheval" (1:507); and on Bassompierre, "Il n'a jamais bien dansé: il n'estoit pas mesme trop bien

Pluvinel's students epitomized not just stylishness on a horse, but attention to style more generally: other masters might be able to train the horse, but only Pluvinel could train the man.[18] Moreover, this style was definitively French. Pluvinel declared that "for civility and for morals, foreign schools are not appropriate for French spirits,"[19] echoing the remarks Thomas Pelletier had made in *La nourriture de la noblesse* (1604):

> It is therefore not an Italian that one has for instruction. It is a French Gentleman that one wishes, the mores, the manner, the grace truly French and not foreign. It is therefore only in France that [the student] will learn to ride, to joust for rings, to dance, to dress according to our style without being judged upon his return more Italian than French, losing in this the time and money that one should spend to erase the tablet and render it with a French air.[20]

By the late seventeenth century, the precedence of style for its own sake gave rise to a culture in which style became the essence of the nobleman—"le style c'est l'homme" was not far off. Much of this book concerns the origins of that transition, which turned French *gens d'armes* (literally, "men of arms") into *gentilshommes,* but it is important to remember that in the early seventeenth century, the education of the nobility was designed with the military profession in mind. Even as politeness, civil conversation, erudition, and skill in social dance began to be expected of nobles and instruction in arms taught some gentle lessons such as attention to detail and self-restraint, fencing, for example, retained its violence, noble careers their bloodshed, and courage its virtue.[21] Pelletier's French gentleman is a knight with grace, manners, and morals. Far from setting the art of war at odds with the profession of style, the military academies stylized martial arts, turning them into ag-

---

à cheval; il avoit quelque chose de grossier; il n'estoit pas trop bien desnoüé" (3:339; all in Tallemant des Réaux, *Historiettes*).

18. Menou, *Practique du cavalier,* 11.

19. Pluvinel, *Instruction du roy,* 195. "Pour la civilité & pour les moeurs l'escole estrangere n'est pas propre aux esprits François."

20. Pelletier, *Nourriture de la noblesse,* fol. 96r–v. "La France ne cede maintenant à l'Italie pour bien eslever nostre Noblesse à tous les exercices dont autrefois elle seule se glorifioit par dessus les autres nations de l'Europe. . . . Ce n'est donc point un Italien qu'on a à nourrir. C'est un Gentilhomme François qu'on desire avoir, les moeurs, la façon, la grace, vrayement à la Françoise & non à l'estrangere. Ce sera donc en la seule France qu'il apprendra à estre à cheval, à courre la bague, à danser, à s'habiller à nostre mode sans estre jugé à son retour plus Italien que François, espargnant en cela le temps & la depense qu'il faudroit faire derechef pour donner de l'eponge sur le tableau & luy rendre l'air de la France."

21. On the merits accorded to violence, particularly in the sphere of the academies, see Brioist, Drévillon, and Serna, *Croiser le fer,* chap. 2, esp. their discussion of the passions at 120–128.

gressive yet mannered sets of behaviors, the acquisition of which doubled as training to stylization itself.

Closely related were the social strategies students came into contact with at the academy. Like the networks of pages, *enfans d'honneur,* and ladies-in-waiting who developed personal fidelities at court from a young age, the academy nurtured political ties, familiarized students with their place in high society, and made them spectators to the social revolutions of *le monde.* As Orest Ranum has shown, the effectiveness of Richelieu's ministry rested not on institutional reform, but on the elaboration of favors, deference and preference, hierarchy, and the elevation of "creatures" to key positions.[22] Loyalty intertwined with authority in clientage systems that many young nobles first entered when they came to Paris to learn to ride. This was another way military academies played a role in the civilizing process, equipping nobles with modes of negotiation that could supplement or replace armed aggression with clientalism, politics, and courtiership.

PLUVINEL's academy was declared the "temple to virtue" that France so needed. In a telling turn of phrase that reveals the synonymy between grace and morality, Alexandre de Pontaymery called the school a "Parnassus to the Muses," explaining that Pluvinel

> not only instruct[s] the Gentleman in the profession of riding, but in the practice of good morals—without which all sciences are only vanity. Also, he who sees the students sees the deportment of Angels and the living images of celestial perfection. Oh happy Nobility! for whom the heavens have caused such an able governor to be born: If one has exerted himself at riding you have gymnastics, fencing, and dance, all under people who the aforesaid Monsieur happily knew to choose, and who are without argument the foremost in their art. You also have Mathematics, painting, and the lute under the most excellent masters one could desire.[23]

According to the schedule of classes outlined in Pluvinel's *L'instruction du roy,* students studied riding and running at the ring each morning, and fencing, dancing,

22. See Orest Ranum, "Courtesy, Absolutism, and the Rise of the French State, 1630–1660," *Journal of Modern History* 52 (1980): 426–451.

23. Pontaymery, *Academie,* fol. 3r. "Sieur de Pluvinel, qui nous fait voir en gros tous les exercices que l'Italie monstre en detail, ayant basty un Parnasse aux Muses, & dressé un temple à la vertu: car à la verité dire, il n'instruit pas seulement le Gentil-homme en la profession du maniage, mais en la pratique des bonnes moeurs, sans lesquelles toutes sciences ne sont que vanité: aussi qui voit ses escholiers, il voit le maintien des Anges & les vives images des celestes perfections. O heureuse Noblesse! pour qui le ciel a fait naistre un tant fortable gouverneur: S'est on exercé au maniage? vous avez le voltigement, l'escrime & la danse, le tout sous des personnages que ledict sieur a sçeu heureusement choisir, & qui sont hors de controverse les premiers en leur art. Vous y avez encore les Mathematiques, la peinture & le lut sous les plus excellens maistres que l'on puisse desirer."

tumbling, and mathematics on Monday, Wednesday, Friday, and Saturday afternoons, when they likely received their instruction in drawing and music as well. On Tuesdays and Thursdays, a "man of letters" addressed the assembled students on moral virtue, politics, and command.[24] This academic program is consistent with that delineated in 1596 by Samson de Saint-Germain, sieur de Juvigny, *gentilhomme ordinaire de la chambre du roy,* captain of the king's light horse, and surely a familiar of Pluvinel's academy. Juvigny advised Henry IV to establish four royal academies in the provinces and made explicit recommendations for the faculty and schedule of classes. Each would employ a principal, an *escuyer* (who was also a gentleman), a master of arms, a dancer, a tumbler, a lutenist, a mathematician, a painter, and a preceptor for letters. Mornings were given over to riding, the handling of arms, and dance; afternoons were reserved for mathematics, the lute, and drawing; and the students spent the hour or two before supper vaulting. Two times a week, the preceptor gave lessons on ancient and modern history, politics, and illustrious personages.[25]

Music, as we shall see, fit generally into Pluvinel's method of dressage, and it was of course central to dance as well. Here, however, we should observe the importance accorded to music as a self-standing discipline with special instructors and separate time for study, something true not just for Pluvinel's academy, but at subsequent academies such as that proposed for Aix-en-Provence in 1611, for which detailed archival records survive.[26] Music was not the first priority in Aix, for after starting off with a permanent faculty of riding and fencing instructors supplemented by individual *maîtres* in arithmetic, vaulting, and dance, only in 1616 was music declared a main subject of study. In the meantime, singing lessons and lessons on the lute could have been covered by adjunct faculty or private tutors in the often tremendous entourages of governors, pages, lackeys, and *valets de chambre* that nobles brought with them to the academy.[27] Ultimately, though, musical facility was the

24. Pluvinel, *Instruction du roy,* 200. Pluvinel describes the four new academies he would like the king to establish: "Sire, Toute la matinee seroit employee pour l'exercice de la Cavallerie, & pour courre la bague l'apresdinee, sçavoir le Lundy, Mercredy, Vendredy & Samedy, pour les exercices de tirer des armes, dancer, voltiger, & les Mathematiques. Et pour les deux autres, sçavoir le Mardy & le Jeudy l'apresdinee, il seroit à propos que celuy que cy-dessus j'ay qualifié homme de lettres, traitast en presence de toute ceste jeunesse assemblee."

25. Saint-Germain, *Advis de l'établissement de quatre académies,* 20–21, 23.

26. O^1 915 (181–201), Archives nationales, esp. 183, 184, 187, and 188. They are analyzed in Schalk, *From Valor to Pedigree,* 188–192.

27. On the households of aristocratic students, see Motley, *Becoming a French Aristocrat,* 155–156. Private lessons on instruments would have been consistent with the possibilities at boarding colleges. See, for example, Thomas Culley, "Musical Activity in Some Sixteenth-Century Jesuit Colleges, with Special Reference to the Venerable English College in Rome from 1579 to 1589," *Analecta Musicologica*

academic ideal, and throughout the seventeenth century, letters patent and other official correspondence relating to the military academies consistently specify that there be a *jouer du luth* on the faculty and *la musique* as part of the curriculum.[28]

The importance of music at Pluvinel's academy suggests a somewhat surprising precursor in the short-lived Academy of Poetry and Music, which had been established by the poet Jean-Antoine de Baïf and the musician Joachim Thibault de Courville in 1570. Though usually taken to have been primarily concerned with the development of *musique mesurée à l'antique,* Baïf's academy was not just literary or musical in a limited sense, but launched an educational program strikingly similar to Pluvinel's. Masters of classical languages, music, and poetry were joined by others specializing in geography, mathematics, and painting, and by "military prefects who taught all those things which are useful for military discipline and for the good of the body."[29] The complex of subjects, the equal attention to body and soul, and the emphasis on military discipline reveal Baïf's academy as a possible model for Pluvinel, who was appointed *premier escuyer* to the future Henry III in the early 1570s and likely knew of its activities. Moreover, the moral claims Baïf and Courville made for musical education suggest one reason Pluvinel's students studied the lute and, likewise, why contemporaries described his academy as a "temple of virtue."

The positions teaching music at the academies seem invariably to have been filled by lutenists. According to Pelletier, playing the lute was counted among the skills of a gentleman, and many fathers believed their sons had learned nothing if they did not take it up.[30] It was an instrument suited to the needs of amateurs, for beginners could produce an acceptable tone on it in just a few weeks without a developed touch, and if the frets were correctly placed (and the strings properly tuned) they allowed the student a good deal of latitude with finger placement on the left hand without marring the harmonies. Lute tablature made learning to read music relatively easy, for it showed the fingerings directly on a graphic representation of the instrument's neck, whereas students of other instruments such as the viol had to "translate" staff notation into fingerings on their own.

---

19 (1980): 1–29, at 4, where Culley relates that students at the German College in Rome "who will know something about singing or about playing will be able to practice" during recreation time.

28. As, for example, at the academy in Toulouse, which was established in 1598 (O[1] 917 [211], Archives nationales), and a Parisian academy proposed in 1669, at which during the hour after lunch "on enseignera à joüer toutes sortes d'instrumens et à chanter" (O[1] 915 [32], Archives nationales).

29. The quote comes from Marin Mersenne's extremely important account of the academy, translated and discussed in Yates, *French Academies,* 23–25. Although she marshaled the sources to do so, Yates never came to the conclusion that Pluvinel's academy was in some sense a direct continuation of Baïf's.

30. Pelletier, *Nourriture de la noblesse,* fols. 88v–89r.

Images of lute-playing cavaliers abound from the period, testifying to the lute's importance as a noble accoutrement (see fig. 2.1).[31] The nobleman in this engraving, clad in fashionable riding boots and spurs, tunes up a ten-course lute of optimal size for the court *airs* that were in vogue, while the text at the bottom of the page suggests the sort of lyrics he might sing in duet with a young lady. The song's articulation of unrequited love encourages *fin amor,* suggestively linking singing, courtly love, and the delicate musical idioms that made *airs de cour* such apt expressions of refinement. Far from negating manliness—and in no ways at odds with the popularity of the lute as an instrument for ladies—such images reinforced a masculine ideal actively cultivated in high society. Louis XIII began lute lessons with Robert Ballard at the age of ten (after having begged to learn when he was as young as six), and when Richelieu began to study with Ennemond Gautier (who taught the queen mother), it created a real fad at court for the instrument. As Tallemant des Réaux recalled, "voylà tout le monde à jouer du luth."[32]

Airs de cour cultivated restrained melodic gestures that fit well with the understatement prescribed by gentle behavior (and the abilities of amateurs), and polyphonic *airs* traded in conservative chordal harmonies that translated easily from part-song to solo song.[33] In many ways, the lute versions had an appeal that the polyphonic ones did not, for the expression of first-person sentiments by a single singer seemed more touching than polyphony and gave singers a chance to shine in the incessantly theatrical world of upper-class society. The royal printer of music, Pierre Ballard, furnished amateurs with *airs de cour* arranged for voice and lute, a repertory that burgeoned dramatically beginning in 1608 with the first book of *Airs de différents auteurs,* arranged by Gabriel Bataille. The series went to sixteen volumes by 1643, its thousand-plus *airs* sustaining a taste for lute songs from the court that greatly benefited the careers of the court composers Pierre Guédron, Antoine Boësset, and Etienne Moulinié. The dramatic acceleration in the production of *airs* in the first decades of the century and the swiftness with which they were arranged for voice and lute goes hand in hand with the deluge of prescriptions that young men learn the lute, the lute instruction at military academies, and the popularity of *airs* for lute and voice at salons and court.

31. On the lute and aristocrats see Michel Brenet [Marie Bobillier], *Notes sur l'histoire du luth en France* (Geneva: Minkoff Reprint, 1973), esp. 50–53. Consider, too, the remarks of Mersenne: "Le luth a pris un tel ascendant sur les autres instrumens à chorde, soit que les honnestes gens luy ayent donné cet avantage, ou qu'il l'aye acquis par son excellence, & par sa perfection, que l'on ne fait quasi nul estat des autres." Mersenne, *Harmonie universelle,* 3:56.

32. Tallemant des Réaux, *Historiettes,* 2:8.

33. For more on this repertory, see James R. Anthony, *French Baroque Music from Beaujoyeulx to Rameau* (New York: Norton, 1978), and Georgie Durosoir, *L'air de cour en France, 1571–1655* (Liège: Mardaga, 1991).

FIGURE 2.1. Abraham Bosse, *Cavalier Playing the Lute.* Bibliothèque nationale de France, Cabinet des estampes, Kd 3 tom. 6

Airs from *ballets de cour* proved especially popular with a public anxious to keep up on the latest goings-on in high society, for ballets were the central cultural events of each season's social calendar. Songs from ballets appeared in Pierre Ballard's series of *airs* for lute and voice in noticeably increasing quantities, and Robert Ballard's two books of lute solos (from 1611 and 1614) featured over sixty *airs* and dances from some twenty-one different ballets.[34] One of the most popular *airs* of the age was Guédron's "Est-ce Mars le grand dieu des alarmes," a song performed at the *Ballet pour Madame* in 1613 and published the same year in two versions, one for four voices and the other intabulated by Bataille for lute and voice (see ex. 2.1).[35] The vocal line begins in a decidedly martial fashion in duple time with attention-getting repetitions of the first note, leaps in the vocal line, a widening range, and emphasis on the fifth and first scale degrees. Guédron captures in a stroke the sound of trumpets and the ambition of the cavalier. But the melody relaxes rhythmically at the end of the first strain with its slow steady rise to the third scale degree, and the second strain goes on to counter the muscularity of the opening with a feint toward the softer sixth scale degree, conjunct motion, and falling lines at the end. These oppositions play up the catchy reversals of the text, which continue in subsequent stanzas ("Is this Mars that I see? . . . judging from his arms, yes . . . but I see the look of Cupid in his eyes . . . no, perhaps it is the Sun . . ." and so forth). While the piece is far more metrical than many of Guédron's *airs,* the range of note values here gives some sense of his keen ear for declamation, and the melodic contours hint at his sensitivity to the affective qualities of large and small intervals and various scale degrees. The harmonic language is restricted and there is no counterpoint to speak of in this melody-dominated song, which communicates expressively without ostentation and the *ports de voix* and diminutions frequently added to *airs.*[36]

Bataille's intabulation provides a light accompaniment that would have appealed to beginners. The C-chords and impressive low C in the bass that open the piece make for a striking beginning with little effort, and the same sequence punctuates

---

34. See Robert Ballard, [*Premier livre de luth*] (Paris, 1611), ed. André Souris and Sylvie Spycket as *Premier livre (1611),* with an introduction and concordances by Monique Rollin, 2nd ed. (Paris: Centre National de la Recherche Scientifique, 1976), xi–xvii.

35. Pierre Guédron, *Second livre d'airs de cour à 4 et 5 parties* (Paris: Pierre Ballard, 1613), fols. 8v–9r; Gabriel Bataille, *Airs de différents auteurs, mis en tablature de luth,* 6 vols. (Paris: Pierre Ballard, 1608–1615), book 4, fols. 6v–7r.

36. On the performance practice of seventeenth-century *airs* see Anthony, *French Baroque Music,* 345–358; Durosoir, *Air de cour,* 305–325; and Théodore Gérold, *L'art du chant en France au XVIIe siècle* (Strasbourg: Université de Strasbourg, 1921). Some *airs* for lute and voice add a few ornaments to the vocal part, which suggests both that these arrangements—rather than the polyphonic *airs*—were aimed at amateurs (who would sing these ornaments rather than devising their own) and that the kinds of elaborate *doubles* shown by Mersenne were not the norm for amateur performance.

the beginning and ending of each strain like a familiar friend (one need only look at the repeating fingerings in the intabulation to get a sense of how elementary it is). Slow chord changes in the first strain (from C to F) and their gentle compression in the second strain leave relatively little new to discover in the piece, and with a couple of other root-position chords in hand (those on D and G), any novice could have gotten through the song with *sprezzatura*. That it was a favorite with amateur lutenists is suggested by its inclusion in Nicolas Vallet's two-volume *Secretum Musarum* (Amsterdam, 1615 and 1616). This collection is of particular interest because Vallet so explicitly aimed it at beginners anxious to learn, as he put it, a quick way to play the lute well. His two "Courantes de Mars" rework Guédron's duple-meter tune into lilting dances in triple time with ornamented repeats, and although they are more difficult than Bataille's accompaniment, they are certainly much easier than the solo lute pieces published by Robert Ballard in the same years.[37] Plus, Vallet included a beginner's crutch that the French prints did not: his tablature shows not just where to press down strings with the left hand and pluck with the right, but exactly which fingers to use to do so. With the "secret" of fingering revealed, amateurs could make the leap from *airs* to solo repertory.[38] Vallet's were not terribly ambitious pieces, but they were recognizably from the same general repertory as those published by Ballard. Armed with a *courante* or two and having internalized Vallet's tips on how to avoid clanging sounds, shaking, and grimaces, a cavalier could toss them off in polite company as a bit of light entertainment.

The lute satisfied those who wished to learn a little music quickly, and *airs de cour* fed the gentle passions of men and women who wanted sweet music and interesting poems to pluck and sing. Nonetheless, many considered the lute to be a frivolous enterprise compared to swordplay. Thomas Pelletier, who played the lute himself, laments in *La nourriture de la noblesse* that "there are those who find it neither necessary nor honorable that they should wish that their children had employed at this study half the time it takes to acquire only some passable ability on this instrument." Detractors of the lute felt that hours playing it led to a reclusive "humeur phantasque" that inclined the lutenist to prefer the company of the instrument above that of friends.[39] Such potential bad effects led Antoine Mathé de Laval, a

37. One need only compare Ballard's tenth Angélique *courante* with Vallet's reduction of it to get a sense of the relative challenge.

38. It should be said, though, that many of Vallet's fingerings are inconsistent and irrational.

39. Pelletier, *Nourriture de la noblesse*, fols. 88v–89r. "Le jouër de luth est conté entre les exercices que doit apprendre un Gentilhomme. Et tel pere y a qui ne croit point que son fils ait rien apprins qui vailles s'il n'a ceste partie. Il s'en trouve d'autres qui de contraire advis ne l'estiment pas si necessaire ny honorable qu'ils voulussent que leur enfant y eust employé la moitié du temps qu'il faut pour acquerir seulement quelque passable suffisance de cest instrument. Leur raison est que ces heures là se peuvent mieux donner ailleurs, & que le plus souvent on voit ceux qui s'y affectionnent estre

Example 2.1. Pierre Guédron, "Est-ce Mars," arr. Gabriel Bataille. In *Airs de différents auteurs, mis en tablature de luth, quatriesme livre* (1613)

D'eſtre auſſi Cupidon, il me ſemble
Qu'il n'a pas
Tant de grace & de beautés enſemble,
Ny d'apas:
C'eſt pluſtot vn ſoleil radieux
Que Cupidon qui n'a point d'yeux.

Le Soleil n'a pas tant de lumiere,
Et ne peut
Prendre l'ame d'vn corps priſonniere
Quand il veut:
Et ces yeux d'Amour meſmes vainqueurs
Prennent les ames & les cœurs.

Inſencé! maintenant je m'auiſe
Que ces yeux,
Sont les yeux de la belle Marphiſe,
Chere aux Dieux:
Sœur de Mars, fille d'vn grand ſoleil
Qui luit icy bas ſans pareil.

Grand ſoleil qui reluis à la France,
Et qui fais
Sous la juſte & la douce aſſeurance
De la paix,
Que les lys ſous ton nom fleuriſſant,
N'iront plus de peur palliſſant,

Grande Reyne de qui l'on adore
La vertu,
Vertu dont la beauté ſe décore
Puiſſe-tu,
Fauorite du Ciel en tout temps,
Voir toujours tes deſirs contens.

young acquaintance of Montaigne's, to alert readers of his *Desseins de professions nobles et publiques* to the case of a *gentilhomme d'honneur* whose childhood indulgences in the lute, music, and poetry kept him from useful service to himself, his friends, and the public.[40] Laval worried that young spirits would lose themselves in a mire of laziness brought on by singing, rhyming, dancing, and playing, to the detriment of their future careers.

Clearly, the lute provoked that age-old question of whether or not music was wholesome. Plato argued in the *Republic* that variety and complexity in music produced licentiousness (in gymnastics, he claimed that it produced disease). Those who mixed rhythms carelessly or mixed modes "panharmonically" were responsible for music's ill effects.[41] On the other hand, he believed that simple songs, like simple exercises, promoted good health by nourishing beautiful rhythms in the soul. Moreover, music was the best means of moral education because it was the least representative of the arts: the plastic arts distracted the soul with superfluous aesthetic pleasures, but music was pure motion that spoke directly to the soul and aligned it with the motion of the universe (*Republic* 3.401d; *Timaeus* 47). And in that other great classical textbook of moral and political life, the *Politics* (8.1340a12–b19), Aristotle attributed to music the ability to improve character by representing in rhythms and melodies the true natures of anger and mildness, and also of courage and temperance and all their opposites and the other moral qualities. By experiencing these states vicariously through music, Aristotle argued, the student would be prepared to judge their ethical natures later in reality.[42]

The balance Aristotle reasoned out between amusement and moral instruction was not unlike the one struck by so many late-sixteenth-century pedagogues, who classified music-making as an *occupation honnête* for idle young minds. We have already heard from François de La Noue, who cited Aristotle when recommending music for young noblemen.[43] Martin Coignet, in his *Instruction aux princes pour*

volontiers d'une humeur phantasque & resveuse, preferant quelquesfois l'entretien de leur lhut [*sic*] à celuy d'un amy, qui viendra chez eux, ou qu'ils devroient aller trouver. Je ne mets point en conte que je l'ay veu quitter par ce que la taille s'y gastoit. Or de repartir que le luth les faict voir & honorer en compagnie, je tiens qu'au contraire il les y fait plutost mespriser, par ce qu'un Gentilhomme de bien bon lieu doit prendre de plaisir d'autruy sans que luy mesme serve de sujet d'en donner aux autres."

40. Antoine Mathé de Laval, *Desseins de professions nobles et publiques* (Paris: Abel l'Angelier, 1605), fol. 5v.

41. Plato, *Republic* 3.399c–400a, 1:249–251.

42. Aristotle, *Politics,* trans. H. Rackham, Loeb Classical Library (London: William Heinemann; Cambridge, Mass.: Harvard University Press, 1967), 656–661.

43. François de La Noue, *Discours politiques et militaires,* 154.

garder la foy promise (1584), likewise saw music as a moral alternative to the cards, dice, gluttony, drunkenness, and whoring that tempted young gentlemen, contrasting music's positive effect on the soul with the shadow of unbridled sexuality that threatened youths with too much time on their hands.[44] So, too, did Pierre de La Primaudaye specifically recommend that students be taught music after puberty.[45] As private tutors well knew, gambling and visiting prostitutes seriously enticed the boisterous teenagers studying at the academies, as did duels.[46] It seems rather optimistic to imagine that the lute could temper the excesses of youth, but at least *airs de cour* directed those who played them toward a social sphere in which moderation and self-restraint were valued.

On a philosophical level, it was the very nature of the lute's design that garnered praise. Many attributed its moral effects to the mathematical properties of the harmonious chords one could so readily play on it. Pelletier, for instance, distinguished between exercises that concentrate on the body alone and the more "serious, useful, and necessary" ones that occupied the body and mind together. To the first class he assigned tennis and other sorts of ball games, while in the second we find riding, dancing, bearing arms, playing the lute, portraiture, and mathematics—precisely the same list Pontaymery gave in his panegyric of Pluvinel's academy, and a list that is, significantly, in almost the same order.[47] Pontaymery and Pelletier both grouped mathematics, painting (representative of geometry), and the lute together, assigning music to its proper place as the culmination of the quadrivium of mathematical sciences. Music was a practical and *practicable* mathematics that occupied both body and intellect, bridging the sensible realm of sound and the speculative realm of number. Tuning and strumming the lute taught harmony, internalizing musical proportions that governed both physical and moral action whether the band was playing or not.

44. Martin Coignet, *Instruction aux princes pour garder la foy promise* (Paris: Jacques du Puys, 1584), trans. Sir Edward Hoby as *Politique Discourses upon Truth and Lying, An Instruction to Princes to Keep Their faith and Promise containing the Summe of Christian and Morall Philosophie, and the deutie of a good man in sundrie politique discourses upon the truth and lying* (London: Ralfe Newberie, 1586), 224.

45. La Primaudaye, *Academie françoise,* fol. 267v.

46. See Motley, *Becoming a French Aristocrat,* 162.

47. Pelletier, *Nourriture de la noblesse,* fols. 84v–85r. Nicolas Faret groups his exercises in similar fashion: dressage, jousting, lute, guitar, hunting, dance, tennis, throwing, jumping, swimming, and target practice (*Honneste homme,* 16–17); in a section on letters, he groups painting and music with poetry (ibid., 31).

THE PYTHAGOREAN BODY

Music, reduced to its numerical properties, constituted a mathematical framework through which a broad array of matter might be construed. From the simple ratios of the harmonic series and the proportions created among them could be generated a universal mathematics that explained everything from the motions of the planets to the proper timing and disposition of a *capriole.* In this way, mathematics traversed the sensible and suprasensible realms in a calculation of vast import that explained the cohesiveness of the universe. Such was the foundation for the acrobat Arcangelo Tuccaro's claim that the dancers in courtly balls imitated the progress of the stars across the sky and the fencer Girard Thibault's assertion that his method moved the body through a series of positions that matched the changing disposition of the heavens and the music created by its motions.[48] Far more than hollow metaphor, this musico-mathematical framework rationalized the correspondence between sounding music and physical motion, and between the music of the spheres and the motion of bodies on earth. And because this mathematics was qualitative, which is to say that some numbers and geometrical figures were "good" and others were "bad," it invested the world with moral import. Freighting the terms "measure" and "proportion" with the weight of social harmony, teachers continually sought to instill *juste proportion* in the actions of their students and in so doing turned lessons in fencing, dancing, and riding into lessons in virtue.

The math at work here is one that equated all manner of proportion with consonant musical intervals. This deserves some explanation. Originating with Pythagoras and Plato, it fed an entire discourse on the nature of the universe that was still relatively intact in 1600. The story—as related in Nicomachus—tells of how Pythagoras was passing a blacksmith's shop when he noticed a pleasing harmony being sounded by the smithies' hammers as they struck the anvil. Upon experimentation, he discovered that hammers in the relative weights of 2:1, 3:2, and 4:3 produced the consonant intervals of an octave, fifth, and fourth when struck together.[49] Having established a rational principle to explain the relationship between solid matter (the hammers) and something as ethereal as sound, Pythagoras then

48. Arcangelo Tuccaro, *Trois dialogues de l'exercice de sauter, et voltiger en l'air* (Paris: Claude de Monstr'oeil, 1599), fol. 36v; Girard Thibault, *Academie de l'espée . . . ou se demonstrent par reigles mathematiques sur le fondement d'un Cercle mysterieux la Theorie et pratique des vrais et jusqu'a present incognus secrets du maniement des armes a pied et a cheval,* 2nd ed. (Leiden: B & A Elzevier, 1628), 1. An edition was published in Paris in 1626.

49. Pythagoras's hammers do not actually produce the perfect intervals he claimed, but it was demonstrable that a plucked string, when divided successively in lengths bearing the ratios 2:1, 3:2, and 4:3, produced an octave, fifth, and fourth.

extended these simple ratios to other sorts of matter, such as the tension applied to strings of like construction. Finally, he projected these "constant" ratios onto the relationship between earth and heaven, using them to explain the unified structure of the cosmos.

Further elaborated by Plato, who accorded to the material of the world soul the rational numbers of music (*Timaeus* 35a–36d; *Republic* 10.616–620), this musical cosmology became a foundation of Greek science, one christianized by Saint Augustine (354–430) and Boethius (d. c. 524) and perpetuated as part of the medieval quadrivium.[50] From Boethius onward, scholastic philosophers accepted the Pythagorean description of universal order, with its basis in the ratios of musical harmony. The spheres in which the sun, planets, and stars were fixed turned according to celestial proportions, producing a music—the music of the spheres—of untold beauty and perfection. This *musica mundana,* as Boethius called it in his *De institutione musica,* could not be heard. The music that people could hear, *musica instrumentalis,* was but an imperfect copy, imitation, or dim reflection of this ideal music unavailable to the senses, but it worked according to the same mathematical principles.

Boethius's research into harmonics is exemplified by the association of his name with the monochord, a scientific instrument with moveable bridges used to divide a string according to various ratios. His studies had many practical results—for example, the improvement of the construction and tuning of instruments—but nonetheless, his project was at heart a mathematical inquiry into the speculative science of music. Even the monochord seems to have been valued most as a calculator.[51] In this tradition, the higher mathematical domain dealing with ratios and the proportions between them was conceptually bound to music and to the method of compounding ratios developed by mathematicians using monochords. Oscar Abdounur has shown this to be true in Euclid's *Elements,* where the compounding of ratios is restricted by the boundaries of the octave.[52] Even Isaac Newton limited himself to this musical terminology in the first edition of his *Principia Mathematica* (1687), a work that, incidentally, began as a proof of the relationships governing

50. A clear and substantial explanation of musical thought among the Greeks and its reception in the Middle Ages and Renaissance can be found in Ann E. Moyer, *Musica Scientia: Musical Scholarship in the Italian Renaissance* (Ithaca: Cornell University Press, 1992).

51. On the speculative and practical sciences of harmonics, see Jamie Croy Kassler, "Music as a Model in Early Science," *History of Science* 20 (1982): 103–139.

52. Oscar João Abdounur, "Ratios and Music in the Late Middle Ages: A Preliminary Survey," in proceedings of the conference "Musique et mathématique à la Renaissance," Tours, 17–19 February 2000, ed. Philippe Vendrix (Tours: Centre d'Études Supérieures de la Renaissance, forthcoming). I thank Professor Abdounur for sharing his paper with me in advance of publication.

celestial motions based on Kepler's third law of elliptical orbits and originally bore the title *De Motu Corporum, or On the Motions of Bodies in Orbit.*[53] Belief in the heavenly substance of musical ratios still drove mathematics at the close of the seventeenth century. Moreover, the ratios of musical intervals and the harmonic proportions among them still informed the mathematics of proportion. Proportions were, de facto, musical ones, which meant that all discourse invoking proportion referred either overtly or unwittingly to Pythagorean harmony. *Juste proportion* or the "just proportions" discovered in things of beauty implied musical *justesse.*

Pythagorean proportions were extended to other matter through geometry. Albrecht Dürer's *Four Books of the Proportions of Man,* first published in 1528 and translated into French in 1557, became a cornerstone in the arguments of French Neoplatonists, who wished to affirm that human bodies were themselves microcosms corresponding to celestial forms.[54] As Pontus de Tyard explained to readers of his music treatise, *Solitaire second,* "admirable is the symmetry with which the members of man are in ratio to one another. And though diverse may be their statures, it seems that in most one discovers the most common proportions (except in the monstrous), the proportions, I say, quadruple, triple, double, half, and others, from which you know musical consonances to be drawn."[55] Tyard called these bodily symmetries *musique humaine* and in so doing equated them with Boethius's *musica humana* or vocal music, finding in the body the Pythagorean concords that Dürer had made visible.

This discourse of a Pythagorean body was not limited to treatises on drawing and music theory. Indeed, the very practical method whereby Dürer laid bare the body and reduced it to a set of graspable geometrical principles appealed to those who wished to claim a similar perfection for riding or fencing. Using the body as a yardstick and geometry as a frame, one could extrapolate the formal outlines of noble practices. Thus geometric formulas of *juste proportion* were applied to physi-

---

53. On the "musical" mathematics of Newton's treatise, see ibid. and Edith Sylla, "Compounding Ratios: Bradwardine, Oresme, and the First Edition of Newton's *Principia,*" in *Transformation and Tradition in the Sciences: Essays in Honor of I. Bernard Cohen,* ed. Everett Mendelsohn (Cambridge: Cambridge University Press, 1984), 11–43. On Newton and music in general see Penelope Gouk, *Music, Science, and Natural Magic in Seventeenth-Century England* (New Haven: Yale University Press, 1999), chap. 7.

54. Albrecht Dürer, *Les quatre livres de la proportion des parties et pourtraicts des corps humains,* trans. Louis Meigret (Paris: Charles Perier, 1557).

55. Tyard, *Solitaire second,* 234. "Admirable est la Symmetrie avec laquelle les membres de l'homme sont rapportez l'un à l'autre. Et, bien que diverses soient les statures, semblables toutesfois en la plus part se treuvent les proportions (sinon aux monstreuses) plus communes; les proportions, dy-je, quadruples, triples, doubles, d'autant et demy et autres, desquelles vous sçavez les consonances Musicales estre tirées."

cal education, particularly in the royal academies. Pluvinel's *Maneige royal* opens with an engraving illustrating the rider's correct posture and the proper relationship between the rider's body and that of the horse (see fig. 2.2). Enumerated as a series of "rules," the precepts are illustrated with a square against which all might be measured.

Apart from the descriptions keyed to the engraving, the treatise offers no explanation of it, leaving the artist to square reality with mathematics. Indeed, it may have been the artist who took it upon himself to include this engraving at the head of a treatise elsewhere materially concerned with *juste proportion* in the shape of turns, the rhythm of gaits, and the graceful arcs drawn by lances as riders lowered them on their approach to the ring. Crispan de Pas taught drawing at Pluvinel's academy and, with his son, engraved the stunning plates illustrating Pluvinel's *Maneige royal* and *L'instruction du roy.* He clearly aimed to teach his pupils an art that, like Dürer's, revealed the essence of things. Unsurprisingly, the students were impatient with the usual slow methods of learning to draw, since they wished primarily to draft plans for battles and fortifications. But de Pas tells that he managed to teach them a clear and easy method of drawing "by means of six or seven mathematical figures."[56] The square in figure 2.2 surely reminded his students both of their lessons and of the universal mathematics he had attempted to convey in visual terms.

Nowhere was Pythagorean geometry imposed more dramatically than in the stupendous treatise of Girard Thibault, who instructed students in his so-called secret method of fencing by having them work within a "mysterious Circle" inscribed on the floor of the studio. All was revealed in his *Academie de l'espée . . . ou se demonstrent par reigles mathematiques sur le fondement d'un Cercle mysterieux la Theorie et pratique des vrais et jusqu'a present incognus secrets du maniement des armes a pied et a cheval* (1626), also engraved by de Pas and his son.[57] Thibault began his treatise in the 1610s, and in 1620 Louis XIII granted him a privilege for the work, which was first published in Paris. We should see Thibault's treatise as a complement to Pluvinel's *L'instruction du roy,* for both were written by masters connected to the court of Louis XIII, both were published in colossal formats in the same years, and both feature engravings by the same artists. Like the sumptuous *Maniement d'armes* (1607), engraved by Jacob de Gheyn at the request of John of Nassau, we should see

56. Yates, *French Academies,* 278 n. 5.

57. On the relative importance of Thibault's treatise and for some information about its publication see Anglo, *Martial Arts of Renaissance Europe.* This work is a valuable guide to the copious period sources on fencing. Brioist, Drévillon, and Serna, *Croiser le fer* (154–159), give a good explanation of how the geometry of the treatise relates to the broader "science" of fencing and physics that developed during the Age of Reason.

FIGURE 2.2. "Rules that the Rider Should Observe." Engraving by Crispan de Pas and Crispan de Pas le Jeune in Antoine de Pluvinel, *Le maneige royal* (1623). Bibliothèque nationale de France, Cabinet des estampes, Ke 7 folio

all of these princely treatises on the martial arts as using the technical excellence of copperplate engraving to further military aims. They are quite simply the bibliographic masterpieces of their age, testaments to the printed art of war.[58] As far as we know, Thibault's "academy" existed only in the two dimensions of his book, but it certainly shared with Gheyn's much-pirated book on weapons drills an effort to notate the warrior's motions pictorially. De Pas's virtuosic plates captured the moving bodies of the fencers upon the grid of Thibault's mysterious circle, step by step and with stunning exactitude (see fig. 2.3). Together with the text, the treatise presents a choreographic description vastly more detailed than anything to be found in contemporary dance manuals or accounts of ballets.[59]

58. Anglo, *Martial Arts of Renaissance Europe*, 75.

59. The same holds true, incidentally, for the descriptions of pyrrhics and equestrian ballets. As Sydney Anglo observes, "The earliest attempts at a systematic depiction of sequences of individual

FIGURE 2.3. The geometric foundation of Thibault's system. Engraving by Crispan de Pas and Crispan de Pas le Jeune in Girard Thibault, *Academie de l'espée* (1626). By permission of the British Library, shelfmark 64.i.3

The *Academie de l'espée* explicitly theorizes both the body of the fencer and its movements in terms of Pythagorean harmony, which makes it extraordinary for its time. Although fencing was closely related to music via dance (dancing masters and fencing masters were often one and the same), only a vivid imagination could bring harmonic theory to bear on the fencer's art. This was an academic project of the sort that took music, in Tyard's words, to be the "image of the whole encyclopedia" comprising world music and human music, and the universal philosophy that is formed of both.[60] Thibault's recourse to Pythagorean theory presents his method as a summa of all knowledge. Beginning with a Neoplatonic *laus musicae,* Thibault reproduces Dürer's illustration of the proportions of man and praises the harmony governing all of matter. "The body [of man] contains a compendium, not only of all the one sees here on earth, but also of that which is in the Heavens itself, representing the whole with a harmony so sweet, beautiful, and whole, and with an affinity of Numbers, Measures, and Weights so just, that relates so marvelously to the four elements and the influences of the Planets that another like it is not to be found."[61] He cites the hands as evidence of the body's perfection because they contain the "first and most excellent Numbers 1, 2, 3, 4, 5, 10, which so many Illustrious Philosophers, such as Pythagoras and Plato, . . . [used] to deduce the greatest mysteries of their doctrines."[62] Thibault's example was particularly apt at a time when lessons in arithmetic began with finger reckoning and choristers learned a solfège system still taught using the Guidonian hand as a device for memorizing the seven interlocking hexachords that made up the musical scale.[63]

---

movements are associated with various types of personal physical combat" (ibid., 44), arenas in which, I might add, the political and mortal stakes were highest.

60. Pontus de Tyard, *Discours philosophiques,* as cited in Yates, *French Academies,* 87.

61. Thibault, *Academie de l'espée,* 1. "Son corps contient un abbregé, non seulement de tout ce qu'on voit icy bas en terre, mais encores de ce qui est au Ciel mesme; representant le tout avec une harmonie, si douce, belle, & entiere, & avec une si juste convenance de Nombres, Mesures, & Poids, qui se rapportent si merveilleusement aux vertus des Quatre Elements, & aux influences des Planetes, qu'il ne s'en trouve nulle autre semblable."

62. Ibid. "Le tres-parfait nombre de Dix luy est continuellement representé devant les yeux, en son entier sur ses propres doigts; & derechef en deux moitiez egales sur ses deux mains, à chascune par le nombre de Cinq doigts; qui sont derechef partis inegalement par le poulce, & par le reste en Un & Quatre, dont l'Un est composé de Deux articles, & les Quatre de Trois: de façon que ceste structure luy met tousjours en veuë les premiers & plus excellents Nombres 1, 2, 3, 4, 5, 10 dont tant d'Illustres Philosophes, comme Pythagoras, & Platon, & tout ceux de leurs Escholes ont fait tant d'estime, qu'ils y ont voulu cacher, & en deduire les plus grands mysteres de leur doctrine."

63. On finger reckoning, Thibault says, "Nous [pouvons] demonstrer [les nombres] clairement sur nos doigts, où nous avons coustume d'apprendre les premieres leçons de l'Arithmetique naturelle"; ibid., 2.

The hand may have been the most overtly musical part of the body, but the whole of it evinced an "Arithmetique naturelle," which led Pythagoras to declare man "the Measure of everything."[64] Although ancient authors had used a variety of regular geometric figures to define physical perfection, overlaying the body with circles, squares, rectangles, and pentagons, Thibault opted for Hippocrates' circular model, which he felt best explained both the construction of the body and its movements. The arms and legs traced smooth arcs as they moved, pivoting upon wrists, elbows, shoulders, knees, and hips that operated like the hinges of a compass—a tool prominently enshrined in the frontispiece to the treatise. In the first figure of his text, Thibault plots the circle that will convene geometry and swordplay by using a sword equal to the height of the navel as its radius. The sword, as a sort of yardstick, is divided into twelve parts, numbered 1 to 12 from the tip to the hilt (see the swords in fig. 2.3 in the center circle and along the bottom of the illustration).

Into the circle fit the fencer's body, steps, and the three positions of attack and defense, both lateral and diametrical (see circle 4 in the lower right-hand corner of fig. 2.3, *prima instantia, seconda instantia,* and *tertia instantia*). Thibault calculated the distance of the second and third positions from the first and that of all the positions from one another using Euclidian geometry and the number 12 (one sword-length) as the radius of the circle.[65] But we should note that ratios do not enter into this discussion, which maps out a static series of geometrical forms.

Ratios and the proportions among them mobilize this geometry only when Thibault treats the force of blade against blade.[66] Moving into the realm of physics, he explains that the force exerted at the tip of the sword (marked "1") is less than

64. Ibid.

65. Ibid., 11–12.

66. Ibid., 21. Georges Vigarello reproduces a number of plates from Thibault's *Academie de l'espée* and discusses the treatise at some length in "The Upward Training of the Body from the Age of Chivalry to Courtly Civility," in *Fragments for a History of the Human Body,* part 2, ed. Michel Feher with Ramona Naddaff and Nadia Tazi (New York: Zone Books, 1989), 149–199. Vigarello argues that in the sixteenth century exercise was not seen as having any corrective function on posture, and any mechanical understanding of the body was fragmentary. Thibault's Pythagorean man, Vigarello emphasizes, is not an ideal to which the body is to be trained—something that would have to attend to the skeleton, the way it supports weight, and the action of the muscles—but a superficial and essentially graphic (rather than physical) illustration of corporeal perfection. I agree that Thibault's conception has been shaped by the draftsman's hand and that its presentation of force is more mathematical than mechanical, but generally speaking, the emphasis on physical activities in the military academies, Montaigne's notion of hardening the young body through exercise, and the repetitive drills developed in training programs at this time do suggest an understanding of the body as something that is formed, even though not in the mechanical terms that will come to predominate.

that at the hilt ("12"), a difference indicated by the numbers assigned to each point.[67] Thibault comes close to theorizing fencing as a series of harmonic relationships between two moving bodies, as a strong octave (2:1) or weaker major third (5:4)—and these are his numbers, not mine. Music was the culmination of the quadrivium precisely because it figured more than the static relationships of geometry and the application of geometry to study of the stars (astronomy). It achieved a science of motion based on proportional relationships, and by explaining motion, it explained the life force of the universe. Like Kepler and Newton, Thibault conceived of physics according to the precepts of *musica speculativa.*

GEOMETRIES OF POWER

At court, artistic programs enlisted this musical geometry in performances structured according to visible symmetries that corresponded to the heavenly source of the king's majesty. In early *ballets de cour,* the dancers arranged themselves in triangles, chains, and circles, then quickly broke apart to form new figures, their bodies gliding from one set of aspects into another. Ronsard spoke of the ethereal circles of the "bal des astres," Agrippa told of the ancient belief that the celestial movements of the stars and planets were a "danse mesuree et bien accordante," and Tuccaro, whose *Trois dialogues de l'exercice de sauter* is steeped in Neoplatonism, maintained that balls here on earth imitated the turning of the sky and the progress of the planets across it.[68] Tuccaro saw in the geometrical figures of ballet the oppositions, sextiles, trines, and squares of planetary aspects.[69] Likewise, Balthasar

---

67. One might calculate the relative force of each cut simply in terms of the difference between the marks, but in reality, the difference in force between the clash of two swords striking respectively at points 2 and 4 is not the same as that of swords striking at points 8 and 10 because the ratios between the pairs of numbers are not the same (they are 1:2 for the first pair and 4:5 for the second, making the second encounter a much closer match).

68. See Ronsard, *Oeuvres complètes,* 2:510, 518, 624; Agrippa, *Declamation,* 89; and Tuccaro, *Trois dialogues,* fol. 36r–v, for examples of the *bal des astres* metaphor; also see Ronsard's "La Charité," 1:576–580, for a beautiful reversal of this metaphor, in which a royal ball is described in heavenly terms. One could easily multiply these examples. See especially Pierre Matthieu, who ended his *Pompe funèbre des pénitens de Lyon* (Lyons: J. Roussin, 1589), with a description of God as "ce sainct & sublime Esprit, qui harmonize les Anges, qui guide le bal des cieux" (24). On Ronsard and Tuccaro see Margaret M. McGowan, *Ideal Forms in the Age of Ronsard* (Berkeley and Los Angeles: University of California Press, 1985), chap. 6, "Dancing Forms," and idem, "The Arts Conjoined: A Context for the Study of Music," *Early Music History* 13 (1994): 171–198.

69. Tuccaro, *Trois dialogues,* fol. 36v. "Ces belles & diverses retraictes, droictes & obliques, qu'on exerce avec tant de grace, sont les mesmes conjonctions & oppositions triangulaires & quadrangulaires, voire sexangulaires qui interviennent quasi tous les jours entre les planettes en leurs spheres celestes."

de Beaujoyeulx, who choreographed the *Balet comique de la Royne* (1581), declared that "Archimedes could hardly have better understood Geometric proportions than the princesses and ladies practiced them in this Ballet," and one panegyric lauds him as a geometer whose artistry resembled that of the great Greek astronomer.[70]

Royal ceremonial, too, placed the king at the center of an orderly universe warmed and propelled by his radiant divinity. The perfect circle of the crown, its golden aurora about his head, the marvel of his bejeweled appearance amidst an entourage of thousands on the court's progresses to far reaches of the realm, the increasingly complex ceremonial actions accruing to his meals, prayers, departures, and arrivals, all created a cycle of rituals symbolically inscribing the king at the center of things. The king was himself described geometrically as a point—a transubstantiated being both God and man, his nature indivisible, somewhere between measurable existence and immateriality. The Bourbon kings became the focal point for a new stagecraft based on perspective and the logical unity in a new statecraft reaffirming their absolute authority, indivisible among the estates.[71] On the battlefield, generals of the royal army used newly invented geometrical compasses to calculate the disposition of troops in a bewildering array of blocks, crosses, shallow lines, and wedges, while students at Jesuit colleges practiced falling into the same formations for royal visits. Pluvinel's cadets performed a geometrical ballet on horseback for the great carrousel of 1612. Geometry was a recurring feature of the performances of power we will study in the coming chapters, a spatialization of *musica speculativa* persistently combined with sounding music.

How was this spectacular geometry commonly understood? By 1600 compasses, T squares, and sextants were used for all sorts of practical reckoning that brought hands-on geometry directly into the lives of merchants, carpenters, sailors, and aristocrats. These tools fostered a measurement revolution that made the speculative science of mathematics more relevant and set the stage for the mathematization of empirical knowledge that was to follow.[72] Carpenters sized up the world with

70. See Balthasar de Beaujoyeulx, *Balet comique de la Royne* (Paris: Le Roy & Ballard, 1582), facsimile ed. and introd. Margaret M. McGowan as *"Le balet comique," 1581: A Facsimile* (Binghamton, N.Y.: Center for Medieval and Early Renaissance Studies, 1982), fol. 56r. "Si bien l'ordre y estoit gardé, & si dextrement chacun s'estudioit à observer son rang & cadence: de maniere que chacun creut qu'Archimede n'eust peu mieux entendre les proportions Geometriques, que ces princesses & dames les pratiquoyent en ce Balet." Also see ibid., fol. E1r.

71. On theatrical perspective in royal spectacles, see Stephen Orgel, *The Illusion of Power: Political Theater in the English Renaissance* (Berkeley and Los Angeles: University of California Press, 1975), esp. 9–16, 35–37.

72. See Alfred W. Crosby, *The Measure of Reality: Quantification and Western Society, 1250–1600* (Cambridge: Cambridge University Press, 1997).

plumb lines, soldiers triangulated the trajectory of cannon fire with sighting instruments, and generals deployed their massed troops in formations derived from tables of square roots that forced officers "to wade in the large sea of Algebra & numbers."[73] Galileo invented a military sector that permitted merchants to calculate compound interest and artillery commanders to figure the amount of gunpowder needed to fire cannonballs of various sizes and densities, even with no education in mathematics. One can easily imagine the objections of seasoned warriors who had won their battles without help from mathematicians. In Shakespeare's *Othello,* Iago jokingly discounts Cassio as "a great arithmetician" and "bookish theoric" who "never set a squadron in the field, nor the division of a battle knows" (act 1, scene 1, lines 19–24), and in *Romeo and Juliet,* Mercutio brushes off the wound just inflicted by Tybalt as one delivered by "a braggart, a rogue, a villain, that fights by the book of arithmetic!" (act 3, scene 1, lines 99–100).[74] Shakespeare's characters express the same impatience with numbers that Crispan de Pas says he encountered in the students at Pluvinel's academy, but whether they liked it or not, advances in measurement and reckoning had begun to develop a consciousness of the underlying structure of the universe among merchants, artisans, sailors, and other ordinary folk.

With the use of quadrants and torqueta, astronomers mapped the heavens with increased precision, adding a new dimension to the already highly political study of the nature and motion of the planets. Galileo's *perspicillum* (spyglass) helped him to discover the moons of Jupiter, which he sagely named for the Medicis, who rewarded him with the position of "Filosofo e Matematico Primario." This and his subsequent verification of Copernicus's heliocentric model of the universe illustrate the high stakes for which court astronomers played as they turned from calculating horoscopes to scientific investigations into the order of the universe and, implicitly, the political and religious orders modeled upon it.[75] It also implicates music in the new science of divine right.

ONE MUST pause to remember that all of these proportional ordines were hierarchical and extraterrestrial. The imposition of them from above and "from without"

73. Quotation from Thomas Digges, *An Arithmeticall Militaire Treatise Named Stratioticos (London, 1571)* (Amsterdam: Da Capo Press, 1968), 70; cited in Crosby, *Measure of Reality,* 7.

74. One wonders if Shakespeare's Tybalt is named for Girard Thibault. For the Shakespeare references I am indebted to Crosby, *Measure of Reality,* 7, and John McClelland.

75. The politics of measurement were as strong as ever at the end of the ancien régime. See Ken Alder, "A Revolution to Measure: The Political Economy of the Metric System in France," in *The Values of Precision,* ed. M. Norton Wise (Princeton: Princeton University Press, 1995), 39–71.

gave them a character of otherness that coincided with the modes of monarchic expression. This is to say that monarchies depended on acknowledgment of the sovereign as an extraordinary being, a living god whose exceptional powers were aptly symbolized by the timeless perfections governing the cosmos.

The external character of sovereignty originated in the forces by which one group, clan, or dynasty came to dominate in society. In his cross-cultural essay on kingship, Valerio Valieri considers the case of acephalous societies that evolve toward unequal forms. Their "headless" society is transformed by the aggressions of a dominant group that imposes inequalities in order to consolidate its power. "However this group was formed, it remains associated, even in its subsequent transformations, with the *external character* of constraint, with the otherness of authority."[76] In France, as well, the sovereign's otherness was keyed to the violence by which dominance was attained. The unequal privileges accorded to France's three estates and the rule of the king over them, however modified by constitutional governance, had at its very origin a conquering king who dominated through military action. War founded monarchies, and continuing military ventures waged abroad helped to restage at home the symbolic victories by which the king had attained his kingdom. A successful foreign war defined kingship, and Charles VIII (in the Italian campaign to Naples), Francis I (in the Battle of Marignano), Henry IV (in the war against Spain), and Louis XIV (in the War of Devolution) all set out on this "royal road to glory" during the first decade of their reigns, with dramatic results.[77] These victories provided the raw materials for panegyrics, ballet librettos, *chants triomphals,* odes, paintings, medallions, statues, entries, and the prologues of *tragédies en musique* that reaffirmed the essential quality of kingship as the triumph of a "conquering" and "foreign" power.[78] During the civil wars, the creation myth of domination from outside replayed itself much more immediately as Henry IV retook his own kingdom, battle by battle and siege by siege.

The symbols of kingship belie a need to repeat the founding event of the monarchy, portraying the king as "other"—one who conquered by extraordinary force and, as in the case of Achilles, with a bloodlust surpassing human bounds. The exotic animals and monstrous men kept by kings are often explained by a desire to inspire wonder and symbolize the geographical reach of the monarch's authority, but

76. Valerio Valeri, "Regalità," in *Enciclopedia Einaudi,* vol. 11, *Prodotti-Ricchezza* (Turin: Einaudi, 1980), 742–771, at 747. Quotations are drawn from the English translation made by Lynn Westwood, forthcoming from Berg, Oxford.

77. See Peter Burke, *The Fabrication of Louis XIV* (New Haven: Yale University Press, 1992), esp. 70–105, on the political importance of the royal military venture.

78. Valeri, "Regalità," 744–747.

the wild lions, leopards, and other fierce creatures also signified the king's uncontrollable—and therefore divine—nature.[79] So too did giants, dwarfs, albinos, and fools remind courtiers and ambassadors of the natural disorders that produced exceptional—and therefore divine—individuals. Like the sirens in court ballets, they symbolized the metaphysical reach of the sovereign's powers to a realm where monstrous men lined the border between man and beast and between man and the devil. Tyard's parenthetical allusion to "the monstrous" in his Pythagorean account of the body was part of the dialectic of the divine, which comprehended powers both beneficent and horrendous. No doubt he had seen the supernatural beings populating the French court: the hairy man from Tenerife, the dwarfs, and misshapen prodigies. We must count music itself as a symbol of the boundless upward reach of princely influence and its ability to convene celestial elements around the person of the king.

The heavens dominated nature, the stars ruled the lives of men from without, and in these unearthly forms we find the particular usefulness of geometry to performances of power. Like the conquering king, foreign and divine, sovereignty found its expression in a geometry that itself seemed to suborn nature to the musical ordines of an otherworldly rule. Moreover, the symbolism that controlled this violent political reality often relied on spatial forms reminiscent of the geography of state. To quote Valeri again on kingship, "this external character and otherness are oftentimes spatialized: and in a world of scattered communities, authority is in fact extraneous, if not foreign." In a country as vast and differentiated as France, royal authority inevitably appeared foreign—this, as we have seen, helped maintain the hierarchy it imposed. Though France benefited from geographic unity, it took six to seven weeks to travel from Calais to Provence.[80] Dialects held fast in provinces, the *langue d'oc* separated south from north, and the king's lieutenant governors ruled in the provinces only by negotiating uneasy truces with local councils. Even the announcement of royal edicts began with trumpet fanfares like those that heralded invading armies.

While my main concern here is to draw out the supernatural character of the king's power, we must note that it is not always the monstrous power of the victor. The violence of the conqueror is domesticated in the coronation ceremony and positively transformed into legitimate power. Hence the geometrical forms of court fete and ceremonial—external, tamed, divine—are doubly productive, reinforcing a political structure positioning the king above and at the center of a world always

---

79. See Lorraine Daston and Katherine Park, *Wonders and the Order of Nature, 1150–1750* (New York: Zone Books, 2001), 100–108, 193–195.

80. See James B. Wood, *The King's Army: Warfare, Soldiers, and Society during the Wars of Religion in France, 1562–1576* (Cambridge: Cambridge University Press, 1996), 59–61.

being wrested into order. At a time when harmony was only just beginning to be theorized as the movement from dissonance to consonance, the mathematical philosophy of the universe suggested similar operations. Like a colossal stage machine, geometry imposed circles upon the bodies of fencers, squares and triangles and chains upon the choreography of ballets, and detailed linear hierarchies on the order of royal processions. These fleeting sixteenth-century experiments in corporal harmony morphed into more permanent forms in the seventeenth century as the gardens of Versailles imposed geometric orders on nature from a rigidly symmetrical palace placing the king's bedroom at its lofty focal point.[81] They also led toward the political orders of absolutism.

HARMONICALL GOVERNANCE: A PYTHAGOREAN BODY POLITIC

At the time of the civil wars, music was drawn to the center of projects to create moral uplift in France, the most prominent of which were pursued by the Academy of Poetry and Music. Instituted by royal letters patent just two years before the Saint Bartholomew's Day Massacre of 1572, Baïf's academy set out to legislate behavior with new music. In an argument very much indebted to Plato, who—in a bit of wordplay—declared songs themselves to be laws (*Laws* 7.799e), the academicians maintained that "it is of great importance for the morals of the citizens of a town that the music current and used in the country should be retained under certain laws, for the minds of most men are formed and their behavior influenced by its character, so that where music is disordered, there morals are also depraved, and where it is well ordered, there men are well disciplined morally."[82]

Music could rehearse the results of virtues both bellicose and pacific, but it could also produce disorder and extreme emotions. One need think only of Pythagoras, who came across a drunken Tauromenitan on the verge of raping a young woman after having become aroused by a Phrygian song; the philosopher counteracts the negative effects of the Phrygian mode with the calming Hypophrygian. Though not yet articulated as such, descriptions of affect and the passions implicitly attributed to music a strong cathartic effect of the sort described by Aristotle (*Politics* 8.1340a–b), an effect that might be risky to initiate. Certainly the academy aimed

81. See Chandra Mukerji, *Territorial Ambitions and the Gardens of Versailles* (Cambridge: Cambridge University Press, 1997). Mukerji's material history of the gardens links their formal structures to the history of fortification and, by extension, to the military ambitions of Louis XIV and the territorial state of the late seventeenth century. The mobile geometries of the court fetes under consideration here perhaps better sustained a monarchy less concerned with territory and its boundaries than with moving among centers of power.

82. The letters patent are reproduced in Yates, *French Academies,* 319–320; I quote her translation, 23.

to purge society of vice, but the published result of its project is rigidly "ethical" music indeed. Le Jeune's *Le printans* and Jacques Mauduit's *Chansonnettes mesurées* are rigorously homophonic, diatonic, and largely syllabic. From their republic, too, "panharmonic" and rhythmically dangerous songs were banished, though in chambers Charles IX enjoyed imported *musica reservata* such as Orlando de Lassus's highly chromatic *Prophetiae Sibyllarum.*[83] And at the magnificent (yet still private) *Balet comique de la Royne,* courtiers were treated to florid solos and a metrically and harmonically diverse Grand Ballet in E and A, with the high clefs and natural signature that represented the Phrygian mode (though F-sharps sometimes compromise the plangent half-step with E that gave the Phrygian its striking quality). Perhaps the music was intended to heighten the warlike affect of the Grand Ballet—a *bataille rangée,* Beaujoyeulx called one portion of it—eliciting and guiding the anger of those packed into the Salle de Bourbon toward a communal contraction of the soul in sympathy with music.[84] But if catharsis was the aim it was bounded by the well-defined context of a supremely royalist narrative and timed to occur only when the king stepped to center stage holding the wand of the vanquished enchantress Circe.

Just as the academy hoped to discipline society with their songs, political theorists argued for princely sovereignty on the basis of harmonic order. After all, Plato's vision of the ideal state culminated with the orphic "Myth of Er," in which he explained the harmonic nature of the universe, and Plato was not alone in likening the well-governed state to a harmonious whole. But in the 1570s the religious crisis precipitated a monarchic crisis so severe that it was no longer enough to make oblique references to harmony as a self-evident metaphor for public repose. As the nature of sovereignty itself was called into question, from the same cultural spheres that produced the republican musical experiments of the academy, a new political theory emerged in defense of sovereignty.

In 1576 Jean Bodin, a royal councilor, argued that sovereignty had always belonged solely to the king, and in so doing he effectively produced the first theory of absolute power. The choice of French instead of Latin for his *Six livres de la République* guaranteed its maximum impact in contemporary political debate. The book went through seventeen French editions between 1576 and 1596 alone; Bodin

83. On the *Prophetiae Sibyllarum* see James Haar, "Orlando di Lasso: Composer and Print Entrepreneur," in van Orden, *Music and the Cultures of Print,* 135.

84. Beaujoyeulx, *Balet comique de la Royne,* fol. 56r. Rabelais's fifth book includes the story of a ball in the form of a tourney that similarly finished off an evening's divertissement: "Et fut la musique serrée en la mesure plus que de hemiole, et intonation Phrygienne et bellique." He ends his description with an account of the spectacle's affect, recalling "que par telle modulation, Ismanias excita Alexandre le grand estant à table et disnant en repos, à soy lever, et armes prendre. Au tiers tournay fut le Roy auré vainqueur." François Rabelais, *Oeuvres complètes,* ed. Mireille Huchon with the collaboration of François Moreau (Paris: Gallimard, 1994), 784–785.

translated it into Latin in 1586, and nine editions of this version appeared between 1586 and 1641 from presses in Paris and Frankfurt; by the end of the century, translations into Italian, Spanish, and German had been made; and Richard Knolles's translation into English was published in a deluxe folio edition in London in 1606.[85] Vastly influential, Bodin's *République* was the founding text of absolutism.

Bodin concerns us here because he used harmonic rules to generate his theory of absolute power. Behind his book stood Plato's own *Republic,* of course, and Bodin was keen to turn Plato's theories of a musical world order to good use. But rather than use musical terms symbolically, as his Protestant opponents had done, or spin off into musical mythology, as Plato does in the *Republic,* Bodin essayed a political theorem based on demonstrable harmonic principles. Harmonic theory became the mathematical paradigm for civil order, and political theory thereby acquired a scientific dimension.

Jean Bodin began his career in Paris as a lawyer in parlement, and in 1571 he became master of requests and counselor to François, Duke of Alençon, the king's younger brother. He was a longtime friend of Guy du Faur de Pibrac, a supporter of Baïf's academy and the organizer of the Palace Academy active under Henry III, where Bodin lectured for thirteen weeks in late 1576 and early 1577 to a group whose normal membership included, among others, the king, Tyard, and Baïf.[86] It was to Pibrac that Bodin dedicated the *Six livres.* In this intellectual milieu, Bodin worked out a theory of "harmonicall governance" that, like the experiments in *musique mesurée* undertaken by Baïf's academy, attempted to redress the crisis of the religious wars.

The battle for religious freedom was not the only one fought in the civil wars, for Protestantism called into question the sovereignty of Catholicism itself and the divine right of the king to rule. After the Saint Bartholomew's Day Massacre, all hope of conciliation between Catholics and Protestants disappeared, and the conflict escalated to a new level. The Huguenots fortified their strongholds, and, in a far more radical move, they produced a republican constitution establishing within France a Protestant state replete with its own militia.[87] Although the Protestants did not officially renounce allegiance to the king, their constitution expanded resistance

85. For a bibliographical overview and an excellent introduction in English see Jean Bodin, *The Six Bookes of a Commonweale: A Facsimile Reprint of the English Translation of 1606,* trans. Richard Knolles, ed. Kenneth Douglas McRae (Cambridge, Mass.: Harvard University Press, 1962). English quotations are from this edition; page numbers are given in parentheses in the body of the text. The standard modern edition in French is Jean Bodin, *Les six livres de la République,* ed. Christiane Frémont, Marie-Dominique Couzinet, and Henri Rochais, 6 vols. (Paris: Fayard, 1986).

86. See Robert J. Sealy, *The Palace Academy of Henry III* (Geneva: Droz, 1981), 65–68.

87. For the essential facts and issues see Mack P. Holt, *The French Wars of Religion, 1562–1629* (Cambridge: Cambridge University Press, 1995), chap. 4.

from the military to the civil domain and threatened the king's sovereignty in an entirely new way. Pamphleteers proclaimed that the people were the true sovereigns of France, and in 1576 they called for the Estates General to convene as a representative body and depose the Valois. Bodin, whose *République* had come out earlier that year, attended the meeting as deputy for the Third Estate of Vermandois, where he led a movement in favor of peace and against the taxes the king would need to levy in order to pursue the war against the Protestants (one matter in which he fought against the king). Radical Catholics denounced Bodin as a *politique*—one who placed political concerns above the Catholic faith—and while it is true that Bodin found in the anarchy of the civil wars a more present danger to the state than that posed by Protestantism, he did not necessarily favor a state founded on religious coexistence. Rather, his primary aim was to restore order to the kingdom, something that could only be accomplished through temporary toleration of the Protestants and by according absolute power to the king. Thus the political theories that bore fruit in the tremendous monarchy of Louis XIV originated in reaction to civil disorder.

Bodin's musical discussion arises first in book 4, where he makes three general points relating to the commonwealth.[88] He begins with Plato's belief that faulty harmony leads to the downfall of republics, which requires him to establish the point at which concord turns to discord. What degree of difference—in terms of proportionality—can the system support? One senses the political import of the Pythagorean series as Bodin explains how harmony is destroyed when one progresses from fourths and fifths to intervals based on more complex ratios. "If therefore choice be had of such proportions as make a sweet consent in the perpetuall course of numbers, the Commonwealth shall so be everlasting."

Leaving aside the question of how social structures might be numerically construed, Bodin moves from trying to educe a universal system of government based on number to the comparative mode more common to legal humanists. He presents examples of music's cohesive force in society, slipping from mathematics to history, where he finds in Polybius's description of Arcadia firm verification of Plato's dictum that music can change a republic. The Arcadians were well known for the honor they bestowed upon music and the regular musical study they required of all citizens up to the age of thirty. Polybius tells us that (to cite Bodin) "the first lawgivers of that people wisely devised [this rule] to quiet and tame them, being by nature rough and barbarous, as commonly all the inhabitants of the mountaines and cold countries be." Indeed, in the republic of Cynethenses in Arcadia, the people once abandoned music, after which they quickly fell into "such sedition

88. The passages discussed here can be found in Bodin, *Six livres de la République,* 4:82ff., and Bodin, *Six Bookes of a Commonweale,* 455ff.

and civill warres, as wherein no kind of crueltie was forgotten." Without the civilizing impulse of music—and Bodin describes the Arcadians as "civill, courteous, and tractable"—barbarism swiftly returned.

The analogy between a destroyed Arcadia and the very real "sedition and civill warres" amid which Bodin drafted his *République* could hardly have been lost on contemporary readers, either in France or in England, where the translation from which I have been citing was published. But Bodin's French Arcadia is only in temporary turmoil. From a once barbarous and fierce people, the French developed into a people "in civilitie inferiour to no people of all Europe, none being more tractable unto their magistrates or obedient unto their Princes than they, as men by nature well, but by instruction better taught, and in the judgement of all their neighbours most skilfull in Musike." Indeed Bodin attributes the civility of the French to their predilection for songs in the fifth and seventh modes, modes discouraged by Plato and Aristotle for their "power to mollifie and effeminate the minds of men" but well suited to those from colder northern climes, who tend to be more savage. Harmony is not irretrievable in France but can be recuperated among an already highly musical people.

What stands out in this line of reasoning is not only its basis on arguments of racial character and national musical repertories (arguments of the sort that would grow stronger under the ancien régime), but the way stories of modal ethos derail a predominately mathematical argument. The hinge between Bodin's demonstration of numerical perfection and his theory of political harmony is missing. Rather, Bodin replaces it with a story about music's ethical effects, stopping the logical gap in the text with a retreat from numbers into history. One senses here the uneasy juncture of *musica speculativa* and *musica practica* so common at the time, and the difficulties engaged by those who struggled to quantify music's effect on the soul. Descartes ran similarly aground at the end of his *Compendium musicae* (1618), admitting that a proper treatise would describe how each mode created particular motions in the mind but explaining that the limited space of a compendium did not permit him to elaborate. The fact that late in life Descartes talked about publishing a work on music, coupled with the fact that his final work was on the passions, suggests that the desire to explain musical affects scientifically stayed with him to the end and was never resolved. Had music not been so easily rationalized in numerical terms, no doubt the quest to decipher its messages to the soul would not have been pursued so keenly. Music remained central to Neoplatonism and early modern science without being fully explained by either.

Bodin essayed a more densely mathematical theory of harmonic government in the last book of the *République,* where he defined the possible relationships among French subjects and their king according to the arithmetic, geometric, and harmonic means.[89] The means had been used by Boethius to establish the proportional relationships between the octave and fifth and the octave and fourth, and in Bodin's own time, Zarlino discovered that by applying the arithmetic and harmonic means to the divisions of a fifth, he could produce a major and minor triad. With these musical equations, Bodin hoped to determine the optimal balance of power among France's three estates and the king.

The relative power of the estates was a matter of great urgency at the time Bodin wrote, and while he generally tried not to mire his treatise in contemporary political frays, he did respond here to the Huguenot François Hotman, whose revolutionary *Francogallia* (Geneva, 1573) attempted to establish a historical foundation for popular sovereignty in France.[90] It asserted that the ancient Francogallican constitution defined a mixed form of government combining democracy and aristocracy with monarchy. In this anterior system, decisions were made by a common council that enjoyed supreme authority of a sort that Hotman explicitly associated with the original powers of the Estates General.[91] Pamphleteers soon called for the Estates General to convene and exercise its rights, making the immediate dangers of Hotman's research all too clear.

In a crucial passage, Hotman likened this mixed form of government to those described by Plato, Aristotle, Polybius, and Cicero. Calling upon these ancient authorities, he quoted at length from Cicero's *Republic:*

> With lyres and flutes, and also with voices raised in song, a certain harmony should emerge from the individual sounds, but a trained ear cannot endure an altered or discordant note. This harmonious and agreeable concord is produced, however, from the regular arrangement of dissimilar sounds. In the same way a commonwealth which is regulated by reason produces harmony through the consent of dissimilar elements, drawn, like the sounds, from the highest and the middling orders, from the lowest and the intermediate estates. What is called harmony in song by musicians, is called concord in a commonwealth. In such a state concord provides the narrowest and best assurance of security, which no general agreement can embody unless it is accompanied by justice.[92]

89. See Bodin, *Six Bookes of a Commonweale,* 769–794. On his "mathematization" of political theory see Philippe Desan, *Naissance de la méthode: Machiavel, La Ramée, Bodin, Montaigne, Descartes* (Paris: Nizet, 1987), chap. 4. On Bodin's life and education, see Simone Goyard-Fabre, *Jean Bodin et le droit de la république* (Paris: Presses Universitaires de France, 1989).

90. See the introduction to François Hotman, *Francogallia* (Geneva, 1573), Latin text ed. Ralph E. Giesey, trans. J. H. M. Salmon (Cambridge: Cambridge University Press, 1972). On the relationship of Bodin's treatise to the *Francogallia* see J. H. M. Salmon, *Renaissance and Revolt: Essays in the Intellectual and Social History of Early Modern France* (Cambridge: Cambridge University Press, 1987), 119–135.

91. See Hotman, *Francogallia,* 68–71, 287–349.

92. Ibid, 295, quoting Cicero, *De re publica* 2.52.69.

Bodin seized upon this passage and its implications of popular sovereignty in his refutation of Hotman's call for shared government. For where Hotman had assumed that the relationships among dissimilar sounds or estates could be structured in only one way, Bodin showed that there were in fact three different musical means by which to rationalize them—the arithmetic, the geometric, and the harmonic (*Six Bookes of a Commonweale,* 757). He then went on to prove that monarchy was equated with the most harmonious mean.[93]

As he explained it, arithmetically ordered sets contain equal differences among the members, but no proportional relationships. That is, in the series 3, 9, 15, 21, 27, the difference between each number is the same (6), but the proportions among them are dissimilar. In political terms, equality without proportion among the members described an orderless society in which aristocrats received no special dispensations. The merit of noble individuals was subordinated to equality in the eyes of the law; in other words, arithmetic government was popular government. Geometrically ordered sets, in contrast, contain members related by the same ratios—where 3, 9, 27, and 81 relate in ratios of 1:3—but between those members there are unequal differences. The geometric mean thus represented a class system based on noble entitlement in which some individuals received vastly greater preference in the form of offices awarded at the king's discretion. Finally, harmonic groupings include elements of the arithmetic and geometric orders, but are more wholesome. The series 3, 4, 6, 8, and 12 comprehends the ratios of musical harmonies, and more importantly, the differences between the numbers are not nearly as great as those in the geometric series, even while they are still unequal.[94] For Bodin, monarchy or "harmonicall governance" did not efface the differences between nobles and nonnobles, but introduced an order of greater equality.

Later, he established the perfection of harmonic government with recourse to the Pythagorean numbers 1 to 4, "which God hath in Harmonicall proportion disposed

93. As defined by Pythagoras, the three means were as follows: arithmetic, $m = (a + b)/2$; geometric, $a/m = m/b$; and harmonic, $m = 2ab/(a + b)$. The most common calculations used relative string lengths; thus, when calculating the arithmetic mean of the octave, $a = 1$ and $b = 2$, such that $(1 + 2)/2 = 3/2$, therefore the arithmetic mean of an octave is a fifth. The harmonic mean of an octave is a fourth. Bodin's most proximate source of information on the three sorts of mean was Tyard, *Solitaire second,* 100–107. Bodin may well have met Tyard at court: Tyard was a member of the Pléiade, he served Henry III as *valet de chambre,* and his theories are amply evident in the musical experiments of Baïf's academy.

94. Bodin does not give a true harmonic progression but conflates overlapping series of three numbers that, taking the largest number to be 1, relate in the form 1, 2/3, 1/2 (unison, fifth, octave). The progressions are 3, 4, 6; 6, 8, 12. This explanation of the harmonic mean comes from Archytas (fragment 2). Also see Plato, *Timaeus,* trans. R. G. Bury, Loeb Classical Library (London: William Heinemann; Cambridge, Mass.: Harvard University Press, 1981), 36; and Tyard, *Solitaire second,* 103–105.

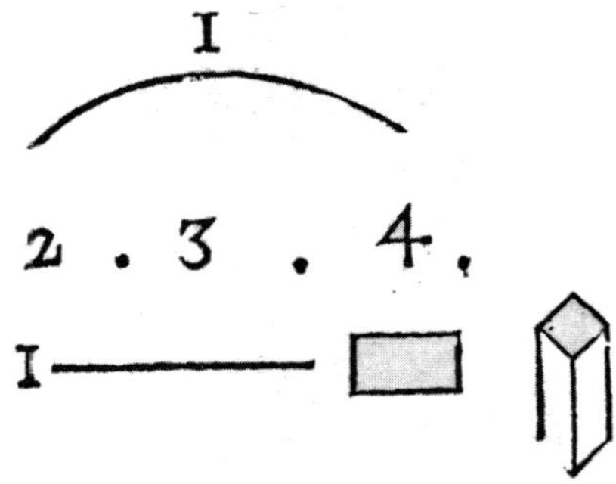

FIGURE 2.4. The Pythagorean series and geometrical figures related to it. In Jean Bodin, *Les six livres de la République* (1576)

to show unto us, that the Royal estate is Harmonicall, and also to be Harmonically governed." The ratios among these numbers produce all harmonic consonances, or as Bodin says, they "containeth the whole ground and compasse of all tunes and concords of musicke." To add the number 5 to this closed system will "marre the harmonie, and make an intollerable dischord" (790). In other words, the "royal estate" of the king (number 1) and France's three estates (numbers 2, 3, and 4) create a harmonic totality to which nothing can be added or removed without causing discord (see fig. 2.4). Of course, Bodin was not the first to liken harmonic orders to political ones, but he was the first to assign Pythagorean numbers to political entities and to use the harmonic mean to rationalize the uneven distribution of political power.

Good jurist that he was, Bodin preserved in his government the checks and balances of constitutional practice, but they were subordinate to the sovereign authority of the king, who stood above all else. Bodin found it impossible to conceive of a "sovereign" of mixed constitution, which theoretically would have been a *persona ficta* composed of all three estates and the king, with all parts sharing in the making of "the sovereign's" decisions.[95] Either sovereignty belonged to the estates in the form of civil laws to which the king was subject, or—as Bodin claimed—it belonged to the king, who was therefore not bound by civil law. Thus Bodin asserted both the superiority of king's authority and its indivisibility. Sovereignty comprised a fundamental unity that could not be divided, just like the number 1 and the "point" in his illustration (fig. 2.4): "Now the soveraigne prince is exalted above all his subjects, and exempt out of the ranke of them: whose majestie suffereth no more

95. Julian H. Franklin, *Jean Bodin and the Rise of Absolutist Theory* (Cambridge: Cambridge University Press, 1973), 26–29.

division than doth the unity itselfe, which is not set nor accounted among the numbers, howbeit that they all from it take both their force and power" (790). Bodin seemed to grant the estates no authority at all, outlining a system in which legal obligation and institutional constraint could not safeguard against tyranny. Only the king's moral respect for God's laws and the "laws of nature" protected the people from despotism.

The king's proximity to God was an essential component of the moral guarantees Bodin foresaw, and he argued that all political authority came from God.[96] Indeed, the king was the lively image of God, and just as God is immortal, so, too, the king never dies. Bodin quotes the French maxim "le Roy ne meurt jamais," referring obliquely to the long-standing belief that the king had two bodies: a mortal body or "body natural," and a royal super-body or "body politic" that was eternal and sacred.[97] In the context of his book, we might think of the body politic in terms of sovereignty itself or, more particularly, as majesty. In Bodin's own words, "Majestie or Soveraigntie is the most high, absolute, and perpetuall power over the citizens and subjects in a Commonweale; which the Latines cal *Majestatem*" (84). Majesty (the power) and *sa majesté* (the individual) were inseparable and indivisible.[98]

The corporate metaphors used to define the body politic helped bind king with kingdom, reaffirming the sovereign's beneficence with anthropomorphic descriptions of the king as a head directing the body of state. We will see how Francis I referred to his subjects as members of his own body, like an arm that could only be severed with great injury, and how royal ceremonial persistently associated the king's body with that of Christ. Bodin likewise called attention to the wholeness of his republic by showing how "the well ordered Commonweale insomesort figured even in the nature of man himselfe" (790).

Understanding, the First unity, intellectual virtue	King
Reason guided by the intellectual virtue Wisdom	First Estate (clergy)
Angry power guided by the moral virtue Fortitude	Second Estate (military men)
Brutish lust guided by the moral virtue Temperance	Third Estate (scholars, merchants, artisans, laborers)

96. On majesty and its divine connotations see Ralph E. Giesey, "The King Imagined," in *The French Revolution and the Creation of Modern Political Culture,* ed. Keith Michael Baker, 4 vols. (Oxford: Pergamon Press, 1987–1994), 1:41–59.

97. On the French saying and this conception of kingship see Ernst H. Kantorowicz, *The King's Two Bodies: A Study in Mediaeval Political Theology,* new pref. William Chester Jordan (Princeton: Princeton University Press, 1997), esp. 409, n. 319.

98. Giesey, "The King Imagined," 1:51–53.

The sovereign prince ruled the republic with divine intelligence, figuring the nature of the human soul and its yearning for knowledge of celestial truths. The king's body politic or majesty thus incorporated the perfection of the world soul from which the human soul (the estates) spilled, fractured and discordant, into the mortal body of man.

Bodin treats the king's divinity in the terms established by a Christianized *musica speculativa,* but his Neoplatonic narrative is not far from articulating political harmony in the practical terms he essayed earlier with his discussion of Arcadia. The commonwealth, as a soul, would be susceptible to musical enlightenment, its horrendous disorders rectifiable with a sounding music that, like the songs Plato called laws, would combat "sedition and civill warres." Bodin does not legislate, as the Arcadians did, regular musical study, but this practical measure is implied. Indeed, we might read *Six livres de la République* as a guide to another sort of musical legislation—the "certain laws" that should be used to retain music in the country, which were proposed by Baïf's academy in the telling reference of its letters patent to Plato's *Laws.*

At the end of the day, our question of Bodin must be, Why music? Why rationalize democracy, aristocracy, and monarchy according to the arithmetic, geometric, and harmonic means? Why assign Pythagorean numbers to the estates and the king? Why relate political macrocosm to moral microcosm in terms of Neoplatonic philosophies that held up music as the means of aligning the soul with cosmic orders? Antecedents such as Plato's *Republic,* the Academy of Poetry and Music, and Hotman's *Francogallia* establish significant contexts for Bodin's musical arguments —we must not underestimate Bodin's desire to improve upon Plato's Myth of Er in the final chapter of his own *République.* But Bodin's systematization of government so vastly exceeds these models and is verified with such exacting attention to harmonic theory that we must recognize both the originality of his musical project and its political legacy. *Six livres de la République* was unquestionably the single most-read and most-cited exposition of sovereignty in early modern Europe, and it defined key terms with substantial musical explanation. The sharing of power in each of the three types of state and the preeminence of monarchy are calculated using musical equations; the relationship between the estates and the king is defined using Pythagorean number, geometry, and Plato's concept of the soul. Justice, like absolutism, is "Harmonicall," for through harmonic theory one can establish "the right division of rewards and punishments, and of that which of right unto every man belongeth." The whole language of political and social relations is shot through with a musical strain: justice is not far from the French *justesse,* or right tuning, just as temperament comprehends the fine tuning of instruments or the soul, or, in the

Francogallia, a mixed or "tempered" form of commonwealth. Accord brought together different tones or interests in concord, a chord, or a harmonious whole.

Broadly speaking, political theory delineated ways to share power unequally among classes or estates. Harmonic terminology supported political discourse precisely because *musica speculativa* was a highly developed science of proportions, and because musical forms were so comprehensive and transmutable. Line lengths could be equated to musical consonances, which in turn related to the weights of hammers, the laws of elliptical orbits, the parts of the body, and so forth. Music brought unlike matter into stable, rational relationships. Performed movingly, calculated numerically, music verified the connectedness of matter and the universal laws that might be deduced from its structures. In a tradition relying on Roman law, which was increasingly seen as inconsistent and historically limited, law needed a new foundation; with the monarchy under attack, sovereignty needed to be redefined; and with civil society torn apart by religious differences that called into question the very nature of faith and heresy, Catholicism could no longer regulate the moral and social life of the country on its own. Of course Bodin reaffirmed that Gallican credo *une foi, un roi, une loi* (one faith, one king, one law), but his mathematical arguments shored up the destabilized traditions of the Catholic state with a musical superstructure.[99]

Bodin's greatest originality lay not merely in defining the state as the supreme political authority, but in founding the state upon mathematics. By securing politics with rational method, Bodin anchored the power claims of the monarchy in the unimpeachable surety of numbers. At a time when syllogistic logic was coming unhinged as a foundation for knowledge, when the data one could draw from history seemed to defy effective utilization, and when the complex customary laws of France were beyond rationalization, Bodin essayed a new set of universal laws based on a comparative historical method and on mathematics. Music theory provided the universals that history could not, offering Bodin a paradigm of subtly balanced relationships that were complex enough to satisfy his design for representation in government and still reducible to the essentials of 1, 2, 3, 4. Bodin's commonwealth certainly benefited from the moral suasion of music, a quality evident in his likening of the harmonically governed state to a well-tuned body and soul, and his love of historical comparison may have yielded a belief that music contributed to the social and political organization of all well-conducted societies, yet musical effect was not his primary concern. Rather, the first architect of absolutism rationalized his

99. In this respect, Bodin's political theory superseded his personal religious convictions, which were outside the conventional norms defined by Protestantism or Catholicism. See Bodin, *Six Bookes of a Commonweale,* A12–A13.

plans according to musical number in order to safeguard the monarchy against the chaos of the Wars of Religion; he restructured government to correspond with universal laws that were as mathematically sound as they were musically sensible.

In all the essays and projects considered thus far, harmonic proportions order more than sounds. Montaigne, La Noue, Pelletier, Pluvinel, and Faret promoted musical education as an ethical practice in the spirit of Plato and Aristotle. In Dürer, Tyard, and Thibault, we see artists and fencing masters attempting to establish paradigms by extending musical proportions to the visual plane through geometry. Finally, in Bodin the civilizing intent of *musique mesurée* achieved a political formulation that was decidedly diatonic in its limited harmonic language. Music is pervasively framed as a governor ruling through a concord that was heard, felt, and "understood." Pierre Charron put it well in *De la sagesse* (1601) when he asserted that "Science, Trueth, and Vertue have no other entrance into the Soule but by the Eare."[100] Music was a moral force that seemed to rectify the relationships between outside and inside, between the body and the soul, between sense and reason, between matter and understanding, and between subject and sovereign.

As Timothy Reiss argued in *Knowledge, Discovery, and Imagination in Early Modern Europe,* it is not just that moral virtue and social justice could be attuned to harmonic proportions, but that this tuning also served to discover what such justice and virtue were.[101] And here we see why Bodin's musical rationalizations were not metaphors, but the result of research. Musical calculations had become a means of discovery, a scientific method whereby the *mathesis universalis* governing the physical world, the senses, morality, and society directed a new search for truth. The turn to musical methodologies in the years around 1600, Reiss posits, is explained by music's unique capacity to speak with moral import in a language that was infinitely more quantifiable than Latin, Greek, or vernaculars. Contemporary attempts to wrest language into some sort of order show the impulse toward the certainty of number: Baïf himself developed an extraordinary system of orthography in an effort to quantify French and "measure" it, as it were. But Reiss shows how such attempts ultimately failed. Earlier, language had expressed meaning and truth

---

100. Pierre Charron, *De la sagesse livres trois* (Bordeaux: Simon Millanges, 1601), trans. Samson Lennard as *Of wisdome three bookes written in French by Peter Charron* (London: Blount & Aspley, [1613]), book 1, chap. 11. On hearing, also see Tyard, *Solitaire second,* 197 (quoting Artistotle), and Montaigne, "Des noms," in *Essais,* 1: xlvi; for the theories of Ficino and their reception see Tomlinson, *Music in Renaissance Magic,* 134–144.

101. Timothy J. Reiss, *Knowledge, Discovery, and Imagination in Early Modern Europe: The Rise of Aesthetic Rationalism* (Cambridge: Cambridge University Press, 1997), 185–186. My debt to this work is profound.

in a system assuming that words were signs of signs of signs in chains reaching back to some secular or divine authority. "So long as the assumption survived that the structure and use of words had in some way been created from objects in nature, however mediated, one could rely on linguistic *'causas'* (as Scaliger still called them in 1540): explicable origins of language and confirmable meanings. Once guarantees of such fixed relations between concept and linguistic sign, between referent and concept, and between donatatum and sign had been removed, then urgent and profound problems arose."[102] Henri Estienne, Petrus Ramus, La Noue, Montaigne, Estienne Pasquier, and other humanists associated with the court and parlement circles blamed the civic catastrophes of the 1570s on "linguistic and conceptual incapacity," finding that social instability resulted from the lack of secure meaning and comprehension in language. Language seemed to break apart as a system of knowledge at the same time as civil society collapsed.

At this juncture numbers became an alternative alphabet, "a genuine language whose order signified and corresponded to mental concepts and judgements."[103] Some decades later, Galileo Galelei would famously put it thus:

> Philosophy is written in this immense book which is always open before our eyes, which is to say the Universe, but one cannot understand it if one does not first commit to understand its language and to learn the characters with which it is written. It is written in the language of mathematics and its characters are triangles, circles, and other geometrical figures, without the means of which it is humanly impossible to understand a word. Without them, it is a vain wandering in a dark labyrinth.[104]

Galileo's concept of philosophy, like Thibault's Pythagorean fencing and Bodin's numerical grounding for the state, asserts the order of numbers as a practical means of discovery that might replace the decaying systems of syllogistic logic he attacks in *Il saggiatore.*

If anything, music became even more important among the arts as measurement created new conditions of possibility for knowledge, for music was made of number. As a contemporary of Le Jeune remarked of his *musique mesurée,* "On apperçoit en sa Musique les secrets de Mathematique" (one perceives in his music the secrets of mathematics).[105] That is, music made the secrets of mathematics perceptible; it sounded a set of first truths.

---

102. Ibid., 98.

103. Ibid., 114.

104. See Galileo Galilei, *Il saggiatore* (Rome, 1623), ed. Libero Sosio (Milan: Feltrinelli Editore, 1965), 38.

105. See the panegyric poem by A. T. seigneur d'Ambry prefacing Claude Le Jeune, *Le printans.*

The remainder of this book explores the social and disciplinary orders established by musical activities (dancing, fencing to music, marching, and riding to music), turning our attention to the actual practices in which the ordering force of musical number was put to work. Royal government was ultimately internalized through music as it operated within the socializing processes of civility, etiquette, ballet, and ceremonial. Less scientifically than Descartes would wish, no doubt, I will pursue a cultural explanation of how music worked on the passions. And while sometimes the practice of, for example, teaching a rank of musketeers to fire simultaneously may seem far off the mark of social concord, my goal is to show the unique ability of music to traverse and connect sense (hearing) and individual action to social and political orders during the period in which the modern state was formed. The universe was written not just in the language of mathematics, but in the language of harmonic proportions, and musical education, even at the basest level of marching or dancing together, taught body and soul the characters of this book, merging them into transcendent orders. In an age that saw the culmination of Neoplatonism and the dawn of the scientific revolution, music traversed the shift from language to number, for it spoke to the soul in a language beyond words.

3

VIOLENCE, DANCE, AND *BALLET DE COUR*

Les Grecs . . . ont donné à entendre qu'elle [la danse] a prins son origine dés le commencement du monde sur le patron des mouvements celestes des astres & planettes, de leurs cours naturels ou retrogrades, des conjonctions, & en somme de l'ordre d'iceux, qui n'est qu'une danse mesuree & bien accordante.

[The Greeks have given us to believe that dance originated at the beginning of the world upon the pattern of celestial movements of the stars and planets, their natural or retrograde orbits, conjunctions, & in sum, upon the order of these things, which is nothing other than a measured and harmonious dance.]

Heinrich Cornelius Agrippa, *Declamation sur l'incertitude, vanité, et abus des sciences* (1582)

Or la plus grande perfection de l'esprit consiste à sçavoir & à contempler les plus excellens ouvrages de la nature, par exemple, les mouvemens des Astres, & des Elemens, & leurs grandeurs, leur lumiere, & leur perfection, & à s'élever par leur moyen à l'Autheur de l'Univers, qui est le grand maistre du Balet que dansent toutes les creatures par des pas & des mouvemens qui sont si bien reglez, qu'ils ravissent les sages & les sçavans, & qu'ils servent de contentement aux Anges, & à tous les Bien-heureux.

[The greatest spiritual perfection consists of knowing and contemplating the finest works of nature, for example, the movements of the Stars & the Elements, & their sizes, light, & perfection, & to elevate oneself through them to the Author of the Universe, who is the grand master of the Ballet that all creatures dance in steps & movements that are so well ordered that they ravish wise men & learned ones, & serve to please the Angels & all the Blessed spirits.]

Marin Mersenne, *Harmonie universelle* (1636)

The notion that the stars turned in heavenly *voltes* to music produced by their astrological ballet came from Plato. His Myth of Er tells of how the nested hemispheres of the world-soul turned upon an axis, while "up above on each of the rims of the circles a Siren stood, borne around in its revolution and uttering one sound, one note, and from all the eight there was the concord of a single harmony" (*Republic* 617b).[1] The human soul had been set in similar motion when the circles of the immortal soul were fastened into it, but when it was plunged into the maelstrom of conflicting motions caused by physical sensations, the proportions between them "produced all manner of twistings, and caused in their circles fractures and disruptions of every possible kind, with the result that, as they barely held together one with another, they moved indeed but moved irrationally, being at one time reversed, at another oblique, and again upside down" (*Timaeus* 43d–e).[2] Ordering the human soul meant restoring the smooth turnings of its inner circles and controlling the influence of the senses upon it.

It is worth repeating this bit of Plato in order to emphasize the importance of physical motion in his cosmography. His universe was not vibrating like the strings of a lute, but whirling in revolutions, which is what made the *bal des astres* metaphor so apropos. It is also worth repeating that while musical consonance bore the greatest burden in discourse about the music of the spheres, Plato did praise rhythm, which instilled "measure and grace" in listeners (*Timaeus* 47). Furthermore, he says that in dance (*emmeleiai*), the body shows itself to be in harmony (*emmelôs*) with music, moving with beauty, agility, and grace (*Laws* 795, 816). No wonder Mersenne and others saw in ballet the mimesis of cosmic harmony, as dancing bodies moved through choreographed patterns in time to the music. Music, in turn, regulated the dancers' steps, measuring them out in repetitive rhythms, consonant chord progressions, and the most audible hallmark of dance music, strong meter. Indeed, for Plato's remarks to make sense, it must have been that for the Greeks, too, the special concord or *emmelôs* of dance arose from a rhythmic harmony between moving bodies and measured music.

Dance drew the body into the moral equation. Setting the soul in motion with steps, leaps, and turns, dance inspired knowledge of first causes and virtue. It was this corporeal edification that led Tuccaro to defend dance in his *Trois dialogues de l'exercice de sauter*. As he says there, when we move, we not only see, feel, and know that which has movement, but we also understand and know virtue and the causes of movement itself, whether material or immaterial.[3] Tuccaro makes evident use here of the *Timaeus*, 47b–e, reminding us that all souls—the world-soul and the human soul—are in motion. When body and soul move in good proportion, we

1. Plato, *The Republic*, 2:503–505.

2. Plato, *Timaeus*, 97.

3. Tuccaro, *Trois dialogues*, fol. 174r. "Nous faisant par le mouvement, non seulement voir, sentir & cognoistre ce qui a mouvement, mais aussi comprendre & sçavoir la vertu & la cause mouvante, materielle, ou sans matiere telle qu'elle est." Also see fols. 16v–19v on bodily movement and its effects on the soul.

experience grace or divine knowledge.[4] And he quotes Homer, who said that the jump made with measure conforming to the proportional times of the cadences of music was a true gift and grace from the Heavens.[5] Dance can enlighten.

The music connected with Baïf's Neoplatonic academy has been studied in depth, as has the academy's involvement in early *ballets de cour.*[6] But the relationships subsequent theorists posited between dance music, dancing bodies, and the social benefits first hoped for by the academicians has—surprisingly—been relatively neglected. In the first part of this chapter, I consider two treatises, each of which attempts to codify theories of rhythmic ethos. Understanding how rhythm was believed to affect the body and its passions makes possible a more pointed analysis of dance as a moral practice relevant to the civilizing process. This is exemplified in the second part of the chapter, which considers the gaillarde and its masculine affects of ambition, physical virtuosity, and even aggression. Finally, we turn to *ballet de cour* as an expression of princely values. Ballets engaged courtiers in the performance of political hierarchies, created opportunities to be seen and to perform, and—just as important—permitted observers to judge those performances with exaction. My desire is to establish the involvement of some of the foremost military commanders of the day in the planning and performance of ballets and to analyze the ballets' treatments of armed violence. In the ballets at court we can observe the continuities between the monarchy's desire to monopolize armed force and the internal social machinery by which this goal was pursued.

FROM NEOPLATONICS TO MECHANICS: RHYTHMIC ETHOS AND THE BODY

In the Renaissance, the speculative science of music was dominated by harmonic theory of the sort considered in chapter 2. Treatises seldom offered more than the most standard explanations of rhythm and mensuration; among the few that did venture more extensive treatments of rhythm are Bartolomeo Ramis de Pareia's *Musica practica* (1482), Franchinus Gaffurius's *Practica musice* (1496), Francisco Salinas's *De Musica libri septem* (1577), and Descartes's *Compendium musicae*

4. Ibid., fols. 45r–46r, 68r; Plato, *Timaeus,* 106–109.

5. Tuccaro, *Trois dialogues,* fol. 42r. "L'incomparable Homere dit que le saut fait avec mesure conforme au temps proportionné des cadences de musique, est plustost un vray don, & grace des Cieux, que de la terre."

6. The foundational studies of early *ballets de cour* are Henri Prunières, *Le ballet de cour en France avant Benserade et Lully* (Paris, 1914; reprint, Paris: Editions d'Aujourd'hui, 1982); Yates, *French Academies,* esp. 60–62, 265–270; and Margaret M. McGowan, *L'art du ballet de cour en France, 1581–1643* (Paris: Centre National de la Recherche Scientifique, 1978).

(1618).[7] Tyard's *Solitaire second* (1555) treats meter only briefly but makes significant allusions to its affects.[8]

Between *Solitaire second* and the *Compendium musicae* the French produced few music theory treatises, which makes all the more significant the manuscript treatise of Nicolas Bergier, *De la musique speculative* (c. 1608), a work entirely devoted to *la musique rithmique.*[9] Bergier, like other theorists before him, understood the science of music to be a science of numbers; his startling innovation was to leave harmonics aside and to treat rhythm exclusively.

For Bergier, rhythm ordered time in a rational way that made it susceptible to the same sorts of mathematical speculation traditionally pursued with the monochord. He extended the methods of Pythagorean theory to the rhythmic domain, demonstrating how a set of numbers produced by the geometric mean (1:2:4:8:16:32:64) comprehends the total possible number of legitimate rhythmic "feet" (combinations of short and long notes grouped in small units based on classical poetic feet).[10] Geometric proportions both complete and enclose a rhythmic system described in terms of ratios, proving that rhythm is a sounding version of geometrical figures (lines and cubes). Moreover, Bergier argues that rhythm is superior to harmony in a metaphor casting harmony as a woman and rhythm as the dominant male. She, harmony, provides only raw material (melody), which rhythm shapes into beautiful forms with beat and measure.[11] Harmony is merely the stuff of the universe, whereas rhythm orders that stuff, disciplining sound, setting it in motion, and regulating the counterpoint of polyphonic music. And just as the motions of

7. See Bartolomeo Ramis de Pareia, *Musica practica* (Bologna, 1482), commentary and trans. Clement A. Miller (Neuhausen-Stuttgart: American Institute of Musicology, 1993), part 3; Franchinus Gaffurius, *Practica musice* (Milan, 1496), trans. and transcribed Clement A. Miller as *Practica musicae,* Musicological Studies and Documents 20 (Dallas, Tex.: American Institute of Musicology, 1968), book 2; Francisco Salinas, *De Musica libri septem* (Salamanca, 1577), facsimile ed. (Kassel: Bärenreiter, 1958), books 5 and 6; and Descartes, *Compendium Musicae,* in *Oeuvres,* 10:92–96. For discussions of the importance of theoretical treatments of rhythm see Reiss on Gaffurius and Descartes (*Knowledge, Discovery, and Imagination,* 163, 193), and Moyer on Gaffurius (*Musica Scientia,* 76–78).

8. Tyard, *Solitaire second,* 79–80, 214–215, 243–244.

9. Nicolas Bergier, *La musique speculative,* ed. and trans. Ekkehard Jost (Cologne: Arno Volk Verlag, 1970). On Bergier's work, see Philippe Vendrix, "Nicolas Bergier: Le dernier théoricien de la Renaissance en France," in *"La musique, de tous les passetemps le plus beau": Hommage à Jean-Michel Vaccaro,* ed. François Lesure and Henri Vanhulst (Paris: Klincksieck, 1998), 369–386. Bergier's concern with quantity in poetry fits with the mathematical trends discussed in McClelland, "Measuring Poetry, Measuring Music," esp. 25–32.

10. Bergier, *Musique speculative,* 154–166. At the outset Bergier established that the etymology of *rime* makes it equivalent to *rythme* (48).

11. Ibid., 98, 130, 194. Bergier is quite insistent in his use of this gendered metaphor (attributed to Martianus Capella).

the spheres are an ideal to which the inner orbits of our souls aspire, so too *la musique rithmique* produces a strong ethic effect upon the soul. "Rhythm has Authority over our passions," Bergier declares.[12]

Bergier's treatise culminates with a system of rhythmic ethos. Tyard had hinted at the relevance of rhythmic "modes" half a century before when he called for a new style of composition approximating poetic meters with musical rhythms based on ancient "modes de chanter," but he never explained precisely how these modes were constituted.[13] Bergier, on the other hand, describes twelve rhythmic modes and their ethic effects, recalling in this respect Glareanus's *Dodecachordon.*[14] In place of Dorian, Phrygian, Lydian, and so on, we have the pyrrhic, spondaic, dactylic, and so forth. Like the harmonic modes, the rhythmic modes can stir the passions and lead men to virtue, and much of Bergier's discourse simply repeats those old yarns about the harmonic modes, revising them to serve his rhythmic theories.[15] The classic tale of Pythagoras and the drunken Tauromenitan exemplifies the pacific affect of the spondaic rhythm in Bergier's retelling. Pythagoras commands a cornettist to play a tune in the regular, slow longs of the spondee, which brings the Tauromenitan to rights.[16] In the same vein Bergier recalls his own encounter with a drunken young man from Lorraine who rampaged about brandishing his sword and swearing madly; Bergier had the presence of mind to grab a lute and play a *passamezzo*—a dance composed, he observes, of double spondees. With just ten or twelve chords the man's fury subsided, and he stopped in his tracks to listen to the lute, after which he recovered his reason and could be led quietly away. Here we should note, along with Bergier, that the *passamezzo* he played was in the Phrygian mode, the deleterious effects of which he attributes to the rhythms usually coupled with it, not its harmonic intervals, which he describes as "soft, gracious, and pleasant by nature."[17] He returns to the locus classicus of modal ethos—

12. Ibid., 190. "Puis donc que la Rime a tant de force et d'Empire sur noz passions que de nous mesme par un mouvement naturel nous nous laissons aller apres elle."

13. See Tyard, *Solitaire second,* 214, 243, and my interpretation of these passages in van Orden, "*Les vers lascifs d'Horace:* Arcadelt's Latin Chansons," *Journal of Musicology* 14 (1996): 338–369, at 361–364. On the pieces that attest to Le Jeune's theories of rhythmic ethos, see His, *Claude Le Jeune,* 278–295.

14. Bergier, *Musique speculative,* 186. The modes are those based on feet of two or three syllables: the pyrrhic foot (⏑ ⏑), iamb (⏑ -), trochee (- ⏑), spondee (- -), tribrach (⏑ ⏑ ⏑), dactyl (- ⏑ ⏑), amphibrach (⏑ - ⏑), anapest (⏑ ⏑ -), bacchic foot (⏑ - -), amphimacer (- ⏑ -), palembacchic foot (- - ⏑), and the molossus (- - -).

15. Bergier, *Musique speculative,* 198–200.

16. Bergier's telling resembles that of Montaigne, who relates how Pythagoras's "poisante, severe et spondaïque" music induced sleep. See Montaigne, *Essais,* 1:277.

17. Bergier, *Musique speculative,* 198. "Et neantmoings Aristote, & plusieurs autres tesmoignent que la mode Phrygienne avoit cette puissance & proprieté d'esmouvoir les Esprits, leur inspirant certaines bouttades et antousiasme: & les poussant jusques a la fureur. Ce que ceste Mode ne pouvoit

stories about the Phrygian mode—in order to show that rhythm trumps harmony in ethic effect.

In contrast to the lugubrious longs of spondees and their soporific affect, the successive shorts of the pyrrhic mode ignited a bellicose vigor in men. Soldiers attack in a pyrrhic rhythm (all short beats), its rapidity eliciting martial anger and courage.[18] The spondaic and pyrrhic hang in balance with one another, as do the *passamezzo* (pavane) and military marching. Thoinot Arbeau chose the same examples in *Orchésographie* (1589) to illustrate the two kinds of "useful" dance described by Plato and Lucian of Samosata: the dances of war (marching and Pyrrhic dance) and the dances of peace (social dances, beginning with the *basse danse* and pavane). Whether Bergier was reading Plato, Lucian, or Arbeau is not a concern, although these other texts do suggest one theoretical background for his own. In the context of my argument here—that dance music exemplified the kind of "sounding number" sought by music theorists at this time—Bergier's chapter on rhythmic ethos is striking first and foremost because its examples are so mundane. He moves from speculation into the realm of everyday life to prove that his theories have practical applications, and suddenly his text comes alive with moving, fighting, relaxing, and dancing bodies.[19] Rhythm is evident not only in sounds, but in dance and in the beating of the heart and arteries. Bodies are pulsing with it, and not just any bodies, but male bodies. The masculine rhythmic property is shown to course through a testosterone-laced world full of unbridled anger (remedied with spondees) and military virtue (inspired by the pyrrhic).

In the *Compendium musicae,* Descartes, too, treated rhythm with visceral examples drawn from the soldier's life. Cannon fire, flutes, and side drums all resound with palpable force in the text, prompting his observation that "sound doth concusse, or shake all circumjacent bodies."[20] Sound sets things in motion through

---

avoir de la part de ses intervalles harmoniques, qui sont doux, gratieux, et plaisans de leur nature. Mais cela luy arrivoit de la part de la Rime, laquelle comme plus forte que l'harmonie, faict changer de forme et condition aux modes."

18. Ibid., 192. The same description of marching and combat rhythms can be found in Thoinot Arbeau's *Orchésographie* (1589).

19. My reading of this chapter contrasts with—though does not contradict—that of Vendrix, who concentrates on the relevance of Bergier's theories to the reform of plainchant (Vendrix, "Nicolas Bergier").

20. Descartes, *Oeuvres,* 10:95; Descartes, *Renatus Descartes Excellent Compendium of Musick: with Necessary and Judicious Animadversions Thereupon by a Person of Honour,* trans. Lord William Brouncker (London: Thomas Harper for Humphrey Moseley, 1653), 6. An excellent modern-language translation with a good introduction is René Descartes, *Abrégé de musique; Compendium Musicae,* ed. and trans. Frédéric de Buzon (Paris: Presses Universitaires de France, 1987). Buzon's edition is keyed to that of Adam and Tannery.

a physical process by which musical accents create a pulse that travels through the ear and impacts the passions.[21] These beats strike the body, making it move: "For this Rule is there kept, that we may distinguish every stroke of the Musick with a single motion of our bodies; to the doing of which we are also naturally impelled by Musick."[22]

Descartes's rather blunt characterization of dancing as a physical response to musical accent accords well with the aesthetic of French Renaissance dance, which employed a basic vocabulary of hops and leaps. Compared to French baroque dance, Renaissance movement was dramatic indeed, and it fit with the clearly accentuated beat patterns of dances such as the pavane, gaillarde, and branle in more obvious ways than the steps of the minuet and gavotte, for instance, which set up a delightful rhythmic dissonance between body and musical meter.[23] Descartes was right to see in the social dance of his era a strong correlation between beat and step.

Moreover, in marching and the rhythmic drills being developed at the time, Descartes would have observed the percussive effect of military drumming directly, which did indeed seem to "smartly and violently concusse or agitate [the] Spirits, by which [men] are excited to motion."[24] Repeating the arguments of Neoplatonists (though stripping these claims of their moral implications), Descartes explained musical affects thus: "a slow measure doth excite in us gentle, and sluggish motions, such as a kind of Languor, Sadness, Fear, Pride, and other heavy, and dull Passions," whereas "a more nimble and swift measure doth, proportionately, excite more nimble and sprightly Passions, such as Joy, Anger, Courage, etc."[25] The ethical ends of

21. Descartes, *Compendium of Musick,* 6; *Oeuvres,* 10:94. Some commentators on the *Compendium* have ignored this passage when trying to situate Descartes's understanding of acoustics at the time relative to Beeckman's fairly advanced corpuscular theories and the subsequent theories of Mersenne. While it is true that acoustics turns around the study of pitch and not duration, Descartes's comments here suggest that he was thinking about sound as a physical phenomena, not strictly in terms of mathematics, as maintained in H. Floris Cohen, *Quantifying Music: The Science of Music at the First Stage of the Scientific Revolution, 1580–1650* (Dordrecht and Boston: D. Reidel, 1984), 161–166. Buzon's nuanced interpretation acknowledges Descartes's discussion of sympathetic vibrations as one that makes an important distinction between Cartesian thought—even if it is less mechanistic than it was to become in 1633—and Zarlinian mathematics (Descartes, *Abrégé de musique,* 12–16). For an analysis of the rhythmic portion of the treatise and its nascent mechanism, see Bertrand Augst, "Descartes's Compendium on Music," *Journal of the History of Ideas* 26 (1965): 119–132, esp. 129–132.

22. Descartes, *Compendium of Musick,* 6; *Oeuvres,* 10:94–95.

23. Descartes specifically mentions the gaillarde in his letter to Mersenne of 18 March 1630. See *Oeuvres,* 1:134.

24. Descartes, *Compendium of Musick,* 6; *Oeuvres,* 10:95.

25. Ibid. The term "proportionately" is key. Descartes did not aim to categorize the ethic results of stirring the passions. Rather, he wished to show that the senses respond most immediately to objects of simple proportions. He specifies the correlation between tempo and relative excitement with

musical education were transformed into a utilitarian mechanism ripe for military exploitation.

From Descartes and Bergier we can mine a very real sense of the threats that music might contain. Masculine rhythm rules the world, bridling irrational anger, regulating action, and enforcing physical order. Both Descartes and Bergier wrote in the wake of the most horrendous of the religious wars, in which nobles chose their own sides in the conflict, amassed armies and bands of marauders, and often acted for independent gains having little to do with spiritual beliefs. Many showed no fidelity to the monarchy, whose own family was torn with factionalism (the youngest of the Valois brothers, François, Duke of Alençon, had to be taken under house arrest in 1572 for siding with the Huguenots). In short, the Second Estate acted out of control, and when the Edict of Nantes officially brought the eighth civil war to a close in 1598, a new conception of noble behavior arose in conjunction with the widespread desire for lasting peace. Valorous action continued to evince virtue, but within a culture of civility that created new modes for its display. Physical activities requiring balance and poise rivaled military action as registers of moral uprightness, and, as we shall see, as expressions of masculine virtues. In this regard, it was all the more important that dancing was not just one among many noble professions, but stood in a direct relation to the tradition of physical valiance that defined the nobleman, much as civility and absolutism belonged to a dialectic initially defined by war.

CIVILITY AND DANCE: *GALANT ET GAILLARD*

At European courts, etiquette and politesse were the lifeblood of the civilizing process.[26] What began to a large extent with Erasmus's slight manual on manners for children, *De civilitate morum puerilium libellus* (Basel, 1530), and Castiglione's *Il cortegiano* (Venice, 1528) had, by the early seventeenth century, grown into a vast discourse vesting gesture, bearing, and physical habit with social import.[27] From the well-thumbed vernacular translations of Erasmus collected by ordinary folk and

reference to meter, explaining that duple meter is "slower and duller" than triple, its basic proportion of 1:2 eliciting less excitement than the more engaging proportion of 1:3, which presents three divisions to the senses rather than just two.

26. Norbert Elias, *The Civilizing Process: Sociogenetic and Psychogenetic Investigations,* trans. Edmund Jephcott, ed. Eric Dunning, Johan Goudsblom, and Stephen Mennell, rev. ed. (Oxford: Blackwell, 2000).

27. On the repertory of civility manuals see Roger Chartier, "From Texts to Manners: A Concept and Its Books; *Civilité* between Aristocratic Distinction and Popular Appropriation," in *The Cultural Uses of Print in Early Modern France,* trans. Lydia G. Cochrane (Princeton: Princeton University Press, 1987), 71–109.

the innumerable French civility texts modeled on Castiglione's to the unspoken laws of conduct at seventeenth-century Parisian salons, the expansion of etiquette created a discourse of action, movement, and practice that aimed to wipe away brutish behavior. Yet whereas Erasmus had envisioned the regularization of manners throughout Europe as a great equalizer that would smooth interaction among people from different countries and social spheres—a common physical currency for cross-cultural exchange—this equalizing strain of civility was ultimately superseded by the creation of manners as a means of distinction. François de La Noue, Antoine de Pluvinel, Thomas Pelletier, and others worried increasingly about educating young Frenchmen abroad, where they would never learn *les moeurs, la façon, la grace,* and *la mode françoise.* French identity—noble identity—relied on education in manners and learning the French way of doing things. Manners were evident, according to Pelletier (writing in 1604), in the way one rode, jousted for rings, danced, and dressed, which identified one as a French nobleman.[28] Here we see the influence of *Il cortegiano* and the Italian courtly strain of civility, in which "culture" is a *marque de noblesse.*[29] "Culture" or style is tied to the martial arts, but their violence has been overlaid with a new necessity to evince grace in the French manner.

Much recent scholarship that sees the imprint of civility on early modern literature and the arts has concentrated on its foundations in oratory and theatricality. As Stephen Greenblatt put it in *Renaissance Self-Fashioning,*

> the manuals of court behavior which became popular in the sixteenth century are essentially handbooks for actors, practical guides for a society whose members were nearly always on stage. These books are closely related to the rhetorical handbooks that were also in vogue—both essentially compilations of verbal strategies and both based upon the principle of imitation.[30]

The improvised conversations of *Il cortegiano* trafficked in word play, dissimulation, and the glossy fiction of *sprezzatura,* which masked artifice with nonchalance. That

---

28. Pelletier, *Nourriture de la noblesse,* fol. 96v.

29. See Schalk, *From Valor to Pedigree,* chap. 8. On the tradition of *politesse mondaine* and its difference from the Erasmian tradition established by *De civilitate morum puerilium* see Chartier, "From Texts to Manners," and Jacques Revel, "The Uses of Civility," in *Passions of the Renaissance,* ed. Roger Chartier, trans. Arthur Goldhammer, vol. 3 of *A History of Private Life* (Cambridge, Mass.: Belknap/Harvard University Press, 1989), 167–205. On the reception of Castiglione, Della Casa, and Stefano Guazzo's *La civil conversatione* in France, see Maurice Magendie, *La politesse mondaine et les théories de l'honnêteté en France au XVIIe siècle, de 1600 à 1660,* 2 vols. (Paris: Librairie Félix Alcan, 1925), esp. 1:305–354.

30. Stephen Greenblatt, *Renaissance Self-Fashioning: From More to Shakespeare* (Chicago: University of Chicago Press, 1980), 162.

Castiglione was read as a rhetorical guidebook to social mobility is evident in the title of Stefano Guazzo's adaptation of it as *La civil conversazione* (1574).[31] The dialogue format in which Castiglione presented his work became a conversational model for social negotiations, one that historians attempting to decipher the codes of French lyric, *air de cour,* dance, and seventeenth-century literature have eagerly adopted as an interpretative tool.[32]

Yet caution is in order, for education in language and letters was not the aristocrat's first priority in late-sixteenth- and early-seventeenth-century France. Conversational finesse aimed—like other manners—to convince others of one's moral character. It required control of the body and its passions and, moreover, brought interlocutors into contacts that allowed them, as Mark Franko has put it, "to observe the bodies of others in order to understand and eventually manipulate them as well as to control one's own appearance to others to the same end."[33] Beginning with tutelage at home, young nobles developed a "civilized" use of the vernacular and an understanding of proper address as but two of a number of manners that were primarily nonverbal.[34] Table manners, etiquette, dance, parlor games, and participation in religious ritual all contributed to a program designed to teach physical and emotional restraint. Indeed, instruction in letters itself aimed to train the body, for learning to write—disciplining the hand—was considered a "grace" that evinced French style. Script was a cipher of a cultural formation taught along with manners and as a manner.

Civility texts, too, suborned speech to the principles regulating carriage and habits, eliding word and gesture in a social economy dominated, as we have already

31. In the 1666 translation by l'abbé Duhamel, Giovanni della Casa's *Galateo* would also advertise this attribute in its title: *Le Galatée, ou l'art de plaire dans la conversation.*

32. For musical studies pursuing the conversational model, see Stefano Lorenzetti, *Musica e identità nobiliare nell'Italia del Rinascimento: educazione, mentalità, immaginario* (Florence: Olschki, 2003), 105–118; Brooks, *Courtly Song,* 227–254; Durosoir, *Air de cour,* 184–188, 331, and, for the slightly later period, Catherine E. Gordon-Seifert, "'La Réplique Galante': Sébastien de Brossard's Airs as Conversation," in *Sébastien de Brossard, musicien,* ed. Jean Duron ([Versailles]: Centre de Musique Baroque de Versailles; [Paris]: Klincksieck, 1998), 181–201; and Lisa Perella, "Bénigne de Bacilly and the *Recueil des plus beaux vers, qui ont esté mis en chant* of 1661," in van Orden, *Music and the Cultures of Print,* 239–270. On dance, see Mark Franko, *The Dancing Body in Renaissance Choreography, c. 1416–1589* (Birmingham, Ala.: Summa, 1986), in which he describes dance in terms of rhetorical codes and the conversational civility epitomized by Guazzo. Also see Franko, *Dance as Text: Ideologies of the Baroque Body* (Cambridge: Cambridge University Press, 1993), in which Franko declares that *ballet de cour* (excluding burlesque ballet) is "textual in every conceivable way short of having dancers speak" (5).

33. Franko, *Dancing Body,* 71.

34. See Motley, *Becoming a French Aristocrat,* chaps. 1 and 2.

seen, by physical expressions of inner balance and decorum.[35] Giovanni della Casa put it thus in *Il Galateo* (1558), a manual that, although inspired by *Il cortegiano,* was widely read outside courtly circles in France and popularized in seventeenth-century reprints: "One finds beauty where there is a proper proportion of one part to another, and also of the parts with the whole, and one can truly call beautiful that thing in which one finds the aforesaid proportion and measure."[36] And later: "It is therefore necessary that the well-instructed person take into account this measure and proportion of which I told you, whether he walks, stands, or sits, whether in his actions or his carriage, in his habits, speech, relaxation, or exercise."[37] By adopting *juste proportion* as a standard measure of perfection, civility texts participated in the Pythagorean discourse discussed in chapter 2, turning it into a philosophical commonplace in society at large, not just in the courtly environs of the Neoplatonic academies. Given these ideals, courtesy books for good reason considered accomplishment in dance a prerequisite for those who wished to enter into civil society. It was not so much that dance was a mute form of rhetoric, as some theorists claimed, but that it externalized self-control in a persuasive form that that fit well with the endless theatricality of the age.

Dance lessons began with the social graces Orest Ranum has distinguished as "courtesy"—the *révérences,* lowered eyes, genuflections, and other signs of deference that equipped the young to enter into society.[38] In aristocratic households, children commenced their study around age five and continued through adolescence. Louis XIII studied dance from an early age, as did all of Henry IV's children, who lived together at Saint-Germain-en-Laye. They learned to walk in a princely fashion and studied the steps of social dances. In 1608 the fourteen-year-old Duke of Vendôme organized a Lenten ballet with the other children; two years later Louis, age nine, took charge of the ballet danced before the king and queen.[39] They

35. Franko assesses this tendency of civility manuals as the reduction of language to "vocal gesture" (*Dancing Body,* chap. 5).

36. Giovanni della Casa, *Le Galatee, premierement composé en Italien par J. de la Case, & depuis mis en François, Latin, & Espagnol par divers auteurs* (Lyons: Jean de Tournes, 1598), 381–383. "La beauté se trouve là ou il y a une convenable proportion d'une partie à l'autre, et puis des parties avec le tout, & peut on veritablement appeller belle la chose, ou ladite proportion et mesure se retrouve."

37. Ibid., 403–405. "Il est donc de besoing que les personnes bien apprises ayent esgard à ceste mesure et proportion dont je t'ay parlé, soit qu'il marche, ou qu'il se tienne debout, ou soit assis, soit en ses faicts ou en son port, en ses habits, en ses paroles, en son repos, ou en son exercice."

38. See Ranum, "Courtesy, Absolutism, and the Rise of the French State," and Motley, *Becoming a French Aristocrat,* 57–58, 148–149.

39. François Bayrou, *Henri IV, le roi libre* (Paris: Flammarion, 1994), 477. On the young Louis XIII also see Durosoir, *Air de cour,* 178–179. On the importance of beginning study early on, consider

were not alone in mounting such useful *divertissements.* In 1593, for example, the Duchess of Rohan presented her five children to Henry IV in a ballet that included her thirteen-year-old son Henry and nine-year-old son Benjamin, both of whom would assist the king in his defeat of the Spanish at Amiens just four years later.[40]

Dance developed strength and in this way began a basic training that was foundational for the lessons in fencing and riding that would follow. Indeed, fencing masters often doubled as dancing masters, which demonstrates the proximity of the two arts. At the Valois court, the Italian masters Pompée and Sylvie taught the young princes both fencing and dance.[41] So too, fencing masters may have given basic dance instruction to college students, though military academies always employed individual *maîtres* in dance alongside the masters of arms, equitation, and lute. Fencing lessons often took place in the same galleries used for dancing, where the geometric patterns on the floors helped students measure the size of their steps. Both arts sensitized the student to the slightest gestures of others, taught them to judge their proximity to dance partners or fencing opponents, and gave them mastery of their physical surroundings. Both employed turned-out feet and steps forward, backward, and to the side, and the arm positions of baroque dance eventually came to mirror those standardized early on for fencing. Of course the psychological conditioning of each differed, for whereas dance instilled courtesy and awareness of social hierarchies, fencing awoke courage and bloodlust, but the two practices did overlap considerably in their physical vocabularies and in the overall schooling of young noblemen.

Balls provided regular occasions for socializing among elites and became a forum for young nobles to take their first public steps into the world of high society. Henry III scheduled dances on Thursday and Sunday evenings during his reign, a tradition continued by Henry IV, while special balls, sometimes *en masquerade,* were held for weddings, entries, and state visits.[42] Students at Pluvinel's academy probably had dance lessons three or four afternoons a week, and at Aix the dancing master taught the students how to walk, greet others, and dance, skills they put to use

---

the recommendations of Pelletier, *Nourriture de la noblesse,* fols. 85v–86r. "Quant à la danse, j'estime que c'est l'exercice qu'un Gentilhomme doit apprendre fort jeune: car d'apporter un grand corps dans la salle d'un maistre à l'âge de dixhuict, à vingt ans, tout barbu, grossier, roide, & lourd comme un Elephant, ce sera bien un chef d'oeuvre si on prend jamais honneur à un tel escholier."

40. Catherine de Parthenay, Duchess of Rohan, *Ballets allégoriques en vers, 1592–1593* (Tours, [1593]), ed. Raymond Ritter (Paris: Champion, 1927), 29–48, 51.

41. Brantôme, *Oeuvres complètes,* 5:277–278, describes a fencing match between the youngsters and their masters in which the teachers are "killed" by their students and dragged off by demons.

42. On the daily routine of the court of Henry III see Boucher, *Cour de Henri III,* 51–53; on Henry IV see Janine Garrisson, *Henri IV, le roi de la paix* (Paris: Tallandier, 2000), 104.

at monthly balls.[43] There is no question that, as François de Lauze put it in his *Apologie de la Danse,* dance could fashion the young and "render them deserving of civil conversation."[44] Thomas Pelletier likewise maintained that of all the exercises a child could study, "none gives access to good company like dance. For to stay planted against the tapestry of a hall like a melancholic, or only to look over the shoulders of others, it seems to me that this is more fitting behavior for a chamber valet than a Gentleman."[45]

Of all the social dances, the gaillarde was the truest test of a dancer's skill. In his 1596 outline for the establishment of four royal military academies, Juvigny lists the gaillarde first among the dances to be taught there.[46] It was performed *à deux* as the gathered company looked on, which made it a natural showpiece.[47] First the man selected a partner, led her to the end of the hall, and, after doffing his hat in a shallow *révérence,* escorted her around the hall in a *basse danse* and its reprise. He then bowed again and took leave of her at the change of meter signaling the gaillarde, at which she went dancing away across the hall in a sprightly triple-time. The man pursued her with showy jumps and flourishes. As described by Arbeau, the dance proceeded as a series of separations, reunions, and professions of love sublimated in mincing steps and capers.[48] Popular at weddings and at court balls, it usually fin-

43. On Pluvinel's academy see his *Instruction du roy,* 200; on Aix see O[1] 915 (183), Archives nationales.

44. François de Lauze, *Apologie de la Danse et la parfaicte methode de l'enseigner tant aux Cavaliers qu'aux dames,* facsimile ed. (Geneva: Minkoff Reprint, 1977), 13–14. "Les façonneroyent & les rendroit dignes d'une civile conversation." On seventeenth-century dancing masters see Wendy Hilton, *Dance of Court and Theater: The French Noble Style, 1690–1725* (Princeton: Princeton Book Co., 1981), 45–55.

45. Pelletier, *Nourriture de la noblesse,* fol. 86r–v. "Je tiens que de tous les exercices que la jeunesse peut apprendre en ce bas âge, il ny en a pas un seul qui la face plus paroistre ne qui luy donne plus d'accés aux bonnes compagnies que faict la danse. Car de demeurer planté contre la tapisserie d'une salle comme un songe creux, où de ne regarder que par dessus l'espaule des autres; il me semble que cela est plus propre à un valet de chambre qu'à un Gentilhomme."

46. Saint-Germain, *Advis de l'établissement de quatre academies,* 30. "Le Balladin monstreroit toutes sortes de dances usitees dans le Royaume, & sur tout la gaillarde, la courante, la volte, la pavane, & les canaries: parce que ces especes de dance acquierent la disposition & font mieux recognoistre la grace."

47. The *courante* was the other common dance *à deux* in the early seventeenth century. As with the gaillarde, this made it important socially, and so de Lauze spends considerable time describing its steps and the *révérences* appropriate to it (*Apologie de la Danse,* 28–36, 65–66).

48. Thoinot Arbeau, *Orchésographie* (Langres, 1589; facsimile of Paris 1888 reprint, ed. Laure Fonta), fols. 38v–63r. With slight revisions, I use the translations in Arbeau, *Orchesography,* trans. Mary Stewart Evans, introd. and notes by Julia Sutton (New York: Dover, 1967), 76–119. The first set

ished off a suite of dances, occasioning the exchange of kisses, favors, or bouquets. Paired with the stately *basse danse* or pavane that usually preceded it, the gaillarde organized a whole system of courtship—and courtiership—based on a conversational model.

At its most elemental, the gaillarde consisted of four steps or hops (*pieds en l'air* or *grèves*), a jump (*saut*), and a pose (*posture*), for which the combination was known as the *cinq pas* or "five steps." This basic vocabulary was the object of infinite variation, and Arbeau provides twenty-some passages derived from it. Dancers embellished the steps by crossing their feet, lifting the feet behind (*ruade*), doubling or tripling their steps, marking with the heel or toe, making coupées (*entretaille*), or combining two quick changes with a kicking hop to the opposing foot (*fleuret*). The jump, meanwhile, might be low (*petit saut*), high (*saut majeur*), or ornamented by beating the feet in the air in the manner of an entrechat. Arbeau calls this leap a *capriole,* and it represents the height (literally) of male social dance, for which Arbeau's fictional student is named Capriole. At the gaillarde's most theatrical, men held long sticks before them with tassels at the end that they kicked to show off the height of their *grèves.* Mastery of this cornucopia of dance steps, like the wealth of *bons mots,* subtle turns of phrase, and the many-faceted discourse synonymous with eloquence, enabled men to dominate the dialogue established by the alternating solo passages. Indeed, compared to the other dances, which were performed side by side or in circles, the gaillarde barely had a set choreography at all, relying instead on a collection of steps and jumps the dancer was expected to arrange on the spur of the moment. Its virtuosity depended not just on the execution of *caprioles,* but on the imaginative deployment of its varied steps. The man was expected to call the tune—literally, to recommend melodies to the musicians, either by name or by singing a few bars—and to introduce new graces with each passage. The gaillarde permitted the lover to "plead tacitly with his mistress" in a conversation waged with ornamental repartees meant to impress her with his "seemliness" and "grace."[49] Love was not always the ultimate aim in this courtship, for the dance inspired the courteous behavior whereby men gained good reputations. This is why Capriole had returned to his old teacher in the first place—to become a more agreeable social companion by adding dance to his skills in fencing and tennis.

The grace demonstrated in the gaillarde depended on more than the balance required to break the forward motion of the hops and jump in order to land lightly in a pose at the end of each series of *cinq pas*—though this good beginning was an

---

of page numbers in subsequent citations refers to the facsimile of the 1888 reprint; the second set refers to the English translation.

49. Ibid., fol. 5v; 16. Arbeau provides a choreography based on the supplications and rejections of courtship for the *courante* as well (fol. 66r; 123–124).

accomplishment in itself. It rested in variety of the sort associated with extempore speech, immediacy, and naturalness. As Terence Cave has shown, Renaissance theories of improvisation and inspiration set up oral utterance as a model for writing in an attempt to "represent writing as arising immediately from the mind, breath, or voice of the speaker."[50] Variation exercises such as those Erasmus provides in *De copia* turned writing into a process for generating a well, storehouse, or cache of figures awaiting release in performance. Writing became a rehearsal for conversation in the same way that memorizing steps and combinations of them rehearsed performances that foregrounded variety as a sign of improvisational brilliance and inspirational wealth. "There are many kinds of steps, and by mingling these a variety of passages are constructed," says Arbeau, who elsewhere laments how often "dancers are satisfied to perform *les cinq pas* and a few passages without any orderly arrangement and are not bothered as long as they keep the rhythm, with the result that many of their best passages go unnoticed and are lost."[51] Discernment is required. And study: Arbeau counsels Capriole to crib passages off good dancers when he sees them, a practice synonymous with those that supported the perpetual evolutions of fashionable behavior, dress, and elocution.[52]

The music for the gaillarde was more complex than that of the other social dances, which tested the musical acumen of the dancer. In the first place, the footwork had to be coordinated with a rhythmic pattern corresponding to the basic set of five steps. As seen in example 3.1a, the jump or *capriole* occurred on a rest or held note and the pose closed off the measure. Dancers needed to hear this rhythm underneath ornamented lines, and so Arbeau provides a sample melody and its reduction along with other examples meant to develop the student's musical judgment. In this respect—its requirements of taste, judgment, and invention—the gaillarde was unique. Arbeau gives predetermined steps for other dances, instructing Capriole to learn their tunes by heart and sing them in his head with the violin. In those cases, music and movement were learned simultaneously, and musical memory acted as an aide-mémoire to recall the steps. But the variability of the gaillarde required an understanding of musical measure, phrase length, repetition, and division. Not only were *les cinq pas* elaborated upon with *fleurets, caprioles, entretailles,*

50. Terence Cave, *The Cornucopian Text: Problems of Writing in the French Renaissance* (Oxford: Clarendon Press, 1979), xii and part 1, chap. 4.

51. Arbeau, *Orchésographie,* fols. 40r, 38v; 79, 77. "Il y a plusieurs manieres d'assiettes, par les meslanges desquelles on bastit les diversitez des passaiges"; "Ceulx qui dancent la gaillarde aujourd'huy par les villes, ilz dancent tumultuairement, & se contentent de faire les cinq pas & quelques passages sans aulcune disposition & ne se soucient pourveu qu'ilz tumbent en cadence: tellement qu'une grande partie de leurs meilleurs passages sont incogneuz & perduz."

52. Ibid., fol. 49r; 93.

and so forth, but the "five steps" might be abridged (to two steps and a jump), diminished (with *fleurets* or other "mincing steps" and a jump), or extended (by delaying the cadential combination of jump and pose for one, two, or three measures, making sets of eleven, seventeen, and twenty-three steps), a kind of passagework Arbeau recommends for endings. The student had to attend to musical diminutions, matching semiminims with *fleurets* (see ex. 3.1b). Finally, hemiolas were commonplace in gaillardes, giving rise to a rhythmic dissonance with the dancer's steps at cadences and the invention of a choreographic hemiola (see ex. 3.1c), which was often applied at cross-accents with the music.

The musical command demanded by the dance's metrical alternations between 6/4 and 3/2, coupled with the advance planning required to extemporize passages of four measures and more, made the gaillarde a true exhibition of musical wherewithal. Its shifting accents and rousing triple time attracted the attentions of instrumentalist-composers such as the lutenist Adrian Le Roy, whose books of tablature for lute, guitar, and cittern always included instrumental gaillardes, not to

Example 3.1. Gaillardes from Thoinot Arbeau, *Orchésographie* (1589)
a. *Les cinq pas*

Tabulature des cinq pas de la gaillarde, de laquelle les mouuements ſont comme au tourdiō, fors qu'ils ſont faicts plus hault & plus virilemēt, et en lieu du pied en l'air, le dāceur fera des coups de pied ou greues

Air de la gaillarde Anthoinette.

Mouuements que le danceur doibt faire en dançant la gaillarde

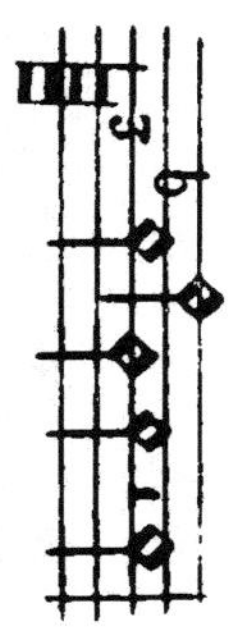

Greue gaulche.
Greue droicte.
Greue gaulche.
Greue droicte.
Sault majeur.
Poſture gaulche.

Example 3.1. (*continued*)
b. *Gaillarde la Milannoise*

Arbeau.

Air de la gaillarde appellée la Milannoiſe.	*Mouuements que le danceur doibt faire en dançant la gaillarde.*	
	Pied en l'air droit, ſans petit ſault. Pied en l'air gaulche, ſans petit ſault. Greue droicte.	Fleuret.
	Pied en l'air gaulche, ſans petit ſault. Pied en l'air droit, ſans petit ſault. Greue gaulche.	Fleuret.
	Pied en l'air droit, ſans petit ſault. Pied en l'air droit, ſans petit ſault. Greue droicte.	Fleuret.
	Pied en l'air gaulche, ſans petit ſault. Pied en l'air droit, ſans petit ſault. Greue gaulche.	Fleuret.
	Pied en l'air droit, ſans petit ſault. Pied en l'air gaulche, ſans petit ſault. Greue droicte.	Fleuret.
	Pied croiſé droit, ſans petit ſault. Reuerence paſſagiere droicte, ou entretaille du droit cauſant greue gaulche. Sault majeur. Poſture droicte.	

Continued on next page

Example 3.1. (*continued*)
c. the *cinq pas* in two steps

Air de la gaillarde appellee Si i'ayme ou non &c.

Cinq pas de deux mouuements que le danceur peult faire en dançant la gaillarde.

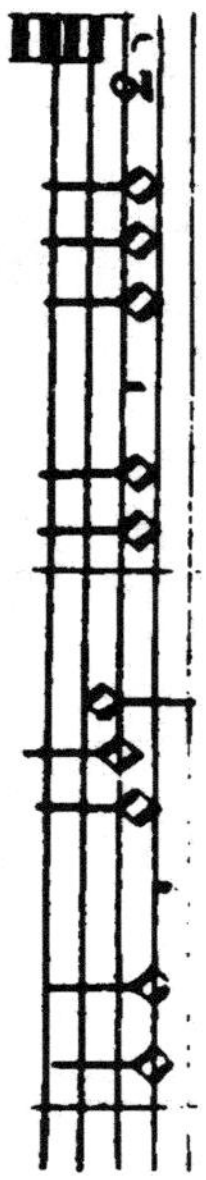

Pieds ioincts ou greue droicte.

Sault majeur auec capriole.

Poſture gaulche.

Reuers.

Piedz ioincts ou greue gaulche.

Sault majeur auec capriole.

Poſture droicte.

Et ainſi continuant en repetant le commencement.

mention many songs based on gaillarde tunes. Nicolas de La Grotte composed a number of songs for court festivities that are essentially gaillardes meant to be sung, and, as Pierre Bonniffet has shown for the measured songs of Le Jeune, most were measured to gaillarde rhythms.[53] On occasion even Ronsard constrained his diction in such a way that his poetry could be sung to gaillarde tunes.[54]

Choreographically, the gaillarde worked similarly to musical divisions, which represented the consummate instrumental art of the age. Just as the divisions on well-known grounds, chansons, and madrigals published by Diego Ortiz showed the student how to "break a ground" or get from one note to the next in a set of

53. On dance songs generally see Heartz, "*Voix de ville*"; on La Grotte see Kate van Orden, "*La chanson vulgaire* and Ronsard's Poetry for Music," in Brooks, et al., *Poetry and Music in the French Renaissance,* 79–109, at 91–97; on Le Jeune see Bonniffet, *Ballet démasqué.*

54. See van Orden, *"Chanson vulgaire."*

"variation exercises," the steps Arbeau provides for the gaillarde convey not the actual dance, but rather one of many paths by which to move through passages.[55] The tabulations were meant only to help until the student had picked up steps from good dancers. "There are an infinity of varieties," Arbeau says, "which you will obtain and learn from those of your own generation."[56] He can only provide a few basic precepts and a handful of brief examples for a dance that turned upon *phantaisie,* combinatorial skill, and great agility.[57]

The solo passages of the gaillarde proved the ultimate test of *sprezzatura.* Arbeau accorded the gaillarde persuasive powers rivaling oratory: "dancing is a kind of mute rhetoric by which the orator . . . can . . . persuade the spectators that he is spirited [*gaillard*] and worthy to be acclaimed, loved, and cherished."[58] But admiration only followed an effortless performance. The prudent shied away from double and triple leaps or *caprioles,* feats that left many dancers on the floor, where they were met with "laughter and jeers."[59] Gone wrong, the gaillarde was a recipe for social disaster.[60] De Lauze, writing three decades later, echoed Arbeau's remarks with even greater accent on the social ramifications of the dance. He devoted an entire chapter of his slight treatise to the *capriole,* which he presents as a *moyen de parvenir* for the socially ambitious even while reminding his students that pride goes before a fall: "A man should never perform the *capriole,* especially in high society [*en lieu de reputation*], if he does not excel at it, unless he wants to serve as an amusement for the company, as some do who cannot represent the carriage and respectability of our nobility, looking to recommend themselves by jumps and other combative movements."[61]

---

55. See Diego Ortiz, *El primo libro nel quale si tratta delle glose sopra le cadenze et altre sorte de punti in la musica del violone* (Rome: Dorico, 1553); idem, *Libro secondo* (Rome: Dorico, 1553); and Christopher Simpson, *The Division Viol, or, The Art of Playing Ex tempore upon a Ground* (London: W. Godbid, 1665).

56. Arbeau, *Orchésographie,* fol. 56v; 107. "Il en est d'une infinité de sortes que vous practiquerez & apprendrez de ceulx de vostre aage."

57. Ibid., fol. 55r; 104. "Il m'est advis que par les meslanges des mouvements divers que m'avez figurez, je composerois bien des cinq pas à ma phantaisie."

58. Ibid., fol. 5v; 16. "La danse est une espece de Rhetorique muette, par laquelle l'Orateur peult, par ses mouvements, sans parler un seul mot, se faire entendre & persuader aux spectateurs, quil est gaillard digne d'estre loué, aymé & chery."

59. Ibid., fol. 63r; 119. "Mais il est advenu maintesfois qu'en faisant les souples saults, ils se sont laissez tumber, dont la mocquerie & les risées s'en sont ensuyvies."

60. By the end of the seventeenth century, gentlemen could destroy a career at court with missteps in balls. See Hilton, *Dance of Court and Theater,* 15–16.

61. De Lauze, *Apologie de la Danse,* 46. "Car en effect un homme ne se doit jamais mesler de caprioler, principalement en lieu de reputation, s'il n'excelle, ou s'il ne veut servir de joüer à la compagnie, comme font aucuns, qui ne pouvant representer le port & la descence de nostre noblesse, cherchent à

Arbeau had described the gaillarde as a means of eliciting acclaim, admiration, and love; de Lauze described it more explicitly as a means of building one's reputation, a difference which relates to the later period in which he wrote and his immediate audience. De Lauze dedicated his book to the Marquis of Buckingham, whose service he entered after teaching in France. He addresses "Cavaliers" throughout, pitching his method at French and English nobles and gentry or those who aspired to imitate their manners. He teaches the evolving graces of social dance, which by his time included specific motions for the arms and hands, as well as the many *révérences* proper to the increasingly hierarchical system of politeness, which demanded various depths of bows according to the quality of the person being greeted and, before the king, had acquired a ritualistic dimension associated with the divinity of his majesty and the godly reverence owed it.[62] Gone are Arbeau's barbarous *canarie,* evocative *bransle des chevaux* and *bransle des lavandieres,* and lascivious *volte de provence,* which Louis XIII banned at court for its grappling turns that exposed the thighs of ladies who failed to hold down their skirts. De Lauze writes with an acute awareness of the ways in which dance affirmed good breeding and mobilized one socially. The difficulty of the gaillarde and especially the *capriole* may explain in part why the gentler (but equally fleet) *courante* seems to have become so popular in the 1610s, at least judging by the dances for lute published by Ballard and Vallet and the remarks of Mersenne.[63] This may also explain why the dance itself became more refined around 1600, when the brusque and athletic *cinq pas* of three quick half notes and a whole note, ornamented with eighth notes, began to acquire ornaments in sixteenth notes (a turn most easily seen in English sources such as the late works of William Byrd and the galliards of John Bull.) By the time of Louis XIV, the dance was much slower and elegant, with extensive ornaments in step, gesture, and music. It still retained the three half notes but had acquired figuration in sixteenths and even thirty-seconds (in the work of Louis Couperin and Jean Henry d'Anglebert), as though the dance itself—formerly as robust as the soldier-courtiers who danced it—had been made to do reverence to courtly and musical demands.

According to de Lauze, a nobleman of ability should perform the *capriole* if he

se recommander par des sauts & autres mouvemens battelleresques." The cautious Pelletier—who always recommended the decorous middle road—also cited the *capriole* as an example of social angling (*Nourriture de la noblesse,* fol. 87r). "Car de sortir de la foule pour venir faire des capriolles au milieu d'un grand bal, avec un petit pennache de quarante sols à vostre chapeau sans sçavoir qui vous estes, ou pour estre trop cogneu, c'est se faire mocquer à credit."

62. On *révérences* see Franko, *Dancing Body,* 31–37, and, for the late seventeenth century, the well-illustrated chapter on etiquette in Hilton, *Dance of Court and Theater,* 268–289. For the performance of rank see Skiles Howard, *The Politics of Courtly Dancing in Early Modern England* (Amherst: University of Massachusetts Press, 1998), esp. chap. 1.

63. Mersenne, *Harmonie universelle,* 2:165.

FIGURE 3.1. Abraham Bosse, *Ball during the Reign of Louis XIII.* Bibliothèque nationale de France, Cabinet des estampes, Kd 5.b tom. 3

excelled at it, even (or perhaps especially) in distinguished company. The phrase *lieu de reputation* (literally, place of reputation) located the dancer in just the sort of social arenas where the ambitious advanced their careers. At the time he wrote, the world of high society, or *le monde* included the court of Louis XIII, where balls and ballets were favored, as well as the new social groups created in Parisian salons. Nobles had moved to Paris in significant numbers after the wars, building large *hôtels* in the Marais and—along with wealthy bourgeois—moving into homes facing the Place Royale and the Place Dauphine. In the salons of these new urban dwellings, nobles and bourgeois alike parlayed wit, brilliant conversation, literary skills, and physical grace into a public reputation.

The most important site of the new *civilité* or *honnêteté* associated with salon culture was the *chambre bleue* of Catherine de Vivonne. The softened gestures and purified speech pursued there became a decisive model of refinement, one distinct from the style of the court. Catherine entertained her guests not only with polite conversation, but also with literary games, comedies, serenades, and concerts that regularly included *airs de cour.*[64] From the lips of social climbers poured *airs* that reflected the polite style of diction in their limpid turns of phrase. Mersenne favorably compared the French *air* to Italian opera arias, dismissing the radical melodic leaps and dissonant harmonies of the Italian style and maintaining that the French represent their passions with less violence: "Airs are composed particularly *and principally to charm the mind and the ear* and to enable us to spend our lives with a little sweetness amidst all the bitterness we meet with, . . . *not . . . to excite anger and several other passions.*"[65] The rejection of excessive passion reached an extreme at the Hôtel de Rambouillet, which by the 1630s had set itself in opposition to the king's court.[66] Catherine found the behavior of Louis XIII indecorous and refused to set foot in the Louvre, which she denounced for its rudeness and ostentation. By contrast, acceptance into her circle could elevate one's social stock and cleanse a sullied reputation.

Conversation, literature, and song prevailed at the *chambre bleue,* though Catherine did hold dances. Certainly balls were central to the activities of subsequent salons, where the physical restraint and decorum enforced by dance aptly reflected the virtue of *honnêteté.* Tallemant des Réaux, a frequent guest at the Hôtel de Rambouillet and one of the keenest gossips of the century, never failed to remark whether the persons he described were able dancers.[67] And when members from the rival salons held at the Hôtel de Créqui and Hôtel de Ventadour found themselves together, all sorts of intrigues calculated to prevent the others from dancing ensued.[68] At subsequent salons as at court, dance created opportunities for self-advancement and social contest, and in this respect it particularly suited those with an ambitious nature. As we have heard from de Lauze, men sought to "recommend themselves" with *caprioles.*

64. On *airs* in the salons, see Perella, "Bénigne de Bacilly," and idem, "Mythologies of Musical Persuasion: The Power and Politics of Song in France, 1653–1673" (Ph.D. diss., University of Pennsylvania, 2003), chap. 3.

65. Marin Mersenne, *Correspondance du P. Marin Mersenne,* ed. Cornelis de Waard, 17 vols. (Paris: G. Beauchesne, 1932–1988), 10:237–238. "Les airs sont faictz particulierement *et principalement pour charmer l'esprit et l'oreille* et pour nous faire passer la vie avec un peu de douceur parmy les amertumes qui s'y rencontrent"; "les airs *ne se font pas pour exciter la cholere, et plusieurs aultres passions* mais pour resjouyr l'esprit des auditeurs." See D. P. Walker, "Joan Albert Ban and Mersenne's Musical Competition of 1640," *Music and Letters* 57 (1976): 233–255.

66. On the salon at the Hôtel de Rambouillet see Magendie, *Politesse mondaine,* 1:120–148.

67. See Tallemant des Réaux, *Historiettes,* 3:339, 2:307–308, on Bassompierre's bad dancing.

68. Magendie, *Politesse mondaine,* 138.

Of course, the ambitious usually aimed for wealth and glory, and, as Faret noted in his *L'honneste homme,* arms were the surest path to "a great reputation and from there to great honors."[69] If military valor brought luster to one's reputation, what of *caprioles?* Here it is important to note the passion expressed in this leap, the ardor required to master it, even its violence. De Lauze described the *capriole* with military metaphors—it is a step "vanquished only with difficulty," older students become exasperated, as if they stood "at the foot of a wall without a ladder"—and remarked that it could appear "une action fort violente," a violence contemporaries associated with the passions.[70] And while some praised the avoidance of too much excitement, military men believed that the pursuit of glory required the cultivation of passions such as ambition, not their repression.[71] It was widely recognized that hops, leaps, and quick steps warmed the body and agitated the passions.[72] In the gaillarde (gendered feminine only because it is *une danse gaillarde*), the male dancer had the opportunity to show his strength, fiery nature, and even aggression. *Un gaillard* was, after all, a strapping fellow, a good, physically strong man—the perfect soldier. Moreover, the gaillarde unfolded at the dancer's whim, independent of a preestablished plan. The deregulated quality of the gaillarde amplified its aesthetic of individual merit and self-creation (or, for Christian moralists, its folly).[73] Like the military career—a risky one with few rules compared to that of a jurist but with vastly greater potential rewards—the gaillarde never followed a set path.[74] Brilliant dancers succeeded by choreographic wit, invention, and agility. The great rose highest, while the inept fell furthest.

For noblemen, dance was not in conflict with the military career but was, rather, a standard complement to arms. Indeed, military men typically passed the seasons

69. Faret, *Honneste homme,* 12. "C'est par les armes principalement que la Noblesse s'acquiert, c'est par les armes aussi qu'elle se doit conserver, et s'ouvrir le chemin à la grande reputation, et de là aux grands honneurs."

70. De Lauze, *Apologie de la Danse,* 47. "Il est grandement necessaire que ceux qui l'entreprendront n'attendent point à un aage trop avancé, autrement ce qui se pourroit quasi acquerir insensiblement par succession de temps, venant à estre precipité ne se pourroit vaincre qu'avec grande difficulté, & peut estre se trouveroient-ils au bout de leur compte, au pied d'un mur sans eschelle," and "et bien qu'il semble que Caprioler ou aller par haut soit une action fort violente, penible, & tres-malaisee à acquerir . . . on se la pourra rendre aussi facile que les moindres actions qui se pratiquent en la danse."

71. Dewald, *Aristocratic Experience,* 17–26.

72. Agrippa, *Declamation,* 88; [Lambert Daneau], *Traité des danses* ([Geneva]: François Estienne, 1579), 10–15.

73. In this respect, the choreography of the gaillarde resembles the essential qualities of burlesque ballet, on which see Mark Franko, "Double Bodies: Androgyny and Power in the Performances of Louis XIV," *Drama Review* 38 (1994): 71–82.

74. On ambition and the uncertainty of the military career, see Dewald, *Aristocratic Experience,* 22–26.

of each year on shifting terrain, with summer military campaigns followed by winter stations at court culminating in balls and *ballets de cour* mounted for Carnival. When Henry IV reprimanded a nobleman for not having learned to dance and the warrior explained that he served his majesty with lance blows in war, the king replied, "'Then I counsel you to arm yourself with a frock-coat in peacetime,' as if he had wished to say that once the furors of war had abated, a Knight could occupy himself with no more noble exercise than one that so advances the understanding of his Court and his world."[75] Clearly, dance—a required subject at military academies—was not just a social prerogative, but a royal one. Like civility more generally, the new accomplishments expected of knights loosened the synonymies between military action and nobility in ways that, while not imposed from above, did further the interests of those who favored an absolutist state.[76] As we shall see in the political study of ballet that follows, it was not lost on monarchs that should nobles relinquish their former violent function, conceivably their independence would diminish as well.

BALLET AS WAR ON ANOTHER STAGE

Catherine de' Medici was the first inspiration of French *ballet de cour,* a genre that emerged from the lavish productions she mounted at moments of extreme political anxiety. Like the Fontainebleau fetes of 1564 that brought Catholics and Protestants together for balls, banquets, and masquerades in the wake of the first civil war, the early ballets are often interpreted in light of Catherine's conciliatory politics as celebrations of political harmony, but here I will essay an interpretation of ballet as ritual strife. The examples of courtly spectacles thematizing peace are many, but the Valois' legendary taste for flamboyance, the luxury goods of musical theater and ballet, and the intellectual refinement of the Neoplatonic academies they patronized did not negate the violence of life at their court. Encounters on the battlefield provoked *querelles* among noblemen at home, and duels resulted from insults, contests over women, vengeance for the deaths of friends and relatives, defenses of honor, and arguments over property.[77] In April 1578 a minor insult sparked a particularly grisly duel between three minions of Henry III and a troop in league with the king's

75. Reported by de Lauze, *Apologie de la Danse,* 18. "Un de nos derniers Roys . . . , luy blasmant un gentilhomme (au reste fort accompli) de n'avoir pas apris à danser, & luy demandant ce qu'il sçavoit faire, je sçay bien, SIRE, dict il, donner en guerre un coup de lance pour le service de vostre Majesté: Je vous conseille donc (repliqua ce brave Prince) de vous armer d'un froc en temps de paix, comme s'il eust voulu dire que les fureurs de la guerre cessees un Cavalier ne pouvoit s'occuper à un plus noble exercice que celuy qui luy donne une grande entrée en la cognoissance de sa Cour & de son monde."

76. Schalk, *From Valor to Pedigree,* 122–124.

77. See, for example, L'Estoile, *Mémoires-journeaux,* 1:243–244, 309, 310, 311, 312, 315, 317.

enemies. The eighteen-year-old Louis de Maugiron and the even younger Chomberg died on the spot; Ribérac died the next day; Livarot eventually survived his gruesome head wounds; and the count of Caylus, who took nineteen blows, lingered pitifully for thirty-three days before dying. Only the Baron d'Entragues left with just a scratch, no doubt owing to the maturity that came with his thirty-eight years.[78] Dueling ravaged the nobility just as surely as battles. Add to it the perils of assassination—Condé, the Guises, Henry III, and Henry IV all died in such attacks—and the violent tenor of court life is obvious. Fencing masters regularly refused to teach with rebated swords, forcing students to confront the cutting edges and sharpened points they would soon enough face in armed brawls.[79] In this way fencing lessons awoke the courage aristocrats needed to survive their bloody age, in which a hot temper counted as something of a virtue.

Ballet de cour did not ignore the politics of armed aggression but often incorporated violence, whence its power came. The same noblemen who faced off in summer military campaigns and fought duels in war's absence worked out aggression in Carnival ballets at court for which artillery men fabricated fireworks and engineers built elaborate stage machines.[80] Indeed, the camaraderie kings and nobles developed in sweating it out together on stage was not so far from the closeness that came from shared blood sports, fencing, jousts, and fighting together in battle. In 1619, for example, the *Ballet du Roy . . . sur l'adventure de Tancrede en la forest enchantée* proceeded as a series of balletic combats as woodcutters with axes, sawyers with silver saws, and centaurs with bows and arrows battled a group of monsters in turn, the whole in time to the music. When these forces failed, Louis XIII himself appeared with two "Chevaliers des Adventures." Armed with swords and silver shields, they vanquished the monsters in "swordplay that was agreeable for being made in

78. Ibid., 1:243–244. On the social importance of dueling and the intersection of its projected military glory with the *ballet de cour* see Brioist, Drévillon, and Serna, *Croiser le fer,* 246–277.

79. See Anglo, *Martial Arts of Renaissance Europe,* 34–35, and Motley, *Becoming a French Aristocrat,* 150.

80. The military theme was so commonplace in ballets that it was parodied as well. The *Fées de forest de Saint-Germain* (1625), danced by royals and courtiers, included a satirical suite of mock jousting, battles, and a tilt for doctors. On this ballet see Franko, *Dance as Text,* 99–102, who suggests that in it, "the courtiers' warrior status is hollow because purely balletic," something which may have been true for burlesque ballets, but not for other *ballets de cour.* On artillery men being enlisted to work on ballets, consider the regular contributions to court ballets made by Horace Morel, *comissaire ordinaire de l'artillerie,* on which see McGowan, *Art du ballet de cour,* 119, 128–129, 138, 140. Morel's *Subject du feu d'artifice fait a l'entree du Roy dans sa ville de Paris Ensemble le Ballet representé sur la riviere de Seine, devant le Louvre* (Nantes: Hilaire Mauclerc, 1629), which included a ballet for "huict hommes couverts de feu, & dançant, leurs figures enflameront quantité d'artifices & grosses fusées" (5), gives some idea of the spectacles he staged for the court.

cadence, for there was not a blow given or received that did not mark a beat."[81] What none could know was that the military supremacy promised by Louis's staged victory would be visited upon one of the courtiers who danced at his side in the grand ballet. Among the dancers were two seasoned warriors—François de Bassompierre and the Duke of Rohan.[82] The year after *L'adventure de Tancrede,* Rohan took command of the Protestant army. In 1621 Louis besieged Rohan's stronghold at Saint-Jean d'Angély; the following season, with Bassompierre promoted to marshal of France, Louis's army pushed south to Montpellier, where Rohan made his last stand.[83] The Protestants were defeated, and the next *ballet du roy* was a bacchanalia.

Even as royal ballets furthered the political agendas of French monarchs with a deliberateness that rivaled their battle plans, nobles fought their own battles through ballet. For instance, the wars of 1620–1622 did not really resolve the conflict between the Duke of Rohan and Louis XIII, and during Carnival season in the spring of 1623, Louis XIII had Rohan thrown in prison.[84] Rohan was an excellent commander, and Louis wanted him in his service, so he offered him the position of constable of France if he would convert to Catholicism. But Rohan refused and sat in prison until the stalemate was resolved by other means: his wife threatened not to dance as planned in a coming ballet alongside the queen. Faced with a potential scandal surrounding the *ballet de cour,* Louis was forced to liberate his prisoner, and a political impasse was decided by a ballet. Like the private armies they mustered, French nobles mobilized artistic resources that challenged *ballets du roy* on their own terms, and Carnival season was marked with as many as a dozen competing ballets.

As Louis Marin so aptly characterized princely fetes under Louis XIV (playing upon the famous line of von Clausewitz), "*Divertissements* . . . represent wars in the times of peace that interrupt them, just as wars are the fetes of the prince on another field and another stage. War is the fete continued by other means, as a *divertissement*

81. *Relation du grand ballet du Roy . . . sur l'adventure de Tancrede* (Paris: Jean Sara, 1619), 24. "Ceste escrime fut agreable pour estre faicte en cadence: car il n'y avoit coup ny donné ny recue, qui ne marquast un temps."

82. Bassompierre, wounded at Ivry, subsequently served as *colonel général* of the Swiss and grand master of artillery, though Tallemant des Réaux had no kind remarks about his military skills (Tallemant des Réaux, *Historiettes,* 3:334); Rohan served as *colonel général* of the Swiss.

83. Others of the dancers fought in these campaigns: the Duke of Luynes participated in the siege of St.-Jean d'Angély, and the young Count of Soissons was charged with the blockade of La Rochelle in 1622.

84. The story is related in Nicole Vray, *Catherine de Parthenay, duchesse de Rohan: protestante insoumise, 1554–1631* (Paris: Librairie Académique Perrin, 1998), 153.

is politics pursued on another register."[85] I would add that before the monopolization of armed force under the *roi soleil* allowed for the creation of a pacified court life, ballets featuring armed combats continued war just as vividly as duels.

THE FUNCTION of combat in ballet is best understood in the context of early *ballets de cour,* which built upon the tradition juxtaposing social dance with martial arts in court festivity. Combat moved *into* ballet, a transformation that illuminates the political relevance of *ballet de cour.* The ability of ballet to reference and incorporate war—something noted even in the geometrical ballets danced by ladies—contributed to the potential complexity of its message. Ballet was not a royal pastime pursued for aesthetic pleasure, but a war machine in its own right.

In 1572 the court performed a combat-ballet entitled *Le paradis d'amour* to celebrate the marriage of Marguerite de Valois, the king's sister, to Henry de Bourbon, the Protestant king of Navarre who eventually ascended the throne.[86] The marriage between a Catholic princess and a Huguenot king promised civil union, though the horrendous interruption of the festivities by the massacre on Saint Bartholomew's Day quickly dashed hopes for peace. Indeed, as Protestants reflected on the massacre, the narrative of *Le paradis d'amour* stirred suspicions that Catherine de' Medici had planned the wedding as a ploy to draw Protestant nobles to Paris and slaughter them. The program featured the Catholic king of France casting the Protestant king of Navarre into hell, and Agrippa d'Aubigné, for one, believed that it foretold of the Huguenot massacre.[87]

Although d'Aubigné's innuendoes ignore the long courtly tradition of festive combat, they rightly identify the politics at the center of the spectacle, which dramatized the religious and dynastic oppositions behind the civil wars. At one end of the hall was a paradise defended by the king, Charles IX, and his brothers, Henry, Duke of Anjou, and François, Duke of Alençon. United, they battled a series of assailants, dispatching them one by one into the jaws of hell at the other end of the hall. Devils and grotesque creatures guarded the inferno, scampering forward to

85. Louis Marin, *Le portrait du roi* (Paris: Minuit, 1981), 241. "Les divertissements, en ce sens, représentent les guerres dans le temps des paix qui les interrompent, comme les guerres sont les fêtes du prince dans un autre champ et sur une autre scène. La guerre, c'est la fête continuée par d'autres moyens, comme le divertissement, c'est la politique qui se poursuit sur un autre registre." See Franko, "Double Bodies."

86. I rely on the description of the ballet in Simon Goulart, *Mémoires de l'estat de France sous Charles Neuviesme,* 2nd ed. (Meidelbourg: H. Wolf, 1578), fols. 268v–269v.

87. Agrippa d'Aubigné, *Histoire universelle* (Maillé, 1616–1620), ed. André Thierry, 11 vols. (Geneva: Droz, 1981–2000), 3:318. Simon Goulart makes even more forthright accusations that the theme of the ballet related to the planned massacres (*Mémoires,* fol. 269r–v).

seize the defeated soldiers and drag them into prison, where a diabolic wheel ringed with bells turned noisily. At the far end of the hall, a heavenly wheel embellished with the signs of the zodiac, the seven planets, and flaming stars spun silently, turning the garden of paradise beneath it and twelve beautiful nymphs therein. The heavenly wheel represented the world-soul carrying the planets and symbolized divine majesty.[88]

If Catherine had contrived a special effect by pitting her three sons against a series of attackers, it likely aimed to show a united family front rather than to presage a massacre. François would soon defect to the Protestant cause and had to be held under house arrest when the bloodshed began, and Henry's alliance with the ultra-Catholic Guises similarly thwarted Catherine's policy of conciliation. That the Protestant groom waged war on the king was not unusual: the king always defended some edifice in mock combats—a convention based on the very real tactics of fortification and siege. Even royal snowball fights were so organized.[89] D'Aubigné must have known that Henry of Navarre's role in this spectacle was not wholly untoward.

What *was* strikingly new about the combat was its resolution with a ballet. After all assailants had been imprisoned in hell, Mercury and Cupid descended from heaven dancing and singing. The famous castrato Etienne Le Roy played the role of Mercury, and after finishing his song he made a speech before the three Valois brothers, who then arose from their seats, crossed into the Elysian fields, and escorted the nymphs onto the dance floor, where they performed a ballet that lasted over an hour. Before *Le paradis d'amour,* masques had often ended with dancing initiated by the king, but here for the first time the ball was replaced with a ballet. It was a ball made into ritualized theater in which the king's ceremonial function was carefully maintained rather than being dissipated by general dancing. Charles IX and his brothers presided over a ballet performed in a space that they had secured against all enemies of the throne. It was a space that would collapse dramatically when the ballet finished, for the evening ended with the prisoners delivered from hell, pikes shattered in combat, and smoke bombs set off in the center of the hall. But for the duration of the ballet, the Valois held court in heaven.

Perhaps, as Frances Yates has argued, the ballet resolved the conflict of the civil wars, saving the Huguenots from their heretical beliefs when they were rescued

88. Heavens abounded in court spectacles—one need only consider of the musical *voûte dorée* in the *Balet comique de la Royne* and the cosmic amphitheater described by Jean Dorat (see Yates, *French Academies,* 272–273)—but wheels sometimes represented the wheel of fortune. See the fete described in Victor E. Graham and W. McAllister Johnson, *The Royal Tour of France by Charles IX and Catherine de' Medici: Festivals and Entries, 1564–6* (Toronto: University of Toronto Press, 1979), 41.

89. See Graham and Johnson, *Royal Tour of France,* 103.

from hell.[90] Or perhaps, as Margaret McGowan believes, the nymphs "have come to bring harmony through their dancing and to consecrate the peace that the king's presence and his actions have brought about."[91] Yates's interpretation correctly draws our attention to the highly charged politics read into the ballets by contemporaries—or at least by d'Aubigné—while McGowan, who fails to account for the spectacle's apocalyptic ending, is still certainly right to emphasize the king's role in the proceedings. Charles does not dance, but plays the Most Christian King, at once a mortal ruler and the mystical embodiment of kingship. In the context of a society that fetishized the king's body even after death, the moment when Charles sets the ballet in motion is crucial.[92] The music he elicits from the *ciel royal* activates the dancing and brings a moving paradise into the hall; with this act he links the dramatic representation of war and peace to the reality of the civil wars in which he also played a part.[93] The combat between good and evil unfolded not in a hermetic world of metaphor, but in a reality intersecting the Salle de Bourbon. Because Charles literally incorporated his realm—for his natural body was conjoined with the body politic—his actions in the spectacle broke the bonds of representation and demonstrated his actual harmonic governance of the state. McGowan's Neoplatonic reading concludes that the ballet manifested a reality in which Charles could transmute civil dissent into concord.

Although Catherine de' Medici's methods were often Machiavellian, her politics were conciliatory. The schism between national identity and religious conviction was the greatest threat to the monarchy, which she fought to save by reasserting the divine right of the king. Why, then, does her ballet deliver its peaceful message so precariously in the midst of armed encounters, and why does the evening end with indoor fireworks that dissipated the celestial harmonies and frightened everyone out of the hall? One answer is that the ballet was inserted into a larger program organized with the conflicting aims of showing off the king's military prowess and the queen mother's refined tastes. The piecemeal character of *Le paradis d'amour* suggests Catherine's inability to control and unite her sons in defense of the monarchy at the same time as it confirms her faith in its ability to withstand attack. This would not last. In the following year, the *Ballet des provinces,* given for the Polish ambassadors, suppressed any such dramatic conflicts, incorporating martial affects into the ballet itself with an *air de guerre,* march, and battle formations. The long and geometrically complex ballet, designed by Jean Dorat, Ronsard, and (possibly)

90. Yates, *French Academies,* 256.

91. McGowan, "Arts Conjoined," 195.

92. See Ralph E. Giesey, *The Royal Funeral Ceremony in Renaissance France* (Geneva: Droz, 1960).

93. Thomas Greene takes a similar analytical approach to the king's role in ballets in his "The King's One Body in the *Balet comique de la Royne,*" *Yale French Studies* 86 (1994): 75–93.

Baltassar de Beaujoyeaulx, took humanism's allegorical tack, asserting the power of the monarchy with a theme of provincial unity and flexing a muscle of military might with the regimentation of the very large troop of sixteen dancers.[94] Nonetheless, it seems to me that in *Le paradis d'amour* as well, firepower and combat gave the celestial scenes of the ballet meaning, and that without violence and ritual strife, *ballets de cour* could resolve little to nothing at all.

Early *ballets de cour* were danced by Catherine's *dames d'honneur* and other women. But by 1600 the queens' ballets, of which the *Balet comique de la Royne* is the most famous example, began to compete with the ballets of noblemen such as Montmorency, Condé, Vendôme, and the king, who not only patronized the new genre, but danced in it as well.[95] For the visit of the Knights of the Garter in 1585, Henry III led a ballet of twenty-four courtiers opulently decked out in matching white doublets and hose, pearl-studded silver hats, and sashes.[96] No doubt many of the dancers were Knights of the Holy Spirit, who had participated in the other ceremonies marking Henry's induction into the English order. Indeed, courtiers quickly adopted ballet as a part of ceremonial, which, given its overriding concern with establishing hierarchy through the ordering of individuals and corporations in processions, religious rituals, at banqueting tables, and so forth, was itself a form of political choreography.[97] Although Henry IV was not as fabled a dancer as the Valois kings who preceded him, he did dance in the ballets staged by Marie de' Medici (to whose dismay he met Charlotte de Montmorency, the last great love of his life, during rehearsals for the *Ballet de Diane* in 1609).[98] Henry furthermore expected his courtiers to dance when the occasion required. In 1597, after Sully refused Montmorency's invitations to participate in a ballet he was planning, the king ordered his faithful battle companion into a frock coat for the performance.[99] Perhaps their ex-

---

94. Yates, *French Academies*, interprets *Le paradis d'amour* as the first result of the interest of Baïf's academy in dance, 62, 267. For a tradition in which balletic combats between Protestants and Catholics continued, see Roy Strong, *Art and Power: Renaissance Festivals, 1450–1560* (Suffolk: Boydell Press, 1973), 124–125. Strong suggests that after Henry IV ascended the throne, ballets in the Protestant holdout of Béarn continued to pursue themes of combat and conciliation consistent with that of *Le paradis d'amour.*

95. See the indispensable catalog of ballets compiled in McGowan, *Art du ballet de cour,* 251–309.

96. See Roy Strong, "Festivals for the Garter Embassy at the Court of Henry III," *Dance Research* 1 (1983): 45–58, at 52–53.

97. English parallels are presented by Howard, *Politics of Courtly Dancing,* and Orgel, *Illusion of Power,* esp. p. 40 on the division of performance roles between professionals (who spoke) and courtiers (who danced).

98. See Bayrou, *Henri IV,* 465–466.

99. See Maximilien de Béthune, Duke of Sully, *Mémoires,* ed. Louis-Raymond Lefèvre, 2nd ed. (Paris: Gallimard, 1942), 172.

change was the one related by de Lauze, in which the king announced that even the most valorous warriors had to dance.

A dozen *galants de la cour* inaugurated the new arsenal with an armed ballet in 1609; French knights amazed the Romans with their dancing of a Carnival ballet there in 1615; a Pléiade of noble *cavaliers celestes* performed in the *Balet du veritable amour* in 1615; and Bassompierre, who Tallemant reports was hardly nimble, nonetheless staged several ballets and danced in those of the king.[100] Louis XIII's closest advisor, Charles d'Albert, Duke of Luynes, seems to have been wholly ineffectual as either a statesman or warrior but furthered his career by planning *ballets du roy* in which he cast himself in starring roles beside his protégé.[101] So astounding and self-assured was Luynes's ascent that in his last ballet he danced as Apollo—having assigning the roles of Castor and Pollux to his brothers, Luynes could only offer Louis a modest commission leading a band of blacksmiths.[102] The *Ballet d'Apollon* consolidated Luynes's stunning political gains in a temporary assumption of kingship before the whole court and with the king's active collaboration.

Louis XIII's political coming of age was marked with a suite of martial *ballets du roy* planned by Luynes, two of them based on Torquato Tasso's *Gerusalemme liberata* (1617 and 1619) and one on Ludovico Ariosto's *Orlando furioso* (1618).[103] The first of these, *La délivrance de Renault,* exemplifies the pursuit of political gains through court ballet. Its story describes the triumph of military virtue over *otium* and the restoration of reason to the hero, Renault, who is rescued from the enchantress Armide. At the outset of the ballet, Renault (played by Luynes) has been

100. See, respectively, "Ballet des Fous Armés dansé par 12 galants de la cour pour étrenner la salle neuve de l'Arsenal," in Collection Michel Henry, F. fr. 24357, Bibliothèque nationale de France (much of the collection is catalogued in François Lesure, "Le recueil de ballets de Michel Henry," in *Les fêtes de la Renaissance,* ed. Jean Jacquot, 2nd ed., 3 vols. [Paris: Centre National de la Recherche Scientifique, 1973–1975], 1:205–219); *Description du balet dansé a Rome par des Cavalliers François* (Rome, 1615); *Description du balet du veritable amour de madame la duchesse de Montmorency* (Béziers: Jean Pech, 1618). Bassompierre's ballets are listed in the appendix of McGowan, *Art du ballet de cour,* under the years 1598, 1600, and 1604. On Bassompierre's dancing see Tallemant des Réaux, *Historiettes,* 3:339, 2:307–308, in which Tallemant relates a fight between Bassompierre and Montmorency over Bassompierre's dancing that almost ended in a duel. On armed dances at the court of Savoy, see Evelyne Samard, "Les danses guerrières dans le ballet de cour en Savoie au XVIIe siècle," in *Le ballet aux XVIe et XVIIe siècles en France et à la Cour de Savoie,* ed. Marie-Thérèse Bouquet-Boyer (Geneva: Slatkine, 1992), 77–115.

101. On the Duke of Luynes see Pierre Chevallier, *Louis XIII, roi cornélien* (Paris: Fayard, 1979), 193–194, and McGowan, *Art du ballet de cour,* 179–181.

102. See the *Ballet du roy faict dans la salle du petit Bourbon le 14 février* (Paris: N. Rousset, 1621).

103. See *Discours au vray du ballet* [*de la délivrance de Renault*] *dansé par le roy* (Paris: Pierre Ballard, 1617), *Le ballet du Roy représentant la furie de Roland* (1618), *Relation du grand ballet du Roy . . . sur l'adventure de Tancrede* (1619), and McGowan, *Art du ballet de cour,* 179–181.

overcome with passion for Armide and listlessly reclines in her bower, which blossoms in an eternal springtime. Her spell has created a state of disorder, and so the first dances in the ballet are performed by strange demons from a devilish underworld. The king made his first appearance in the ballet as the demon of fire, leading Renault and the demons of water, the hunt, folly, war, and so forth in a series of dances. After these *entrées,* two Roman soldiers come to Armide's garden in search of Renault. A beautiful nymph tries to seduce them with an *air* praising love over war, but they have stopped their ears and are unmoved by her siren song. Next, Armide's demons attack the soldiers, but to no avail. Renault is delivered from his enchantment, and, full of shame, he flees Armide and returns to camp. Reunited with Godefroy of Bouillon, leader of the Christian army (played by the king), Renault's return to reason prepares for the triumph of Christian faith predicted at the end of the ballet, as the army stands ready to continue its crusade to the Holy Land.

No expense or effort was spared in mounting the production.[104] As early as late December the king attended rehearsals in the apartments of Luynes and his mother, which continued until the performance on 29 January.[105] The libretto was written by the *poète ordinaire* of Marie de' Medici, Étienne Durand, with some verse for the *airs* supplied by René Bordier and Pierre Guédron, who wrote most of the vocal music.[106] Antoine Boësset and Gabriel Bataille also contributed musical settings to the large collaborative project, which, according to the libretto, was performed by an ensemble of sixty-four voices, twenty-eight viols, and fourteen lutes led by Jacques Mauduit; a second ensemble of sixteen musicians costumed as Roman soldiers led by Guédron; and trumpets and other extras that combined forces at the

104. An excellent description of the production, the ballet's creators, and the political tensions between Louis and his mother that inflected the ballet are given in McGowan, *Art du ballet de cour,* 101–115.

105. Héroard, *Journal,* 2:2430, 2433, 2437.

106. The music for the ballet survives in several sources. Ballard published the polyphonic airs and lute songs in a luxurious edition of Durand's libretto that also includes a lengthy description of the ballet and engravings of the sets and costumes (*Discours au vray du ballet* [*de la délivrance de Renault*]). The Philidor manuscript compiled at the end of the century contains two-part versions of the dance music; Collection Philidor, Département de la Musique, Rés. F494, 137–139, Bibliothèque nationale de France. Ballard simultaneously published the *airs* in smaller formats for solo voice (*Second livre d'airs de cour* [Paris: Pierre Ballard, 1617]) and lute (*Airs de différents auteurs, mis en tablature de luth, septiesme livre* [Paris: Pierre Ballard, 1617]). Guédron and Boësset included the *airs* they had composed in single-author prints: Pierre Guédron, *Quatriesme livre d'airs de cour à 4 et 5 parties* (Paris: Pierre Ballard, 1618), and Antoine Boësset, *Airs de cour à 4 et 5 parties* (Paris: Pierre Ballard, 1617). On the *airs* see Durosoir, *Air de cour,* esp. 100–102, and Charles T. Downey, "Musical-Dramatic Productions Derived from Ariosto and Tasso in the City of Paris, 1600–1800" (Ph. D. diss., Catholic University of America, 1998), 210–220.

end to create a massed ensemble of ninety-two voices and more than forty-five instruments.[107] Even given the usual hyperbole of such accounts, it is not difficult to believe that Europeans had truly never heard anything so ravishing. The staging, too, was spectacular, with flamboyant costumes (at one point Armide's monsters metamorphose into snails, turtles, and crayfish) and five changes of scenery engineered by one of the Francini brothers, who brought the Italian art of theatrical machines to the French court.[108] As the last scene swiftly appeared—two magnificent palms laden with trophies and armor sprang up at either side of the stage as a golden pavilion with Louis enthroned beneath a canopy of stars turned into view—the audience gasped and applauded as at the apparition of some miracle.[109]

Tasso's story of the crusade to redeliver the Holy Land into Christian hands fit neatly with Louis's growing aspirations to reinstate Catholic worship in the southern principality of Béarn. He began to build up the royal army in 1618, and in 1620—at age eighteen—he embarked on the first of three campaigns to the south. In one sense, then, the trilogy of heroic *ballets du roy* mounted in the seasons leading up to his first military venture prepared the king and his court for war. Even more immediate was the king's desire to assert his authority at court. In the very month the ballet was being planned, his mother's Italian minister, Concino Concini, reached the apogee of his power.[110] Louis's unusual appearance in the ballet coincided almost exactly with his initiatives to take control of his kingdom.[111]

La délivrance de Renault cast the king in contrasting roles representing *otium* overcome by *virtus.* At the beginning of the ballet he appeared as a fiery demon in a radiant costume covered with reflective enameled flames that struck an affect somewhere between black magic and Apollonian majesty. His was a kingdom turned upside down, a reversal rectified by his return at the end of the ballet as Godfrey of Bouillon.[112] The narrative's upward-bound trajectory from the demonic to the divine and from secular to Christian was reinforced by his last entrance at the top of the magnificent pavilion capped with its heavenly canopy. Both times he led a ballet performed by the barons, counts, generals, and other nobles who made up his most immediate court (fig. 3.2). Half of the gentlemen were knights of the

107. *Discours au vray du ballet* [*de la délivrance de Renault*], fols. 3r, 20r, 22v.

108. McGowan, *Art du ballet de cour,* 81.

109. *Discours au vray du ballet* [*de la délivrance de Renault*], fol. 24r.

110. See the excellent overview of Concini's ascent and demise in René Pillorget and Suzanne Pillorget, *France baroque, France classique, 1589–1715,* 2 vols. (Paris: Robert Laffont, 1995), 1:124–144.

111. Before *La délivrance de Renault,* Louis had danced in only one other ballet since reaching his age of majority in 1614. It was a couple of months before. See Héroard, *Journal,* 2:2423.

112. For an analysis of the staging of this reversal with particular emphasis on Louis's demonic appearance at the start of the ballet, see Mark Franko, "Jouer avec le feu: la subjectivité du roi dans *La délivrance de Renault,*" in Careri, *La Jérusalem délivrée du Tasse,* 159–177.

FIGURE 3.2. The Grand Ballet. In *Discours au vray du ballet* [*de la délivrance de Renault*] *dansé par le roy* (1617). Bibliothèque nationale de France, Cabinet des estampes, Qb1-1617

Order of the Holy Spirit; two had been raised with Louis as *enfans d'honneur,* their parents sending them to court at untold expense to be raised with the dauphin and establish a future at court; and Blainville and Luynes were *premiers gentilshommes de la chambre,* of which there were only half a dozen at any given time.[113] In short, the courtiers who danced with Louis in the ballet had grown up with him, guarded him, shared his interests, and been rewarded with knighthood, and their collaboration in the ballet strengthened its affirmation of Louis's sovereignty.

At the end of the drama, the gentlemen stand ready to fight (see fig. 3.2). But they are not armed. They wear ruffs and feathered headdresses, while what armor might be associated with them hangs decoratively at either side of the royal pavilion from which they have just descended. Nor do they sing or dance to warlike music.

Earlier sixteen musicians had appeared in the camp of the Christian army clad as "ancient knights" in Roman breastplates and helmets (see fig. 3.3). Not the courtiers, but the *cavaliers antiques* in the orchestra sing "Allez, courez, cherchez de toutes pars," a call to arms in F that stands out tonally from the rest of the ballet (see ex. 3.2). It begins with imperatives set to trumpeting arpeggios and the same iambic figures Le Jeune exploited in the battle cry from *La guerre,* "Bataille compagnons" (see ex. 1.6), a choice associated there with inflamed passions that Guédron reaffirms here. Guédron composes out the rousing effect of the iambic mode, which is first announced in the top voice alone: "Come on, run, look everywhere." The five-part choir echoes and amplifies the command of the superius, singing, "Let's go, let's run, let's look everywhere [for Renault]" to the same music.

But Renault, the proud vanquisher of Mars, is a slave of love, as the stately second half of Guédron's *air* reminds us. The rhythm evens out, taking up the smooth anapests that Le Jeune had likewise juxtaposed with his iambic air, a rhythmic turn that makes the opening of the song all the more striking (and all the more like Le Jeune's *La guerre,* which had finally been published posthumously in his 1608 *Airs*). Such rhythmic sensibilities are evident throughout the *airs* of the ballet, which, like the dance music itself, trade imitative textures for homophony's ability to show off complex rhythmic progressions. The unique rhythmic vocabulary of each *air* emphasizes this as a site of meaning.

The soldiers in the chorus are not the only characters with bellicose music, but neither were the other aggressive roles played by courtiers. The soldiers sent to rescue Renault—Charles and Ubaldo, in Tasso's telling—were played by two more anonymous *cavaliers antiques,* presumably professional fencers or dancers.

113. On Louis's *enfans d'honneur* see Eugène Griselle, *État de la Maison du roi Louis XIII* (Paris: Éditions de Documents d'Histoire, 1912), 49; the *premiers gentilshommes de la chambre* are given in idem, *Supplément à la Maison du roi Louis XIII* (Paris: Éditions de Documents d'Histoire, 1912), 80.

EXAMPLE 3.2. "Allez, courez, cherchez de toutes pars," by Pierre Guédron. In *Discours au vray du ballet* [*de la délivrance de Renault*] *dansé par le roy* (1617). Bibliothèque nationale de France, Cabinet des estampes, Qb1-1617

Llons courons cherchons de toutes pars Ce super-
be Renault, ce fier vainqueur de Mars, Dont le cœur genereux En vn loin-
tain sejour, Par l'effort d'vn bel œil, Est exclaue d'amour.
Llons courons cherchons de toutes pars Ce super-
be Renault, ce fier vainqueur de Mars, Dont le cœur genereux En vn loin-
tain sejour, Par l'effort d'vn bel œil Est exclaue d'amour.
be Renault, ce fier vainqueur de Mars, Dont le cœur genereux En vn loin-
tain sejour, Par l'effort d'vn bel œil Est excla- ue d'amour.

FIGURE 3.3. "Cavaliers antiques" with Peter the Hermit. In *Discours au vray du ballet [de la délivrance de Renault] dansé par le roy* (1617). Bibliothèque nationale de France, Cabinet des estampes, Qb1-1617

The libretto describes them as "plains d'ayse & d'ardeur" (full of joy and ardor, fol. 16r), passions no less fiery than those that caused Renault's lovesick collapse in Armide's bower, and certainly less debilitating. They dance their first entrée to trumpets (the only time trumpets are said to have played during the ballet), and their battle with the burlesque monsters—the one bit of forward-moving action in the drama—brings the victory upon which the whole ballet turns. The trumpets, swordplay, and vanquishing of *otium* produced an extremely heroic scene, thus all the more suggestive that it was danced by *anonymes.*[114]

In the Grand Ballet, Godefroy and the leaders of his army celebrate a victory won by others. They do not dance Pyrrhics, as Xenophon tells us Greek warriors did. Indeed, the music for their ballet shifts into a gentle triple time with hemiolas that work in an almost precious counterpoint to the meter (see ex. 3.3). This is the only dance music for the ballet in three, a time signature commonly reserved for Grand Ballets like this one, which ended *ballets de cour.* Later in the century, the rocking circularity of these 3/2 ballets inspired the chaconnes that ended *tragédies en musique* with stabilizing messages of sovereign authority.[115] These dancers are neither ancient knights, Roman centurions, nor crusaders, but French courtiers. Louis leads a pacified nobility, and herein lies the true power of the ballet. *La délivrance de Renault* is not a story of war made into peace, for the army Godefroy assembles at its end has yet to engage in its greatest battle. Rather, by excluding the nobles from actual military encounters, the ballet redistributes the control of armed force in a hierarchical and centralized form that differed significantly from contemporary reality.

The musicians and professional dancers have become an infantry under the control of nobles who command from above, something emphasized by the physical placement of the orchestra in the hall (see fig. 3.3). Louis, who enters atop the golden pavilion, reigns from the center of a peaceful court established by a military base now pushed to its margins. Indeed, the vastness of the large orchestra and chorus is always partially hidden from view, by the foliage of Armide's enchanted garden, the trees in the woods, or the shadows at the edge of the hall. Like the backdrops used for each scene, whose perspectives disappeared into the boundless orders of the universe, the king's musical army seems infinite.[116] The ninety-two voices

---

114. *Discours au vray du ballet* [*de la délivrance de Renault*], fol. 13r. "Ces Monstres plaisans & difformes tout ensemble, attaquerent les deux Cavalliers, comme ils entroyent dé-ja dans le Jardin, & eux leur resistant par les armes, & par la puissance de la baguette, leur contraste donna lieu a un Ballet de bouffonnerie & de gravité entre-meslée, qui n'eust pas la derniere place en la loüange de ceux qui les regarderent."

115. Geoffrey Burgess, "The Chaconne and the Representation of Sovereign Power in Lully's *Amadis* (1684) and Charpentier's *Médée* (1693)," in *Dance and Music in French Baroque Theatre: Sources and Interpretations,* ed. Sarah McCleave, Institute of Advanced Musical Studies, Study Texts 3 (London: King's College London, 1998), 81–104.

116. We should consider the size and placement of the orchestra in light of the politics of theatrical perspectives in court entertainment, on which see Orgel, *Illusion of Power.* Owing to the rhythmic discipline convened by the *batteur de mesure,* metaphors of the orchestra as an army and of its organization as a mirror of absolutism became prevalent in the late seventeenth and eighteenth centuries. John Spitzer and Neal Zaslaw, *The Birth of the Orchestra: History of an Institution, 1600–1815* (Oxford: Oxford University Press, 2003), 515–519.

EXAMPLE 3.3. The Grand Ballet. *La délivrance de Renault* (1617). From the Philidor Manuscript, Bibliothèque nationale de France, Rés. F. 494

and forty-five instrumentalists who performed the final chorus likewise left listeners with a sense of the orchestra's limitlessness: "[the ensemble] made such a sweet noise that it hardly seemed to come from the four parts in which [the music] was composed."[117]

The nobles form an officer corps at a higher level of command. At the beginning of *La délivrance,* Louis and his *fidèles* had danced entrées in groups of two or four, each strikingly costumed as a different demon. The uniqueness of each character and the separate entrées reinforced the proud individualism of the noble dancers,

117. *Discours au vray du ballet* [*de la délivrance de Renault*], fol. 22v. "Le nombre de quatre-vingt douze voix & de plus de quarente cinq Instruments, estant joinct ensemble faisoit un si doux bruit qu'il ne sembloit point revenir au quart de ce dont il estoit composé."

with whom Louis danced as a relative equal. But the Grand Ballet at the end revises the relationship between the king and his court. His transformation from a fiery demon to the leader of the Christian army is emulated by his closest followers, who relinquish their fiendish powers and swear on as "les chefs de son Armée." They are costumed uniformly and so bejeweled that (as the *livret* tells us) it was hard at first to discern individual faces for all of the light reflecting off the stones. The brilliance of the enameled flames attached to Louis's demon costume now consumes the scene as his fiery majesty achieves its divine manifestation.

WHEREAS *Le paradis d'amour* dramatized the military supremacy of king in the face of armed opposition from within his immediate circle of nobles, *La délivrance de Renault* stages the consolidation of armed force in the hands of the king. The contrasting roles played by courtiers in each ballet mirrored a very real shift in the strategies of court life that was fundamentally connected to the aspirations of absolute monarchs. As we know, civility would banish violence and rude behavior, but politesse depended upon keeping war at bay. As Norbert Elias put it, "Through the formation of monopolies of force, the threat which one person represents for another is subject to stricter control and becomes more calculable. Everyday life is freer of sudden reversals of fortune. Physical violence is confined to barracks; and from this store-house it breaks out only in extreme cases, in times of war or social upheaval, into individual life."[118] It follows on from this, Elias argues, that "the concentration of arms and armed troops under one authority makes the use of violence more or less calculable and forces unarmed people in the pacified social spaces to restrain their own violence through foresight or reflection: in other words it imposes on people a greater or lesser degree of self-control."[119] The policing of society made the civilizing of it possible.

We might see in *La délivrance de Renault* a structural demonstration of the civilizing process, yet it was one sooner aspired to than achieved. In 1617 Louis had no significant standing army or royal police force and was obliged instead to rely upon a system in which the private armies raised by France's *grands* formed essential components of royal forces. Even at that time he worried that Vendôme and Mayenne seemed to have more direct control over the towns and forts they garrisoned than he did.[120] Rohan, La Rochefoucauld, and Guise all mustered substantial armies in those years, but not necessarily—as the case of Rohan shows—in support of the king. And even when these armies were created in response to royal appeal, their officers' loyalty ran to the lord who engaged them and often no further. This would

118. Elias, *Civilizing Process,* 372.

119. Ibid., 373.

120. Lynn, *Giant of the* Grand Siècle, 284–287, 601.

change. By century's end, the royal army had expanded enormously: the wartime strength of Louis XIV's army numbered as many as 340,000, while he maintained peacetime forces of some 150,000 men, as compared to the 10,000 or so Henry IV maintained between 1600 and 1610.[121] But the transition to a huge army of state took place in fits and starts.

Moreover, nobles continued to exercise private violence through dueling, which presented just as profound an obstacle to sovereign authority as did private armies. Not only did aristocrats kill each other in alarming numbers—estimates for the years 1598 to 1608 run to eight thousand—dueling symbolically asserted an aristocrat's freedom to wage war on his own terms. In 1610 one commentator lamented that so great was the number of duels that it seemed civil war continued despite the peace. "In the absence of legitimate war, we consume ourselves by a fatal rage, and we cannot breathe this sweet air nor take possession of this happy territory without doing violence to ourselves and burying each other."[122] Certainly the most compelling dramatization of the politics of dueling is the case of the young Count Montmorency-Bouteville, who devoted himself to dueling beginning at age fifteen. By the time he was twenty-eight, he had twenty-two duels to his credit.[123] In 1627, having already been forgiven by the king for earlier transgressions, he engaged in yet another duel in, of all places, the Place Royale. Such an affront was impossible to ignore, and Louis XIII had him tried and beheaded. A stunned nobility objected that he should have been spared based on his eminent lineage and noble character, but the king asserted his prerogative to determine when and where violence could be exercised. *Raison d'état* won out alarmingly over the chivalric values that Louis and his nobles once seemed to share. Nonetheless, dueling did not diminish until after the aristocratic rebellions known collectively as the Fronde (1648–1653), when the aristocracy itself became more willing to relinquish this form of violence for other privileges.

Even though the Grand Ballet seemed to manifest a French state freed from the threat of aristocratic rebellion, the king cajoled allegiance from the *grands* by playing upon their desire for royal favor. The staging emphasized this narrative with reflective substances and floods of light that were multiplied by the king's luminous costumes, and the *livret* pointedly asserted that the king's majesty was shared by all.

121. Ibid., 7–8, 32–64, 599–602.

122. P. de Boyssat, *Recherches sur les duels* (Lyons: Irenee Barlet, 1610), 4. "Il advient qu'en pleine paix se fait une guerre civile, & qu'à faute de guerre legitime, nous nous consumons par une rage fatale, & ne pouvons respirer parmi ce doux air, ni posseder cest heureux terroir, sans nous violenter & enterrer les uns les autres."

123. For a fine political analysis of the Montmorency-Bouteville affair, see Billacois, *Duel dans la société française,* 247–275.

The manipulation of nobles' desire for special favors was to become a hallmark of Richelieu's ministry, as Orest Ranum has shown, and the acceleration of desire and reward that intoxicated absolutism's ranks of seemingly peaceable minions, *créatures,* and *fidèles* only recreated violence in a new form. In a slightly sinister turn of phrase, the *livret* of the ballet hints at the violence imminent in all princely relationships:

> The grand Ballet was danced with such order and disposition that none before it could lay claim to the same beauty. Not a single Frenchman can refrain from blessing the Heavens for the kindness of his king. The Majesty that seems contrary to such actions always preceded his steps. If those who had accompanied him had not sometimes captured it, to make themselves admired by imitating him, grace would have belonged to him alone. But all assembled felt the power that his Majesty had over their minds, for those who had no good fortune acquired it, and those who already had it could now never lose it. In this way the Ballet finished.[124]

The force exerted in *La délivrance* transcended the clash of sword against sword that created the conditions of its possibility, continuing war on a new field. The royal coercion demonstrated in the order and disposition of the ballet, the power that His Majesty had over them, was not strictly beneficent. For majesty was expressed in clemency that might be revoked with horrifying consequences, as the Montmorency-Bouteville affair would make so terribly clear. Majesty seemed contrary to kindness. And no one learned this lesson more surely than the man who penned those words, the librettist of *La délivrance,* Étienne Durand. Just three months after the ballet, Louis revolted against his mother, a development that resulted in the execution of the marshal of Ancre and set in motion a purge that included having Durand drawn and quartered for assisting in her plots against the king.[125] Louis exiled his mother to Blois, throwing off her regency in a coming to his senses more than one observer likened to Renault's escape from Armide's enchantments.[126] And as for her minister, on 24 April 1617 Concini was arrested and

124. *Discours au vray du ballet* [*de la délivrance de Renault*], fol. 25r. "Ce grand Ballet fut dancé avec tant d'ordre & de disposition, qu'aucun autre devant luy ne se peut vanter de la mesme beauté, un seul des Françoys ne se peut tenir de benir le Ciel en la gentillesse de son Roy, la majesté qui semble contraire a telles actions estoit tousjours au devant de ses pas, & la grace n'eust esté que pour luy seul, si ceux qui l'accompagnoyent ne l'eussent par fois derobée pour faire admirer ce qu'ils faisoyent en l'imitant: mais tous ensemble se sentirent de la puissance que sa Majesté eust alors sur les esprits: car ceux qui n'avoyent point de bonne fortune, en aquirent, & ceux qui en avoyent les mirent en point de ne pouvoir estre perdues. Ainsi le Ballet se finit."

125. The details are recounted in McGowan, *Art du ballet de cour,* 113–115, and Downey, "Musical-Dramatic Productions," 165–169.

126. René Bordier wrote one *air* for Armide that was cut from the ballet before the performance ("O Dieux! quel est le Sort dont je suis poursuivie?"). Downey believes that the omission of Armide's

assassinated at the entrance to the Louvre. Luynes, Richelieu, and the rest of the court looked on as Louis's arrival at the scene was met with cries of joy and he exclaimed, "Merci! grand merci à vous! à cette heure, je suis roi!"[127]

---

love song for Renault indicates the growing mistrust of Marie de' Medici at court. Detractors likened her influence over her son to Armide's sorcery ("Musical Dramatic Productions," 165–169).

127. Quoted in Pillorget and Pillorget, *France baroque, France classique,* 1:143.

4

THE CROSS AND THE SWORD

Toulouse, late August, 1576. For three nights men and women process to a series of stations, robed and barefoot, carrying torches and crosses. Their six-hour marches ignite devotion in a hundred thousand onlookers and drive off two armies of sackers in the region. Antoine de Bertrand experiences an epiphany, renounces his "chansons impudiques," and drafts a collection of airs spirituels *in just a few days. He is martyred for writing Catholic songs.*

Lorraine and Burgundy, summer, 1583. White-robed penitents crisscross the countryside singing antiphons, hymns, and litanies. Those who hear them are moved to tears, take whatever white cloth they find at home, tie it about themselves, and join the processions against heresy. By summer's end, 72,000 penitents have converged on the region's great churches.

Paris, Feast of the Annunciation, 25 March 1583. Robed in white, the king, courtiers, and singers of the chamber march through Paris chanting the litany in fauxbourdon. At Notre-Dame, they kneel and sing a polyphonic Salve Regina. Days later, the king's pages and lackeys stage a mock procession in the Louvre wearing handkerchiefs over their faces with holes cut through for their eyes that scare the children. The king has 120 of them whipped.

THE REIGN of Henry III (1574–1589) was the great era of religious processions in France. Henry himself probably participated in more religious processions than any previous king in French history.[1] He walked behind the consecrated host in Corpus Christi processions, made his way to the Place de Grève to light the Saint John's Night bonfire, and walked in the feast day processions of his three royal orders, the Order of the Holy Spirit (est. 1578), the Confraternity of the Annunciation (est. 1583), and the Congregation of the Oratory of Our Lady of Vincennes (est. 1584). In addition to these annual events, he made extraordinary pilgrimages, walking to Chartres and Cléry in 1584 with forty-seven of his penitential confreres to

1. Holt, *French Wars of Religion,* 125.

pray for an heir, a pilgrimage he had made the year before with Queen Louise and from which they returned with blistered feet.[2]

Certainly the religious wars pressed everyone to affirm their sectarian convictions, and processions had the advantage of joining individuals together in their public proclamations of faith. Like the corporate metaphors of the Catholic Church, in which believers became one with Christ by consuming his body, processions gave material form to shared beliefs.

Yet it was not just going about in groups that manifested faith. Music breathed spirit into processions and asserted the proximity of the penitents to the divine center of things. Their singing declared an unverifiable truth: God was with them. All who told of the *processions blanches* in the summer of 1583 commented on the perpetual singing of the "white" penitents, who tried to wash sin from the soul of the country with *Aves* and *Salves*. Of course, sacred music was very transportable, and just as relics and the Corpus Christi were increasingly put on public display during the wars, processions brought the sacrament of ecclesiastical song to the streets. Even more important, sacred music touched the laity deeply. They "owned" psalms and hymns and litanies with a completeness that could never extend to relics or even the Corpus Christi, which Catholics generally took only once a year. Apart from a Book of Hours or psalter, a rosary, and some small objects of veneration, the chants and psalms the faithful knew from catechisms, Mass, and vespers were the most densely meaningful religious symbols they might possess.

In the processions described at the beginning of this chapter, music is more than a declaration of faith—it is used to proselytize. Bertrand is converted by the penitents and writes spiritual songs to support their cause. Singing swells their ranks by tremendous proportions. The Jesuit father who organized the processions in Toulouse, Edmond Auger, was explicit about music's efficacy, for he considered Huguenot psalms a grave threat to Catholicism. At the outset of the first religious war in 1563, he wrote to his General in Rome suggesting that only Catholic songs in French could "quench these psalms of Clément Marot."[3] "For the French love singing very much," Auger continued, "and with this would be a battle like that in the time of St. Chrysostom against the songs of the Arians." One can almost see the companies of Christian soldiers singing vernacular hymns as they clash with psalm-singing Protestants (who did sing on the march and, like Catholic martyrs

2. L'Estoile, *Mémoires-journaux,* 2:121, 149–150; the pilgrimages are discussed in Yates, *Astraea,* 182–183.

3. On Auger's request, see T. Frank Kennedy, "Jesuits and Music: Reconsidering the Early Years," *Studi Musicali* 17 (1988): 71–100, at 82, from which this translation is drawn. For a good biographical sketch, see Jacqueline Boucher, "Auger," in Jouanna et al., eds., *Histoire et dictionnaire des guerres de religion,* 683–686.

before them, even at the stake).[4] By the 1580s penitential fervor had levied armies of crusaders, and the church militant took to marches that were threatening enough to drive off brigands, as Auger's penitents did near Toulouse.[5] In short, Catholic hymns *were* sent into battle against Protestant psalms. While processions established physical dominion over streets, parishes, towns, and provinces, their sacred songs echoed across the territories they captured, blessing them in the name of God.

Much of this chapter charts the broad terrain of religious beliefs upon which this battle was waged and takes stock of the musical arsenal deployed by the faithful. My purpose is not to set aside those events staged at the Catholic court, to which I will turn soon enough, but to include *le menu peuple* in my history. Indeed, without this broader perspective, it is difficult to comprehend fully the religious issues at stake in the wars and the meaning of the music and processions under consideration.

The research of Nancy Roelker, Barbara Diefendorf, Denis Crouzet, Larissa Taylor, and Natalie Zemon Davis has shown that for the common folk above all, the civil wars were about religious difference.[6] Among elites, constitutional concerns, noble factionalism, dynastic wrangling, and even personal vendetta spurred on the wars, and, viewed from the perspective of those in power, the wars often appear to have consisted of the sorts of conflicts that had always characterized European rule. But for the great majority of the populace, the violence of those years erupted from a panic to rid their world of moral pollution. Protestants decried the clergy as filthy, made unclean by their concubines, sodomy, and idol worship, while Catholics feared contamination from fiendish heretics whose sessions of psalm-singing ended in orgies.[7] Such antipathies erupted during religious processions, when columns of the devout turned to "warriors of God" and crowds to mobs.[8] French towns and villages became sectarian battlegrounds. It is true that, during the religious wars,

4. See Diefendorf, *Beneath the Cross,* 136–144.

5. Edmond Auger, *Metanoeologie, sur le suget de l'Archicongregation des Penitens de l'Annonciation de nostre Dame, & de toutes telles autres devotieuses assemblées, en l'Eglise sainte* (Paris: Jamet Mettayer, 1584), 188–189.

6. Nancy Lyman Roelker, *One King, One Faith: The Parlement of Paris and the Religious Reformations of the Sixteenth Century* (Berkeley and Los Angeles: University of California Press, 1996); Diefendorf, *Beneath the Cross,* Denis Crouzet, *Les guerriers de Dieu: La violence au temps des troubles de religion, vers 1525–vers 1610,* 2 vols. (Seyssel: Champ Vallon, 1990); Larissa Taylor, *Soldiers of Christ: Preaching in Late Medieval and Reformation France* (New York: Oxford University Press, 1992); and Natalie Zemon Davis, *Society and Culture in Early Modern France* (Stanford: Stanford University Press, 1975).

7. See, among others, Davis, *Society and Culture,* chap. 6.

8. Ibid., and Crouzet, *Les guerriers de Dieu.*

FIGURE 4.1. Procession of the King's White Penitents (1583). Bibliothèque nationale de France, Cabinet des estampes, Qb1-1583

peasants in the Dauphiné revolted against noble oppression and taxes, and that grain riots and other protests sometimes fractured civil society along class lines. Nonetheless, the wars in France between 1562 and 1629 were wars of religion, no matter how loud the rhetoric of haves against have-nots became, and no matter how dramatically the concerns of nobles shaped them.

My interest in sacred music and how it animated religious conflicts ultimately contributes to the second and larger object of this chapter: to show how the liturgical music of the Catholic church was foundational to royal ceremonial. We should see Henry's keenness for processions, his establishment of three new orders, and his creation of a Grand Master of Ceremonies in 1585 as parts of a ceremonial offensive designed to assert his divine right to rule.[9] It was a campaign of public image derived in part from religious ritual. The case of Henry III is revealing not only because the musical ceremonies of his reign attained enduring forms, but because—like the *ballet de cour*—they evolved in reaction to religious strife. Yet unlike the *ballet de cour,* which was by definition a courtly genre, the ceremonies I will examine here attempted to seize for monarchic expression religious music and practices shared by all.

Even in the thumbnail sketches above, we can see how the Confraternity of the Annunciation made its debut with a penitential procession not at all unlike the *processions blanches* in Burgundy and Lorraine. Only the singing of the *Salve Regina* in "very harmonious polyphony" distinguished the royal penitents from others (see fig. 4.1).[10] The king's singers added sumptuousness to the ceremonies, but polyphony was virtually the only *marque de noblesse* not foregone by the king and courtier-penitents, who, hooded and swathed in white linen from head to toe, marched on foot with little to signal their rank and station. The engraving Pierre de L'Estoile pasted into his scrapbook depicts the singers with music books that match the size and oblong format Le Roy et Ballard used for their motet prints and contain chant notation on four-line staves. They seem to refer simultaneously to the singing of *fauxbourdon* and motets. The other penitents carry what must be little prayer books, which, along with the motets, hint that the procession was not like others.

What we also know from L'Estoile—whose colorful and biased chronicle is a primary source of information on the reign of Henry III—is that Henry's pious dis-

9. For a pointed overview of ceremonial and absolutism see Cosandey and Descimon, *Absolutisme en France,* 75–82.

10. L'Estoile, *Mémoires-journaux,* 2:110. "Les Chantres du Roy et autres marchoient en rang, vestus de mesme habit, en trois distinctes compagnies, chantans mélodieusement la litanie en faux-bourdon. Arrivés en l'église Nostre-Dame, chantèrent tous à genoux le *Salve Regina* en très-harmonieuse musique."

play backfired.[11] Preachers condemned the royal penitents as hypocrites and called their hoods a mask behind which they mocked God, Parisians reacted badly to the sight of their king in sheets, and we have already read of how the king's household servants mocked the procession with a Carnivalesque reenactment in the Louvre. Whether it was intended to or not, the lackeys' counterprocession challenged the king's profession of faith in its own terms, terms he had appropriated from the penitential confraternities that attracted scores of avid believers in those same years.

In these ceremonies, then, we discover a hotly contested musical language. My design is to move back and forth between court ceremonial and the sacred rituals staged by others, examining the struggles between a king and his subjects for control of religious symbols. At stake was power itself. Religious rituals employed a symbolic "magic" to establish the dominion of God, and what we see in the contests of penitential devotion described above are assemblages of Catholic symbols—hymns, hosts, robes, candles—that empowered those who deployed them. They placed the penitents closer to God. This had long been the method of monarchic ritual, which borrowed from Christian ceremonies that had themselves often been aligned to the needs of Roman emperors for divine affirmation.

Ritual forged relationships between temporal and divine power. As Clifford Geertz put it in his essay on "Centers, Kings, and Charisma," we should "look for the vast universality of the will of kings . . . in the same place as we look for that of gods: in the rites and images through which it is exerted. More exactly, if charisma is a sign of involvement with the animating centers of society, and if such centers are cultural phenomena and thus historically constructed, investigations into the symbolics of power and into its nature are very similar endeavors. The easy distinction between the trappings of rule and its substance becomes less sharp, even less real; what counts is the manner in which, a bit like mass and energy, they are transformed into each other."[12] Geertz reminds us to see in the mystique of court cere-

11. Ibid., 2:109–112. Nothing is known of the professional life of L'Estoile (1546–1611), whose detailed journal and collection of political broadsides chronicle happenings in Paris during the reigns of Henry III, Henry IV, and the first year of the reign of Louis XIII. His milieu seems to have been that of Parisian magistrates and members of parlement, which gave him access to information about goings-on at court. Though written in a tone of reportage, his politics shine through in the journal. Roelker has characterized his bias as "royalist, *politique,* anti-ultramontane to the nth degree" (*One King, One Faith,* 484). He was certainly more approving of Henry IV than of Henry III. See the overview in ibid., as well as that in Pierre de L'Estoile, *Registre-Journal du règne de Henri III,* ed. Madeleine Lazard and Gilbert Schrenck, 6 vols. (Geneva: Droz, 1992–2003), 1:7–50, esp. 10–13, 26–28.

12. Clifford Geertz, "Centers, Kings, and Charisma: Reflections on the Symbolics of Power," in *Local Knowledge: Further Essays in Interpretive Anthropology,* 3rd ed. (New York: Basic Books, 2000), 121–146, at 124.

monial the actual machinery of power. But we might see it in the other *processions blanches* as well, as penitents attempted to resacralize their country and combat unruly soldiers. This chapter, then, takes sacred song to be a religious symbol of tremendous power and charts the struggle between Catholic extremists and their king over the quintessential hymn of divine authority—the *Te Deum laudamus.*

ROI-DIEU

When Henry III created the Grand Master of Ceremonies in 1585, he stressed that all things in his court should be conducted with order. Only in this way could the dignity and splendor befitting his royal grandeur be acknowledged.[13] This was not the idle fancy of a king so often accused of ignoring his subjects in favor of revels with his minions. Rather, ceremonial allowed Henry to assert his authority at a critical moment in his reign. *Royale grandeur* was a quality that he alone possessed, a radiance, brilliance, or éclat emanating from the sacredness of kingship, and ceremonial reified it.

Henry's attention to ceremonial may seem frivolous at a time when challenges to his rule were so serious. His barren marriage and the death of his brother, François, in 1584 left the Valois without an heir, placing the Protestant Henry of Navarre next in line to the throne. Panicked at the prospect of a Protestant king, nobles and commoners alike swelled the ranks of the Catholic League, which insisted that the Cardinal de Bourbon was Henry's rightful successor. The Duke of Guise headed the crusade to save France from the certain damnation that crowning a heretic would bring, and in addition to becoming a popular hero, Guise negotiated a treaty with Philip II that filled the League's war chest while the king's coffers hemorrhaged. Pierre de L'Estoile noted in December of 1585 how the League inflamed hatred of Henry, inciting all of Paris to murmur that the king drained his purse "en plaisirs, vilanies, bombances et dissolutions."[14] League cells sprang up across the kingdom, the most notorious one in Paris, where committees of public safety were formed in each of the sixteen *quartiers* of the city. Led by local magistrates, clergy, wealthy merchants, and lawyers from the Parlement of Paris, the League quickly gained the support of the bourgeoisie and lower classes, forming an organization with the political clout of grass roots. This war would be fought not just on battlefields, but in the streets of Paris and provincial capitals, where *colporteurs* flogged seditious pamphlets, placards papered corners, and militants sang League anthems. It was in the

13. Théodore Godefroy and Denis Godefroy le Jeune, *Le cérémonial françois, recueilly par Théodore Godefroy et mis en lumière par Denis Godefroy,* 2 vols. (Paris: Cramoisy, 1649), 1: [unnumbered opening folios]. Similar impulses gave rise to the demands of courtesy, on which see Ranum, "Courtesy, Absolutism, and the Rise of the French State."

14. L'Estoile, *Mémoires-journaux,* 2:221.

face of urban dissidence that Henry launched his ceremonial offensive, which was designed to meet the League on its own ground.

THE MOST concentrated affirmation of the king's sacredness was the *sacre et couronnement,* a ceremonial whose symbolic vocabulary would be heavily mined during the religious wars. The very name of the coronation ceremony ("consecration and coronation") indicated that for the French, kingship was a state simultaneously divine and temporal, embodied by a king who was himself both God and man.[15] As Ernst Kantorowicz demonstrated in *The King's Two Bodies,* the medieval "political theology" inherited by early modern kings maintained that the king was a twinned being in whom a natural, mortal body and a royal superbody conjoined.[16] To paraphrase Kantorowicz, it was as though the "soul" or immortal part of kingship migrated from one mortal king to the next, reincarnating the body politic in a king of flesh, doing away with the human imperfections of his body natural, and conveying a sort of immortality to him.[17] Like Christ, the king was by nature a man and by grace a *Christus*—an anointed one or God-man. Thus the cries of "Vivat Rex in aeternum" at coronations and "Le Roy est mort—Vive le Roy" at French royal funerals made clear that the immortal body of the king lived on even though his mortal body had perished. In life, too, the king possessed Christlike abilities, laying hands upon the sick in order to cure victims of scrofula with his "royal touch." So certain was the king's divinity that in 1625 the General Assembly of the Clergy would declare that "kings are ordained by God, but not only that: they are themselves gods."[18]

Of course, this claim issued from a clergy far more secure in the Catholicism of the crown than it had been in the 1580s, when Henry III refused to drive the Huguenots from the kingdom, or in the early 1590s, when a Protestant claimed the throne. This uncertainty was made all the more acute by the sacerdotal (rather than constitutional) nature of kingship in France, where the coronation included an oath to protect the country from heresy. The king was not only a *Roi-Dieu,* but a *Roy-Prestre* as well, whose defense of the faith counted among his primary responsibilities.[19] Unlike any other European monarch, the French king bore the title of *Rex*

15. Richard A. Jackson, *Vive le Roi! A History of the French Coronation from Charles V to Charles X* (Chapel Hill: University of North Carolina Press, 1984), 3.

16. Kantorowicz, *The King's Two Bodies.* On the consecration of kings, see especially chap. 3.

17. Ibid., 13. See Giesey, *Royal Funeral Ceremony,* on the delicate period of interregnum, when this transfer of kingship between mortal bodies was believed to occur.

18. Cited in Jackson, *Vive le Roi!,* 218.

19. See ibid., 3, on the king as *roi-prêtre,* especially the taking of both species of communion during the Coronation Mass.

christianissimus or Most Christian King, making him a *Rex regum* in the Catholic world. The French took these as signs that they were God's chosen people, to whom he delivered the Holy Ampulla containing the balm used to anoint French kings and the fleur-de-lis, sent from heaven as a mark of God's special care for France.[20] The English might suffer Protestant monarchs, but not the French.

At the outset of the religious crisis in France, the crown dealt firmly with Protestant sedition, the case in point being the Affair of the Placards in 1534. Parisians on their way to Mass had been greeted one Sunday morning with copies of a placard bearing the title "True Articles on the Horrible, Gross, and Insufferable Abuses of the Papal Mass Invented Directly Contrary to the Holy Supper of Jesus Christ."[21] It attacked the doctrine of transubstantiation, denying that Christ's body and blood were in any way actually contained in the consecrated bread and wine, and argued furthermore that by this devilish doctrine Catholic priests led their poor flocks astray like bewitching wolves, destroying everything and disinheriting kings, princes, merchants, and everyone imaginable. Francis I, to whose bedroom door one of the placards had been nailed, had not been not slow to respond, for its denial of transubstantiation and announcement that the king had been disinherited from the kingdom of heaven attacked his authority at its foundation. As the coronation ceremony made so clear, the king was himself a transubstantiated being. To deny Christ's real presence in the Eucharist negated not only the sacrament of communion, but the king's divinity as well. The placard had attacked the essence of a *Rex christus,* in this way committing a crime against the royal body politic. Suspects were imprisoned and six were burned, presumably with spectacular sermons for their souls and opportunities to convert at the stake, but the king's most striking response came in the form of an unprecedented procession featuring the Corpus Christi, with Francis walking just behind it.[22]

The procession included relics from across the city, which were brought in separate processions to the royal parish church of Saint-Germain-l'Auxerrois, where the general procession began. It included the rarely seen crown of thorns from the Sainte-Chapelle and reliquary caskets of the city's patron saints, Saint Marcel and Sainte Geneviève. Not only did the city give over all of her relics (many of which had never left their parishes) for the event, but the whole of corporate Paris participated as well: the monarchy, law courts, university, monks, clergy, magistrates, and

---

20. Ibid., 33. Also see Ralph E. Giesey, "Models of Rulership in French Royal Ceremonial," in *Rites of Power,* ed. Sean Wilentz (Philadelphia: University of Pennsylvania Press, 1985), 41–64, at 43–44.

21. This exposition is based on the overviews in Holt, *French Wars of Religion,* esp. chap. 1, and Mark Greengrass, *The French Reformation* (Oxford: Basil Blackwell, 1987).

22. On the procession see the sources and analysis in Diefendorf, *Beneath the Cross,* 46–47; and the relation in Godefroy and Godefroy le Jeune, *Cérémonial françois,* 2:939–945.

guilds. At the procession's center, the bishop of Paris carried the host in a jeweled monstrance beneath a royal canopy held by four princes of blood, and Francis, out of reverence for the Eucharist, walked behind it in the manner of a simple Christian, bareheaded, dressed in black, and carrying a burning torch. He prayed at four stations en route to Notre-Dame, impressing onlookers with his great piety. The king's singers sang "devout songs and motets in the concord of polyphonic music" during the procession and "anthems" at each of the stations.[23]

After Mass, the king retired to the bishop's palace to dine with the dignitaries from the procession. He took the occasion to reiterate the Oath of the Kingdom he had sworn during his coronation, this time embellishing it with the physical metaphor of a diseased body. "If one of the arms of my body was infected with this corruption, I would cut it off, and if my children were tainted with it, I would myself offer them in sacrifice."[24] As Barbara Diefendorf observes, the speech articulated the need to purify the social body even as the king's reference to his arm reminded listeners that all of his subjects were members of the body politic.[25] Of course, the organization of the procession around the body of Christ further stressed the likeness of the king to Christ and his duty to protect the sacraments by juxtaposing royal and heavenly majesty. If processions often kindled religious riot, the inverse was also true—that they could resacralize the spaces through which they passed. In Francis's own words, the 1535 procession had been made "in order to invoke the grace of our Redeemer by a common consensus of all."[26] By focusing the entire corporate and sacred wealth of Paris upon the Corpus Christi, the procession unified her citizens and restored Christ's purifying grace to the city.

Had the hard line against heresy laid down by Francis I been possible to maintain, the crown's role in the religious wars would have been far different, but as it was, during the reigns of Henry II and his sons (Francis II, 1559–1560; Charles IX, 1560–1574; and Henry III, 1574–1589) the relationship between the crown and the Huguenots became increasingly complex. Henry II dealt firmly with Protestant "rebels," setting up a special *chambre ardent* in the Parlement of Paris for the prosecution of heretics and banning Protestantism entirely in 1551. Although the psalms

23. See Georges Guiffrey, ed., *Chronique du Roy François* (Paris: Renouard, 1860), 113–126, at 113. Discussed in John T. Brobeck, "Musical Patronage in the Royal Chapel of France under Francis I (r. 1515–1547)," *Journal of the American Musicological Society* 48 (1995): 187–239, at 220. On the Royal Chapel in this procession and that of 1528 (also against heresy), see Christelle Cazaux, *La musique à la cour de François Ier* (Paris: École des Chartes and Programme "Ricercar," 2002), 175–179.

24. Guiffrey, *Chronique du Roy François,* 125. "Si ung des bras de mon corps estoit infecté de ceste peste je le vouldrois coupper, et si mes enfans en estoient entachez, je les vouldrois moy mesme immoler."

25. Diefendorf, *Beneath the Cross,* 46–47.

26. Guiffrey, *Chronique du Roy François,* 124. "Pour invoquer, d'ung commung constantement de tous, la grâce de nostre Rédempteur."

of Clément Marot had been popular with Catholics in the 1540s and even at Henry's court, at this time psalm-singing in public was banned.

During her regency (1560–1563), Catherine de' Medici attempted to reconcile religious differences—and to distance herself from the equally dangerous Guises—with a moderate approach. The Edict of Toleration of 1562 accorded limited rights to Huguenots. Still forbidden to worship inside any towns, to congregate at night, or to raise arms (restrictions aimed, of course, at preventing seditious activities), Huguenots were permitted to preach openly in the countryside as long as it was by day and peaceably.[27] Tragically, the crown's recognition of the Protestant faith precipitated the civil war Catherine wished to avoid, and fighting broke out before the edict could be signed into law.

Henry III pursued his mother's tolerationist policies, which placed him at odds with Catholic militants and cast doubt upon his Catholicism, for by allowing Protestants to remain within the kingdom, he broke the covenants of his coronation. Leaguers argued that one could legitimately oppose a king who defied the Oath of the Kingdom, and by 1589 regicide was condoned from Catholic pulpits and street songs preached deposing the *roy hérétique.*[28] In the public outcry following Henry's assassination of the Duke of Guise (December 1588), "there was not a son of a good mother in Paris who did not vomit abuse and gibes against the King, whom they called *Henri de Valois, bugger, son of a whore, tyrant* . . . in the streets and crossroads of Paris."[29] Leaguers inflamed public opinion with stories of Henry consorting with the devil and other unnatural acts, and while much of this slander played upon what seemed to be evidence of Henry's deviant sexuality—his childlessness, his taste for lace, pearl earrings, and other finery, his appearance in travesty during court festivities, and his affection for his minions—it is important to note that these reports of sexual disorder, gender reversal, and diabolic monsters were themselves evidence of religious disorder.[30] His was a body turned back to front, male to female, a body incorporating the topsy-turvy reversals of creed brought by

---

27. See Holt, *French Wars of Religion,* 45–49, on the edict and reactions to it.

28. Ibid., 131. For one song see *Les belles figures et drolleries de la Ligue,* in L'Estoile, *Mémoires-journaux,* 4:70–72.

29. L'Estoile, *Mémoires-journaux,* 3:242. "Car aussi n'y avoit-il fils de bonne mère à Paris qui ne vomist injures et brocards contre le Roy, qu'ils apeloient *Henri de Valois, bougre, fils de putain, tiran,* estant tous les jours crié et deschiqueté par les rues et quarrefours de Paris."

30. See, for example, the pamphlet *Les choses horribles, contenues en une lettre envoyee à Henry de Valois, par un Enfant de Paris, le vingthuitiesme de Janvier 1589* (Paris: Jacques Gregoire, 1589) and the analyses of images of Henry III in Guy Poirier, *L'homosexualité dans l'imaginaire de la Renaissance* (Paris: Honoré Champion, 1996); David Teasley, "The Charge of Sodomy as a Political Weapon in Early Modern France: The Case of Henry III in Catholic League Polemic, 1585–1589," *Maryland Historian* 18 (1987): 17–30; and Katherine B. Crawford, "Love, Sodomy, and Scandal: Controlling the Sexual Reputation of Henry III," *Journal of the History of Sexuality* 12 (2003): 513–542.

the Reformation, a body poisoned by heresy that aptly evinced the troubles of the body politic.

Henry's own actions exacerbated this situation. Certainly his spectacular appearances in ecclesiastical robes, as a flagellant, or as a penitent were easily construed as masquerades not unlike the other fetes that exemplified the moral dissipation of which he and his court were accused. And despite these zealous professions of faith, he rejected the politics of Catholic extremism, refusing to expel the Protestants in favor of a moderate path to civil union. Though to say so risks drawing the story lines of the wars too boldly, we must stress that Henry's politics damaged his reputation as a good Catholic. And in the end, confessional politics trumped civil ones. Leaguers drove the king from Paris in May of 1588, and as he attempted to regain the city the following year, a young Jacobin monk assassinated him in his camp. In the minds of many, the kingdom was better off rid of its heretical head. To the tune of a song called "Tremblez, tremblez Huguenots," Parisians celebrated his murder with the words "He is dead, this traitor King, He is dead, O hypocrite, He is dead in disarray, Clothed with his iniquitous deeds."[31] In light of the traditional cry of "Le Roy est mort—Vive le Roy," this song's reiterations of "Il est mort" sound all the more like nails in the coffin of a king long dead to some of his people. The Te Deum ceremonies of December 1587, in which he paraded triumphantly through Paris (and to which I now turn), may fairly be seen as his last stand.

THE CREATION OF THE TE DEUM CEREMONIAL UNDER HENRY III

As Henry saw it, ceremonial should celebrate his royal grandeur. His greatness was not in question, nor was the dignity and splendor befitting it; rather, his charge to the new master of ceremonies in 1585 was to order all things in such a way that the dignity and splendor owed to his greatness was acknowledged. The new Te Deum ceremonial that took shape in 1587 responded directly to the royal desire for affirmation by reworking a highly significant moment in the coronation, at which the king sat enthroned and the *Te Deum laudamus* registered the people's acclaim. During the long *sacre et couronnement,* the candidate was consecrated with holy oil, crowned by the twelve peers of France (six ecclesiastical and six lay), and enthroned. Finally the officiating archbishop or cardinal bowed before the king, kissed him, and cried out in a loud voice, "Vivat Rex in aeternum!" At this cue the crowds erupted with "Vive le Roy," musicians played, and, after a time, the officiant began the *Te Deum laudamus* hymn, which was finished by choir and organ. For those in the nave of the cathedral, who would have seen and heard little before the king was brought to the throne erected for him on the rood-loft and the "Vivat Rex" was said, the

31. L'Estoile, *Mémoires-journaux,* 4:218–223. "Il est mort, ce traistre Roy! Il est mort, ô hypocrite! Il est mort en desarroy, Vestu de ses faicts iniques!"

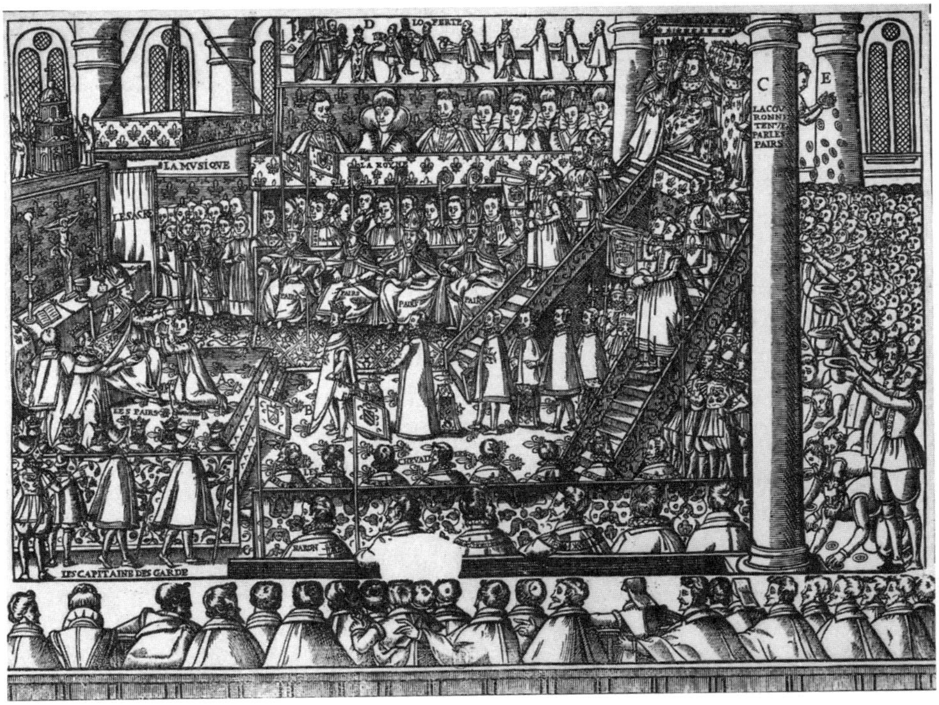

FIGURE 4.2. The Coronation of Louis XIII in Reims (1610). Bibliothèque nationale de France, Cabinet des estampes, Qb1-1610

Te Deum capped the most magnificent part of the ceremony (see fig. 4.2).[32] Heralds threw money and coronation *jetons,* clarions, shawms, trumpets, and drums sounded, and for the coronation of Louis XIII in 1610, hundreds of sparrows, goldfinches, and other small birds were released from the choir, delighting the nine-year-old king as they flew and sang about him.[33] Only as the Te Deum progressed did the cries of "Vive le Roy" both inside and outside the cathedral run their course,

32. The elevation of the throne was quite important. René Benoist begins his ceremonial with a discussion of the throne (*L'ordre et les ceremonies du sacre et couronnement du tres-Chrestien Roy de France* [Paris: Nicholas Chesneau, 1575], fol. 4r), and even though Henry IV was crowned in Chartres, his throne, too, was on the rood screen. See [Nicolas de Thou,], *Ceremonies observees au sacre & coronement du Tres-Chrestien et Tres-valeureux Henry IIII. Roy de France & de Navarre* (Paris: Jamet Mettayer & Pierre L'Huillier, 1594), 8–9.

33. Godefroy and Godefroy le Jeune, *Cérémonial françois,* 1:432; this account is a reprint of *Le voyage de Reims* (Paris: E. Foucault, 1610).

the artillery fire subside, and the sonorities of polyphony take over, bringing the first half of the ceremony to a close before the Mass began.

The Te Deum hymn featured in other royal ceremonial as well.[34] Entries usually ended with a Te Deum and Mass, and even those towns which could not greet the king with triumphal arches did welcome him with the sacred hymn. These royal associations were certainly consistent with Catholic uses of the hymn more generally: it was sung at the end of matins on Sundays and feast days as a festal praise of God; churches and cathedrals across Europe sang it in thanksgiving for significant events in the lives of local congregations; Te Deums greeted visiting church dignitaries and local princes; and the hymn of Saints Ambrose and Augustine, as it was called, was used for the consecration of popes, from which it had likely made its way into the French *sacre*.[35]

The Te Deum is related to the Eucharist of the Roman rite, a similarity that is immediately evident in their shared text "Sanctus, Sanctus, Sanctus, Dominus Deus Sabaoth. Pleni sunt coeli et terra majestatis gloriae tuae" (verses 5–6).[36] The Sanctus represents early Christian hymnody at its most essential and is one of the few texts that can properly be called a hymn in the original sense of "praise-words." Fittingly, the threefold acclamation "Holy, Holy, Holy" was sung by the congregation, and while in early Christian worship it expressed the people's joy upon hearing the bishop tell the story of Christ's life, death, and resurrection, cries of "Sanctus, Sanctus" peppered the rites of Greco-Latin cults as well, who used it to praise various deities and emperors who had declared themselves divine. Although early Christians sought to distance their rituals from pagan ones by choosing a version of the Sanctus from the Hebrew Old Testament, the Sanctus still bore the imprint of ancient ruler worship and imparted it to the Te Deum, something equally evident in the first verse of the hymn and the signature praise-word "laudamus."

The verses surrounding the Sanctus (3, 4, and 7–10) employ the kind of utterance common to the early Christian "Hymn of All Creation." This was a song to the Creator sung by all things, each giving praise in its particular way. First the celestial hosts (angels, cherubim, seraphim, the heavens, and the universal powers) proclaim God's glory, then the terrestrial ones (apostles, prophets, the white-robed army of martyrs, the church). There follows a Trinitarian doxology with the requi-

34. For the earlier part of the century see Cazaux, *Musique à la cour de François Ier,* part 3.

35. Sabine Žak, "Das Tedeum als Huldigungsgesang," *Historisches Jahrbuch* 102 (1982): 1–32. On the history of the hymn in French ceremonial, see Michèle Fogel, *Les cérémonies de l'information dans la France du XVIe au milieu du XVIIIe siècle* (Paris: Fayard, 1989), 154–163.

36. See Dom Paul Cagin, *Te Deum ou Illatio?* (Solesmes: Abbaye de Solesmes, 1906). I thank Richard Crocker for sharing his insights on the Te Deum hymn and eucharistic devotion. On hymnody more generally see Crocker, *An Introduction to Gregorian Chant* (New Haven: Yale University Press, 2000), esp. 90–93.

site acknowledgement of the Father, Son, and Holy Ghost (verses 11–13), the salvation story beginning "Tu rex gloriae, Christe" (verses 14–21), and a closing section of petition and exit (verses 22–end). Richard Crocker has spoken of the Te Deum as an "ideal" hymn, composed in the fourth century at a time when Christian leaders sought to control the language of hymns. Of course one way to ensure orthodoxy—particularly in the Mass—was for the bishops to make up and sing the hymns themselves, which is what they did in the case of the eucharistic prayer, the anaphora. Crocker sees in the Te Deum an idealized anaphora, with its praise-words, salvation story, and Trinitarian doxology. Unconstrained by theological or liturgical requirements, the Te Deum redacts the Eucharist, casting its elements in a lyric form shaped first by artistic considerations. By paralleling the ritual enactment of the Last Supper, the Te Deum invites reflection on the eucharistic presence of Christ that redeems the faithful and elicits the outburst of thanksgiving at the heart of the Mass—"Holy, holy, holy." In the context of a coronation, entry, or other royal ceremony, the Te Deum cut both ways, reminding those who sang and heard the hymn that the king before them was a *rex Christus* and his body politic a *corpus mysticum* they joyfully shared. "Tu rex gloriae, Christe": you are the king of glory, O anointed one.

TEXT

Te deum laudamus:
te Dominum confitemur.
Te aeternum Patrem
omnis terra veneratur.
Tibi omnes Angeli,
tibi Caeli: et universae Potestates:
Tibi Cherubim et Seraphim
incessabili voce proclamant:
Sanctus, Sanctus, Sanctus:
Dominus Deus Sabaoth.
Pleni sunt caeli et terra
Majestates gloriae tuae.
Te gloriosus
Apostolorum chorus,
Te Prophetarum
laudabilis numerus,
Te Martyrum
candidatus laudat exercitus.
Te per orbum terrarum
sancta confitetur Ecclesia,

Patrem immensae majestatis;
Venerandum tuum verum et unicum Filium;
Sanctum quoque Paraclitum Spiritum.
Tu Rex gloriae, Christe,
Tu Patris sempiternus es Filius.
Tu, ad liberandum suscepturus hominem,
non horruisti Virginis uterum.

TRANSLATION

We praise you O God,
we acknowledge you to be the Lord;
All the earth now worships you,
the Father everlasting.
To you all angels cry aloud,
the heavens and all the powers therein;
To you cherubim and seraphim
cry without ceasing:
Holy, holy, holy:
Lord, God of armies,
Heaven and earth are full
of the majesty of your glory.
The glorious company
of the apostles praise you,
The admirable company
of the prophets praise you,
The noble army
of martyrs praise you.
The holy Church does acknowledge you
throughout all the world:

O Father whose majesty is infinite!
Your adorable, true, and only Son,
And the Holy Spirit, the counselor.
You are the King of glory, O anointed one.
You are the everlasting Son of the Father.
When you took upon yourself to deliver man,
you humbled yourself to be born of a virgin.

Tu, devicto mortis aculeo, aperuisti credentibus regna caelorum.	When you had overcome the sharpness of death, you opened the kingdom of heaven to all believers.
Tu ad dexteram Dei sedes, in gloria Patris.	You sit at the right hand of God in the glory of the Father.
Judex crederis, esse venturus.	We believe that you will come to be our judge.
Te ergo quaesumus, tuis famulis subveni, quos pretioso sanguine redemisti.	We therefore pray you help your servants, whom you have redeemed with your precious blood.
Aeterna fac cum Sanctis tuis in gloria numerari.	Make them to be numbered with your saints in glory everlasting.
Salvum fac populum tuum Domine, et benedic hereditati tuae.	O Lord save your people and bless your heritage.
Et rege eos, et extolle illos usque in aeternum.	Govern them and lift them up forever.
Per singulos dies, benedicimus te,	Day by day we magnify you,
Et laudamus nomen tuum in saeculum saeculi.	And we worship your name, world without end.
Dignare Domine die isto sine peccato nos custo dire.	Vouchsafe, O Lord, to keep us this day without sin.
Miserere nostri, Domine, miserere nostri.	O Lord have mercy upon us, have mercy upon us.
Fiat misericordia tua, Domine, super nos, quemadmodum speravimus in te.	O Lord, let your mercy be upon us, as our trust is in you.
In te Domine speravi non confundar in aeternum.	O Lord, in you have I trusted, let me never be confounded.

The eucharistic tone of the Te Deum may have particularly appealed to the French, with their Christ-centered ideas of kingship, but the use of the hymn for ceremonies of power was hardly limited to France.[37] In coronations and imperial investitures the hymn functioned initially as the people's approval of the elected one brought before them and, eventually, as a sign of their adoration of the new ruler. This turn can be witnessed in the evolution of the *sacre et couronnement.*[38] In the 1250 *ordo,* or order of ceremonies, the Te Deum was sung early on in the ceremony before the enthroning: the king-to-be swore to protect the church and its rights,

37. Žak, "Das Tedeum als Huldigungsgesang."

38. See Fogel, *Cérémonies de l'information,* 133–188, in which she presents an indispensable history of the Te Deum hymn in French ceremonials from the *sacre et couronnement* to the Te Deum ceremonial that emerged in the late sixteenth century. On the changing coronation *ordines,* see Jackson, *Vive le Roi!,* 24–40. Alain Guéry, in his study of the symbiosis of Catholicism and political power in Europe, also notes the difficulties some contemporaries had balancing the idea of popular election with divine right. See Guéry, "Le Roi est Dieu, Le Roi et Dieu," in *L'état ou le roi: les fondations de la modernité monarchique en France, XIVe–XVIIe siècles,* ed. Neithard Bulst, Robert Descimon, and Alain Guerreau (Paris: Maison des Sciences de l'Homme, 1996), 27–47.

then the bishops asked the people if they would accept the candidate as their king. At this point the Te Deum was sung while the candidate knelt, after which he rose and swore the Oath of the Kingdom. In the 1365 *ordo* of Charles V, the Te Deum still holds this place, but the writer grumbles that in other kingdoms the Te Deum follows the enthroning, "and one thinks that it is better there than here." For placed after the enthroning, the Te Deum announced the people's acclaim, but placed before it, the hymn registered their consent. By the coronation of Charles VIII in 1484, the Te Deum celebrated royal power unconditionally: the king sat, consecrated and enthroned, bearing all the devices of his royal majesty, while the cries of "Vivat rex in aeternum" gave way to "Vive le Roy" and the supreme acclamation of the *Te Deum laudamus.*

Despite the hymn's ceremonial transformation into a song of popular acclaim, the Te Deum's service to the crown in the sixteenth century remained rocky. Royal historians, noticing the discrepancies between earlier coronation *ordines* and the one then in use, wished to restore the request for consent from the people to the service and move the Te Deum back to its former position. In Jean Champagne's description of the revised ceremony, the king lay prostrate on the ground during the Te Deum to show his mortality. "You are Gods, but you will die like men and fall to the earth like one of the Princes," Champagne declared.[39] In an effort to forestall such scenes, the royal counselors cut both the people's consent and the Te Deum from the ceremony. No Te Deum was sung for the coronations of Charles IX or Henry III, and perhaps not for Francis II.[40] When it was sung, as at the end of royal entries, its authority was often compromised by the entry's themes of power shared between crown and city or lessons on the civic obligations of the king.

In fact, cities enjoyed proprietorship over a festivity not unlike a royal entry that included processions, fireworks, bell-ringing, temporary monuments, and the Te Deum, but—significantly—not the king. In September 1574, the mayor and aldermen of Paris organized *feux de joye* to celebrate Henry's safe return from Poland to take up the crown after the death of Charles IX, and although this particular

39. Jean Champagne, *Discours du sacre et coronnement . . . en forme d'Epistre* (Reims: Jean de Foigny, 1575), fol. I4r. "J'ay dit que vous estes Dieux, mais vous mourrez, comme hommes, & tomberez par terre, comme un des Princes."

40. There are significant discrepancies between printed commemorations of Henry III's coronation—several of which were printed up ahead of the fact—and what seems to have actually taken place. Ibid., and Benoist, *Ordre et les ceremonies du sacre et couronnement,* both give the order of service following the 1365 *ordo* (Benoist even repeats the remark made there that the Te Deum is better said after the enthroning than following the Oath of the Kingdom), but Fogel's research shows that the Te Deum was cut altogether. See Fogel, *Cérémonies de l'information,* 160–162.

event had been ordered by the king, it is important to see how much more inclusive it was than an entry.[41] After a procession to Notre-Dame for the Te Deum, the church bells began to peal and the clock at the palace to chime, and the attention shifted to the Hôtel de Ville, to which the aldermen and mayor had repaired for a meal. A portal erected at the entrance showed the Holy Spirit in the figure of a dove extending its wings protectively over Henry's arms of France and Poland. Verse by Jean Dorat explained the significance of the tableau, but even the illiterate would have been able to decode its basic message: God had bestowed the Polish crown upon Henry on Pentecost of 1573 and the French crown on Pentecost of 1574; the downward-plunging Pentecostal dove—which dramatized the descent of the Holy Spirit upon the apostles fifty days after the resurrection of Christ—symbolized the special favor God showered on Henry's ascent to these European thrones. During the interregnum, Pentecostal symbolism reaffirmed Henry's divine election, and it remained a favorite emblem of the king.[42] He eternalized it in the Order of the Holy Spirit, with its regalia featuring the dove and the fiery tongues of the Holy Spirit.[43]

Whether or not they could even see the tableau, the people cramming the Place de Grève in 1574—an "almost infinite" number—were still treated to a providential spectacle. Forty pieces of artillery, forty mortars, and ranks of harquebusiers fired joyfully, and trumpets and shawms added to the cacophony as a large bonfire burned. Elsewhere throughout the city fires of joy burned into the night, the abbots of Sainte-Geneviève broke open barrels of good wine and tossed sugared almonds to those at their fire, and the following Sunday general processions coursed through all the parishes in the city. In contrast to the inward pull of an entry, which located the king at its center with every perspective painting, every theater, every pavilion, and the hierarchical organization of the procession, this celebration aimed itself outward into all corners of the city. If it was not wholly decentered, at least the theater created by the bonfire in the Place de Grève required no single perspective, and the neighborhood fires and the multiple processions created local manifestations of the fete that engaged countless others. As it traveled away from the Hôtel de Ville, the message of the tableau dissipated into the revelry kindled by the bonfires like those built for the Feast of Saint John and into the general spirit of thanksgiving motivating religious processions for other occasions.[44]

41. *Les feux de joye faicts a Paris pour l'arrivee du Roy en France, Avec l'Ordre tenu à son entree et reception en la ville de Lyon* (Paris: Denis du Pré, 1574).

42. See Jacques de Bie, *La France metallique contenant les actions célèbres tant publiques que privées des Rois et Reynes* (Paris: Jacques de Bie, 1634), 74.

43. See ibid., 75, and Frances A. Yates, "Dramatic Religious Processions in Paris in the Late Sixteenth Century," *Annales Musicologiques* 2 (1954): 215–270, at 215–217.

44. In at least some years, the king processed to the Place de Grève in order to light the St. John's night fire himself. See L'Estoile, *Mémoires-journaux,* 2:199–200.

From the bonfires, parish processions, and the Te Deum hymn, Henry III and his Grand Master crafted a new ceremonial that reassigned the meanings of these popular forms to acclaim for his *royale grandeur.* The historian Michèle Fogel has described Henry's recuperation of the Te Deum for royalty as a stroke of genius, for the new ceremonial cut the ties between the ambiguous meaning of the hymn in the coronation and its role in the negotiations of power conducted through entries, making of it a ceremonial that was the king's alone.[45]

As Fogel has shown, during Henry's reign the singing of the Te Deum in conjunction with *feux de joye* and other civic events increased gradually along with the general escalation of public devotion. Yet it was only in December 1587 that the new ceremonial achieved its full formulation, very much in response to the crisis of the League. To a certain extent, it redressed Henry's flagging public support at a time when Parisians were more in love with Guise than ever. Guise promised action in place of the king's pacifism, and he had come through most dramatically in late November 1587, leading Catholic forces to victory against the treacherous German horsemen known as *reîtres.* The king's army had suffered a brutal defeat just the month before at Coutras, where Henry's "archminion" Joyeuse and several other courtiers were killed. Thus Guise's rout of the *reîtres* should have been sweet revenge for Henry, but he was "very unhappy," "irritated," and "envious" of Guise, seeing that "giving this laurel to the League would only wither his own."[46] Moreover, the clergy—"who loved heavy shells more than church bells"—were elated at Guise's triumph, which defeated the king just as surely as Navarre. "The victory of Auneau became the hymn of the League," says L'Estoile, and despite the care with which Henry described his own subjection of the defeated when he wrote to Paris with orders for a Te Deum to be sung at Notre-Dame, in the end the cathedral "resounded more [with] praises of the Duke of Guise than those of God."[47]

Rhetoric meant little without heroism to back it up, and the king quickly threw himself into what was left of the fray, chasing down the defeated Germans, forcing them to capitulate to harsher terms, and gaining political leverage on Guise if not much military advantage on Navarre. Whereas November's pamphlets had announced Guise's "defeat" of the *reîtres* (and some titles reminded readers that Guise operated "by the commandment of the King" or, in a rare case, assigned the victory to the king alone), December's pamphlets reported the "total surrender" of the *reîtres* and the king's "entire and perfect victory" over them.[48] Triumphal hymns

45. Fogel, *Cérémonies de l'information,* 163.

46. L'Estoile, *Mémoires-journaux,* 3:75.

47. Ibid., 3:75–76.

48. At least forty "news" pamphlets relating these events and others containing hymns and triumphal songs were published at the end of 1587 and are conserved at the Bibliothèque nationale de

sang of Henry's fierce protection of the kingdom even as they warned troublemakers at home that they would be next if they did not acknowledge Henry as true king, protector of France, and defender of the church.[49] Hastily printed pamphlets revived Henry's once-brilliant reputation as a warrior by recalling his victories from 1569 at Jarnac and Moncontour, where his horse was killed beneath him, victories eclipsed only by his triumph at Auneau. Other pamphlets described the battle as something straight out of a *roman de chevalerie:* cutlass at his side, lance in his right hand, Henry rode like a Mars, conducting troops with a valiant heart.[50] Descriptions characterizing him as a Godfrey cast the victory as another canto in the *Gerusalemme liberata,* while medallions struck for the occasion depicted Henry as an Alexander in ornate armor, ribbons flying from his crown of laurels and jaw set as he looked sternly to the future (see fig. 4.3).[51] Guise had a medallion struck for his victory as well.[52]

On 14 December artillery fire sounded for joy, bonfires burned throughout Paris, and the whole of the court, parlement, and city government gathered with the queen mother and the queen in Notre-Dame to hear the Te Deum sung "en musicque & bien solemnellement."[53] On 20 December the queens, courtiers, and the parlia-

---

France. See the *Catalogue de l'histoire de France,* 16 vols. (Paris: 1855–1895; reprint, Paris: Bibliothèque nationale, 1968), 1:315–318, nos. Lb34-371–Lb34-413. A good account of the battles from the Protestant perspective is given in d'Aubigné, *Histoire universelle,* 7:164–171. Also see Jacques-Auguste de Thou, *Histoire universelle,* 16 vols. (London, 1734), 10:37–65. See Roelker, *One King, One Faith,* 331–384, for an account of the League's agitation against the crown in 1587–1588 that pays close attention to the mainstream—and often underemphasized—interests of the Parlement of Paris.

49. *Action de graces a Dieu et chants de triomphe au Roy, pour sa victoire* (Paris: Pierre Ramier, 1588).

50. See, respectively, *La victoire entiere et parfaicte obtenue nouvellement par le Roy à l'encontre des Reistres. Plus les feuz de joye: faicts à Paris, & un Te Deum solennel chanté en l'Eglise nostre D'ame, le lundy quatoriesme jour de Decembre* (Paris: Hubert Velu, 1587), and *Resjouissance Chrestienne des vrais et naturels François, pour la reddition generale des Reistres à l'obeissance du Roy. Avec le Te Deum chanté en l'Eglise nostre Dame, & Feux de joyes faits, ce 14 Decembre, 1587* (Paris: Pierre Chevillot, 1587). See also Pierre Poisson, sieur de La Bodinière, *Harangue au peuple de France sur les louanges des anciens François, & de nostre Treschrestien Roy de France, & de Poloigne Henry 3. n'agueres empesché à repousser les estrangers, pour la manutention de l'Estat, & resjouissance de son heureux retour en sa ville de Paris* (Paris: Denis Cotinet, 1588), which took the occasion to recount the victories of all French kings and to admonish Parisians to be grateful for Henry's protection. We should remember that songs were written to commemorate victories as well, such as Ronsard's "Tel qu'un petit Aigle sort" for Henry's victory at Jarnac, set to music by Nicolas de La Grotte.

51. *Resjouissance Chrestienne des vrais et naturels François,* and Cabinet des estampes, QB1-1588, Bibliothèque nationale de France. Significantly, the name given Henry at birth was Edouard-Alexandre.

52. This is the supposition of Jacques de Bie, *Les familles de la France illustrées par les monumens des medailles anciennes et modernes* (Paris: Jacques de Bie, 1634), 87, 90.

53. See *Les feuz de joye, faicts a Paris, touchant la reddition des Reistres: au Roy Tres-Chrestien de France et de Pologne. Avec un Te Deum solennel chanté en l'eglise nostre D'ame lundy quatoriesme Jour de*

FIGURE 4.3. Medallion struck for the victory of Henry III over the Reîtres (1587). Bibliothèque nationale de France, Cabinet des estampes, Qb1-1588

mentary officers of the court, dressed in the red robes of joy usually worn only in the king's presence, processed to Notre-Dame for a Mass announcing the king's imminent return from the *champs de Mars* that Auneau had become.[54] This dizzying array of media focused consistently on specific aspects of the king's victory—Henry's physical valiance and personal risk, his defense of the kingdom, his complete rout of foreign infidels, and God's hand in his victory, "si sainct & belliqueux"—monumentalizing the event and presaging the seventeenth-century media campaigns that built the image of Louis XIV through themes coordinated across paintings, engravings, statues, monuments, palaces, medals, histories, poems, plays, operas, ballets, processions, ceremonies, military reviews, and carrousels, which registered the king's éclat in multiple forms.[55] Power—the king's ability to vanquish enemies—was quickly transformed into symbols of power: the swords, laurels, crosses, crowns, and so forth that in turn renewed the mystical power of his

Decembre (Paris: Hubert Velu, 1587), 8–9, and *Victoire entiere et parfaicte obtenue nouvellement par le Roy,* 10–11.

54. L'Estoile, *Mémoires-journaux,* 3:78.

55. See Burke, *Fabrication of Louis XIV,* and Christian Jouhaud, "Printing the Event: From La Rochelle to Paris," in *The Culture of Print: Power and the Uses of Print in Early Modern Europe,* ed. Roger Chartier, trans. Lydia G. Cochrane (Princeton: Princeton University Press, 1989), 290–333.

majesty.[56] So vastly did the publication of the king's glory exceed the military importance of his acts that we must see in it some cognizance of where a king's unique power lay—in rites and images.

Henry's entry into Paris on 23 December was staged as a triumphal return from battle.[57] He arrived by the Porte Saint Jacques to the south, forgoing the standard ceremonial route from the north and in this way arriving at Notre-Dame as if directly from Auneau.[58] Furthermore, he wore the simplest of black velvet doublets, making the procession seem all the more spontaneous, as if the city officials, bourgeois, courtiers, and parlement, again in their red robes, had turned out to meet him unbidden. He dismounted before the cathedral, the organ sounded as he made his way toward the altar, the choir sang the *Te Deum laudamus* accompanied by organ, followed by the psalm *Laudate Dominum omnes gentes* set *en musique* and several prayers for the king. The succinct imperatives of the psalm—"O Praise the Lord, all ye nations: praise him all ye people. For his merciful kindness is great toward us: and the truth of the Lord endureth for ever. Praise ye the Lord"—aptly extended the ruler worship established by the Te Deum, adding fullness to the proceedings. The musical setting may well have been the magisterial twelve-voice version of Orlande de Lassus, which had been published by Le Roy et Ballard in 1573. In number of voices it was the largest work published in France to that date and majestic indeed. As Henry left the cathedral, the choir sang "Vive le Roy," also *en musique.* The king then remounted and made his way to the Hôtel de Ville, where a large bonfire was lit. In other parts of the city similar fires burned, and as Henry passed through the streets the people (many of whom had been paid to do so) cried loudly, "Vive le Roy!"

THIS WAS the Te Deum ceremonial in its new form, born of the king's desire to recapture the hymn for the expression of royal majesty by expanding it into a self-standing event. While very much like an entry, the Te Deum ceremonial differed in the king's control over it and the more central position of the hymn, which stood

56. Geertz, "Centers, Kings, and Charisma."

57. See the descriptions in *Congratulation a la France sur les victoires obtenues par le Roy contre les Estrangers, & son heureux retour en sa bonne ville de Paris le 23. Decembre. Avec les magnificences qui ont esté faictes à son arrivee* (Paris: Pierre Chevillot, 1587); L'Estoile, *Mémoires-journaux,* 3:78–79; Fonds Godefroy 426, fols. 49Ar–52Av, Bibliothèque de l'Institut (for procession); and Fonds Godefroy 398, 1–6, Bibliothèque de l'Institut (for ceremony in Notre-Dame). For a synthesis of the sources, see Fogel, *Cérémonies de l'information,* 167.

58. On the rue Saint Denis see Victory E. Graham and W. McAllister Johnson, *The Paris Entries of Charles IX and Elisabeth of Austria, 1571* (Toronto: University of Toronto Press, 1974), 23–24.

at its heart, paired with a psalm. From the famous Catholic hymn of praise, the ceremonial built a daylong celebration lauding the king and his military victories. Battles in this way precipitated processions and bonfires and brought the *Te Deum laudamus* to the street in what could fairly be called a Catholic version of the Roman triumph. And—unlike the royal entry—a Te Deum ceremony could be ordered by the king whenever and wherever he pleased. Its instigation marked a new degree of royal control over the king's appearance before his subjects.

The Te Deum also engaged the populace in celebration in a way that entries never managed: the *feux de joye* at the Hôtel de Ville gave the ceremonies a more civic feeling than entries, which, though funded by the city, circumvented this geographical seat of the mayor, aldermen, and city *bureau.* And the fires effectively spread the good news into all corners of Paris. The more detailed accounts of subsequent Te Deums specify that fires were to be lit in the 150 *dizaines* of the city's sixteen *quartiers* and *faubourgs,* which not only ensured the joyful illumination of the entire city, but also involved scores of locals, who built them and came to watch them burn. Henry's ceremonial cleverly engaged Parisians on the turf of the *quartiers* where the League cells of "The Sixteen" operated, taking back the city with his lengthy progress through its streets and with flames in his honor.[59]

The Te Deum ceremonial never lost its tendency to press the populace into the expression of joy. Indeed, during the reign of Henry IV, this aspect of it became more and more elaborate. Theodore Godefroy's *Le cérémonial françois* (1649) relates detailed accounts of many early-seventeenth-century Te Deums, of which the Te Deum celebrating the birth of the dauphin in 1601 offers an instructive example, not least because Godefroy's accounts of the ceremonies of Louis XIII's lifetime made them models for future Bourbons. The dauphin was born on 27 September 1601, and the Te Deum celebration in Paris took place over the following two days:[60]

28 September 1601

2:00 P.M.: Palace clock began to ring and continued until evening
3:00 P.M.: Procession from the Hôtel de Ville to Notre-Dame with the archers, harquebusiers, crossbowmen, and sergeants of the city, clerk of court, mayor, and aldermen, other city officials, a few nobles, councilors, quarteniers, and bourgeois
Procession met at Notre-Dame by the Messieurs of the Sovereign Courts
Te Deum sung in Notre-Dame *en musique,* with the psalm *Domine in virtute tua*
Te Deum sung in all the parish churches and monastic churches

59. On the administrative division of Paris into *quartiers* and *dizaines* see Jean-Pierre Babelon, *Paris au XVIe siècle* (Paris: Hachette, 1986), 511–516. The figure of 150 *dizaines* is from 1571.

60. Godefroy and Godefroy le Jeune, *Cérémonial françois,* 2:161–168.

Procession back to the Hôtel de Ville for
  Fire: pyramid of wood with "fusées, petards, et lances à feu"
  Royal Artillery discharged in the Place de Grève
  Trumpets, clarions, shawms, and drums
  Cries of "Vive le Roy!" from the crowd
  Distribution of two hogsheads of wine (200–300 gallons in total) and a large quantity of bread
Into the evening, fires burned in each street with cries of "Vive le Roy!"

29 September 1601

8:00 A.M.: Procession of the same officials to Notre-Dame, where they were joined by the cathedral clergy in a procession that visited all the parishes. Likewise joined by the parish clergy, the procession went around the Cité, through the courtyard of the palace, and returned to Notre-Dame
Mass and the psalm *Domine in virtute tua* at Notre-Dame

Early October 1601

Te Deums, processions, and fires in other French cities[61]

The ceremony was reproduced in this form all across the city and country, where it reached a massive public. In Paris the Te Deum hymn was sung in parish churches, a religious procession involving all the parish clergy supplemented the civic procession, and hundreds of fires across the city elicited cries of "Vive le Roy!" in every quarter as the city itself seemed to burn with passion.[62] At the Hôtel de Ville, wine and bread were distributed to allow crowds to partake in the social communion launched by the Te Deum sung at Notre-Dame. The noise from the cannon fire, trumpets, drums, and shawms and the shouts of "Vive le Roy!" left many crying for joy.[63] For so many, the smell of gunpowder, the din of a shouting mob, and the spectacle of explosions, inky smoke, and flames evoked the visceral terrors of war; the containment of these horrors—their enlistment, even—made the feting of royal might all the more impressive.

Within the thirty-five parishes of the city, thousands of souls gathered for the hymn praising God and king.[64] Parish churches were the spiritual homes of most

---

61. See, for example, Jean Pussot, *Journalier ou mémoires de Jean Pussot, Maître-Charpentier en la Couture de Reims,* ed. Édouard Henry (Reims: P. Regnier, 1856), 100; in Reims the birth was celebrated with the Te Deum hymn on 3 October and a procession on 7 October.

62. *Les feux de joye faits a Paris sur la resjouyssance de la publication de la Paix, le 12 Juin, 1598* (Lyons: Jean Jullieron, 1598), 4.

63. Ibid.

64. On the parishes in Paris see Adrien Friedmann, *Paris: ses rues, ses paroisses du Moyen Âge à la Révolution* (Paris: Plon, 1959).

Parisians, who prayed there, venerated their patron saints, participated in saints' day processions, and ended their lives with testaments asking to be buried there and endowing priests to say Masses for their souls. Certainly confraternities provided artisans and professionals with other chapels for devotion, as did convent churches, but nothing supplanted the role of parish churches. In fact, parochial worship received particular emphasis during the late sixteenth century as a means of keeping Parisians under the watchful eye of their pastors: Counter-Reformation admonishments to attend Mass at least once a week in one's parish church thus worked in tandem with more obvious efforts to police the city and maintain public order.[65] Ordering the Te Deum to be sung in the parishes guaranteed that its message would be heard by playing the Catholic insistence on parochial duties to the king's advantage. In addition, the general processions of the following day welled up from individual parishes, turning out the entire city in an extraordinary act of devotion to rival the processions for Corpus Christi.[66]

In 1600 France was the most populous country in Christendom and Paris the most populous city. Both required extensive communication networks. Paris's systems of parishes, *dizains,* and *quartiers,* and even the topography of main intersections and squares used by criers, were all nodes in the conveyance of information throughout the city. As we have seen, the Te Deum ceremony utilized these networks to communicate with the city's 300,000-some souls.[67] The king took possession of his capital by physically dominating the entire urban area with bonfires, artillery salvos, processions, hymns, and church services, using these media to represent his sovereignty in the public sphere. These media were, moreover, highly suited to the traditional modes of oral culture, in which news was disseminated by bell-ringing, public crying, street song, preaching, and artillery fire. At the most essential level, then, Henry's Te Deum ceremonial addressed his subjects in a rich and familiar language.

The Te Deum viscerally exerted the king's dominion in ways that no longer required his physical presence or that of a royal agent, so that whereas Henry III made a point of combining royal entry with his Te Deum for victory over the *reîtres* in 1587, Te Deums ultimately worked without the king, who ordered them to be celebrated concurrently throughout the realm for events of national importance.[68] That

65. Diefendorf, *Beneath the Cross,* 35–37.

66. Ibid., 39.

67. For estimates—always debatable—of the city's population see Babelon, *Paris au XVIe siècle,* 159–166.

68. For example, in March a Jubilee was celebrated throughout France, Italy, and Spain for the defeat of the German horsemen, in which, at least in Beauvais, clergy and monastics chanted for forty hours and the people held one hour of prayers for the eradication of heresy and to thank God for the

the king presided over these distant ceremonies in the provinces is nowhere more evident than in the Te Deum held in Toulouse in 1596 at the city's surrender to Henry IV. After a procession to the Cathedral of Saint Etienne, where the Te Deum was sung "avec une excellente musique," the officials set off a large pyramid of fireworks while the loud band played, then made their way to the church of the Augustins for another Te Deum and another pyramid of fireworks, this one decorated with trophies and all sorts of ancient and modern arms.[69] The fire was lit, the emblems of war went up in flames, and from a covered theater an effigy of the king looked on as the musicians sang a motet for fourteen voices before the theater, singing "Vive le Roy!" at the end of each verse. Like the royal effigies that presided at court during the interregnum, the simulacrum symbolized the sacred and immortal body of the king. In giving shape to the king's eternal power, the effigy ensured the smooth reassimilation of Toulouse and gave form to the body politic its people rejoined.

Musically, the fourteen-voice motet trumped even the polyphonic Te Deum sung at Saint Etienne, not only by its magnitude, but also by prompting the chronicler to exclaim, "One should not forget the traditions of magnificence and largesse at the ancient games and spectacles of the Romans and at the crowning of Emperors."[70] Largesse came in many forms, not least in the barrels of wine that flowed in the square, but we should also hear the motet as an example of princely magnificence of the sort associated with coronations. What motet could this have been? No motets for fourteen voices survive from sixteenth-century France, though there is an excellent candidate for twelve voices, Guillaume Boni's *Quaesumus omnipotens Deus.* Boni was choirmaster at Saint Etienne from 1565 until his death in or after 1594, and he dedicated this expansive setting of the collect of the Coronation Mass to Henry III in 1582, calling the motet the *Oratio Regia* in the title of the print.[71] Its sheer amplitude is extraordinary—the only other piece with as many voices published in France before 1600 was Lassus's *Laudate Dominum omnes gentes*—and by it Boni surely hoped to impress the king's great power upon listen-

---

victory. See Nicolas Ricquier, *Recueil mémorable d'aulcuns cas advenus depuis l'an du salut 1573 tant à Beauvais qu'ailleurs,* ed. Victor Leblond, vol. 2 of *Documents pour servir à l'histoire de Beauvais et du Beauvaisis au XVIe siècle* (Paris: Honoré Champion; Beauvais: Société Académique de l'Oise, 1909), 26.

69. *Discours veritable des ceremonies, feux de joye, solemnitez, et autres resjouissances publiques, faictes en la ville de Tolose, sur sa reduction à l'obeyssance du tres-Chrestien Henry quatriesme Roy de France & de Navarre,* (Lyons: Thibaud Ancelin, 1596), 13–17.

70. Ibid., 16. "Ne faut oublier la magnificence & largesse accoustumee d'estre faicte aux anciens jeux & spectacles des Romains, & au coronnement des Empereurs."

71. Guillaume Boni, *Psalmi Davidici novis concentibus sex vocibus modulati cum oratione Regia 12. voc* (Paris: Le Roy & Ballard, 1582). Transcription my own.

ers.[72] Its scoring for double choir, moreover, marked it stylistically as a ceremonial motet suitable for the occasion, and the fact that Henry III was the original dedicatee of the print would have mattered little under Henry IV, given the continuity of music for royal ceremonial.[73]

Singing the motet before the effigy reenacted part of the *sacre* in the square and, coupled with the trophies of war flaming nearby, reminded listeners of the king's military might. Indeed, armed combat is not far off in the coronation collect, which was added to the coronation *ordines* for the crowning of Charles V in 1364. At that time, English claims to the throne caused the French to reinforce the coronation with assertions that enemies of the king were enemies of God.[74] The image of the Deus Sabaoth or God of armies could not have been more relevant as Henry IV struggled to retake his kingdom, nor the tenets of the *sacre* more valuable to a king whose Catholicity was still in question.

> Quaesumus omnipotens Deus, ut famulus tuus Rex noster Henricus qui tua miseratione suscepit regni gubernacula, virtutum etiam omnium percipiat incrementa quibus decenter ornatus & vitiorum monstra deuitare, hostes superare, & ad te qui via, veritas & vita es gratiosus valeat peruenire.
>
> Deus qui corda fidelium sancti spiritus illustratione docuisti, da famulo tuo Regi nostro Christianissimo in eodem spiritu recta sapere, & de eius semper sancta consolatione gaudere. Per dominum nostrum.[75]
>
> [We beseech thee, Almighty God, that Thy servant our King Henry, who by thy mercy hath taken upon himself the governance of the kingdom, may also receive every increase in all virtues wherewith fittingly endued he may be able in grace both to avoid the monstrosities of vices and overcome his enemies, and to attain to thee, who art the way, the truth, and the life.
>
> God who hast instructed the hearts of the faithful by the illumination of thine Holy Spirit, grant that thy servant our most Christian king may have true wisdom in the same Spirit, and ever rejoice in the holy consolation thereof. Through our Lord.]

Boni's *Oratio Regia* effuses majesty from the start, opening with a headmotive full of semiminims that sweep downward over and over again as all twelve voices

72. Neither motet is in "verses" as the description relates—Boni's is in two parts, Lassus's is through-composed. If instruments doubled the bass lines, as was not uncommon, this could explain why it was said to be for fourteen voices.

73. See [Matthieu Giron], *Au roy sur le sacre de sa Majesté* (Paris: Pierre Chevillot, 1594), 32.

74. Jackson, *Vive le Roi!,* 28–29.

75. The collect text is taken from [Nicolas de Thou], *Ceremonies observees au sacre & coronement,* fol. 15r. Thanks go to Leofranc Holford-Strevens for graciously translating this prayer and the poem by Jean Dorat cited below.

announce the invocation to pray (see ex. 4.1). The point of imitation cadences only weakly on G in m. 5, with plagal motion in the bass and d′ in the superius, which allows the voices of the second choir to prolong it, establishing a static, meditative quality. By contrast, the text that follows, "omnipotens Deus," is set with a syllabic figure alternating between the two choirs in loose homophony and rising with each repetition to the striking superius outburst on a″ in measure 13 that marks the upper range of the top voice. A strong authentic cadence on G with a ringing g″ in the superius arrives on "Deus" at the end of the phrase (m. 18). This sort of textural variety and combinatorial writing for the two choirs characterizes the whole motet, as does the remarkable textual clarity Boni achieves by parsing his text into aphoristic fragments. "Rex noster" receives an undulating D–C♯–D figure in the bass, the memorable voice-leading turning it into an emblem, to which Boni adds a very linear and monumental "Henricus" in block chords at the end of the exordium. The repetition of these musical devices creates an effect by turns litanylike and more pressing. For instance, the first part ends with an eightfold repetition of "superare" in each voice, its pointed rhythm (two semiminims, two minims) and straightforward harmonies imitating a call to arms that reifies the very military supremacy promised in the prayer. The army of singers, the juggernaut of repeated motives, and the breadth established by the massive scoring for twelve voices and their spatial separation into two choirs creates an aesthetic of grandeur, of colossal mass, volume, space, and determination appropriate to the expression of royalty.

In the dedication of Boni's *Quaesumus omnipotens Deus,* the king's poet, Jean Dorat, addressed his majesty on the subject of music and kingship.[76] He opens by encouraging Henry III to patronize sacred songs because it is right for kings to praise God, "who is the greatest King and constantly favors those Kings who praise him." Praising God is a royal affair, and who better to praise God the King than Henry, "miniature image of God the immense"? Singing brought more than spiritual pleasure, for the *Oratio Regia* Boni offered his king was, as Dorat put it, part of the "pious rites by which the French state remains secured." Dorat hit squarely upon music as a source of political certainty in a French state founded on doctrines that were not made manifest in laws or constitutions, but in the "pious rites" of kings. Boni's motet captured the essence of the *sacre* in a form designed for repeat performance, creating an effigy of the coronation and extending the king's presence into one of the many palpable artworks through which his subjects might venerate his ritual transformation from man to *Roi-Dieu.*

76. Boni, *Psalmi Davidici,* fol. A3r. "Regia res Regi, qui sit laudabilis ipse, Laudet ut & dignum laude perenne Deum. Et laudare Deum Regem, quis dignior est te, HENRICE, ingentis parua figura Dei? . . . Tu successor eis in Regia sceptra, piósque; Ritus, res per quos Gallica salva manet."

Example 4.1. "Quaesumus omnipotens Deus," mm. 1–19. In Guillaume Boni, *Psalmi Davidici novis concentibus* (1582). Superius and quinta pars of chorus 1 reconstructed

Example 4.1. (*continued*)

Example 4.1. (*continued*)

Example 4.1. (*continued*)

THE HYMNS OF GOD'S SOLDIERS

I have thus far focused on the political and civic aspects of the new ceremonial, its paths through French cities, and its place in systems of monarchic representation. Yet the Te Deum ceremonial amalgamated religious practices as well, for the hymn appeared in several places in the liturgy, as well as being a devotional song, not to mention one put to use in libels and sectarian conflicts. The king's ceremonial did not erase the meanings that had accrued to the hymn in this way, but enlisted them.

In the pages that follow, I establish the ways in which French people came to know the hymn, its history in the pious lives of the laity, and its role in the general revival of hymnody by Catholic reformers, who hoped to combat heresy with spiritual songs. Yet this is not the whole story, for through this material history of the Te Deum, I should like to establish the nature of its symbolic power in a culture where kings were gods and where Catholic ritual invested the kings with authority. Ceremonial did not just express the king's power, it created power with the same magical rituals religions used to anthropomorphize divinity.[77] Clifford Geertz took the royal progress as the object of his study of charisma, yet the Te Deum ceremonial benefits perhaps even more from an interpretation that moves between the abstract and concrete, not only because the hymn reaffirmed the symbolic conflation of royal and divine majesty, but because its eucharistic magic called the symbolic powers of sacred music into royal service. This is to say, I do believe that singing hymns fostered piety and that Catholics who sang hymns together enjoyed an enhanced sense of community and even spiritual empowerment—for the more who sang, the better God heard the prayer—but it is equally important to acknowledge the mystery attributed to the ancient chants of the church, their charisma, if you will. The Te Deum was written by not one saint, but two, and like the story of the holy dove that sang melodies into the mouth of Saint Gregory, the attribution of this hymn to Saints Augustine and Ambrose consecrated its strains. For Catholics and Protestants alike, it glowed with a particular chrism: it became an emblem of Catholic authority early in the religious struggles in France, and this contributed to the meaning of the ceremonial based upon it. But let us begin with some concrete observations about the hymn's currency in sixteenth-century France.

Most Catholics knew the Te Deum from its use outside the ordinary liturgy. It was sung at the end of matins on Sundays and feast days, but unlike Mass, vespers, and compline, matins was not generally attended by the public. Rather, faithful from the middling and upper classes would have known the text from their Books of Hours, in which it concluded the matins for the Office of the Virgin. Essentially,

77. Geertz, "Centers, Kings, and Charisma," 124.

the Book of Hours was a book of common prayer, and layfolk often said their hours during Mass.[78] Through the *Heures* and *Horae,* the Te Deum became a part of daily life. One Book of Hours even counseled its owner not to sing the Te Deum every day but only in times of joy, a proscription that tries to maintain the celebratory nature of the hymn even as it attests to its popularity.[79]

The Book of Hours is useful for understanding the devotional practices of the laity, for although it is a nonliturgical compilation, it made elements from the liturgy available to all parishioners, serving at once as a hymnal, a catechism, and a repository of litanies, psalms, and useful prayers. When Catholic reformers, desperate to reach their flocks, sought spiritual songs to counter the frightening popularity of Protestant psalms, these common hymns proved a sound place to start. It was as "un remede et contrepoison" to the Huguenot Psalter that Guy Le Fèvre de La Boderie wrote his *Hymnes Ecclesiastiques,* a deluxe hymnal in French verse that naturally included a translation of the Te Deum.[80] Hymns were sent to war against psalms, with French translations for private devotion and communal singing of the Latin originals for parish processions, pilgrimages, and confraternity rites. The Te Deum was deployed on all of these fronts.

Take, for example, the *processions blanches* of the summer of 1583, in which some 72,000 Catholics set out from their homes to save the country from heresy.[81] Accompanied by clergymen transporting the Corpus Christi in a ciborium, the processions swelled with faithful, not only men, women, and children, but rich and poor alike (though gentlemen sometimes brought up the rear on horseback, their wives in coaches).[82] All the pilgrims sang. Young boys served as choristers, leading antiphons and litanies as the rest of the group sang refrains; young girls sang the

78. A point made by Mary Kay Duggan, "Books of Hours: Devotional and Didactic," paper read at the conference "The Culture of Books," Berkeley, April 2002.

79. Abbé Victor Leroquais, *Les livres d'heures manuscrits de la Bibliothèque nationale,* 2 vols. (Paris: Bibliothèque nationale, 1927), 2:266.

80. Guy Le Fèvre de La Boderie, *Hymnes Ecclesiastiques,* 2nd ed. (Paris, 1582), fol. A2r; the Te Deum appears on fols. 91r–92v. See Franco Giacone, "Le catholicisme dans les *Hymnes Ecclesiastiques,*" in *Poésie encyclopédique et kabbale chrétienne: onze études sur Guy Le Fèvre de La Boderie,* ed. François Roudaut (Paris: Honoré Champion, 1999), 99–167, which gives a complete list of its contents.

81. Hubert Meurier, *Traicté de l'institution et vray usage des processions tant ordinaires, qu'extraordinaires, qui se font en l'Eglise Catholique, contenant un ample discours de ce qui s'est passé pour ce regard en la Province de Champaigne, depuis le 22. de Juillet jusques au 25. d'Octobre, 1583* (Rheims: Jean de Foigny, 1584), appendix, "Ensuivent les litanies, hymnes." See Crouzet, *Guerriers de Dieu,* 2:297–310. On the penitential fervor in Normandy under the League, see Philip Benedict, *Rouen during the Wars of Religion* (Cambridge: Cambridge University Press, 1981), 190–208.

82. L'Estoile, *Mémoires-journaux,* 2:134.

Ave Maria in alternation with men and women; and when they arrived at regional churches, Te Deums were "sung with great devotion and joy by all the people."[83] Hubert Meurier, a canon from Reims Cathedral—one destination of the pilgrims—tells us that everyone took pains to sing everything, with those who knew the chants teaching others.[84] The pilgrims' progresses witness how thoroughly Catholics had internalized the music of the church, for "not only those of the Church . . . sang infinite praises to the host day and night, but also laypeople of every quality, including women and small children, who knew most of the Mass as if they had been brought up from the start among ecclesiastics."[85]

The pilgrimages also demonstrate the faith laid in the music of the church. On a symbolic level, the days of walking barefoot reenacted Christ's entry into Jerusalem and his march with the cross, allowing the penitents to participate in his passion, but all was not solemn. For by singing as they progressed toward the towers of the great regional churches, transporting the Eucharist to Reims or Paris, they represented the glorious and triumphant resurrection of the Lord.[86] Meurier tells us how that summer the emotion of God's people was so great and so widespread that it washed away sin for a time: children who could barely talk ran about singing *Ave Maria* in their homes and the streets, making their own processions in imitation of adults; spiritual songs in praise of God edged out dance and lascivious songs; and the simple purity of white sheets effaced the vanity of sumptuous dress. Multitudes raised their voices as one, and the clamor excited by their common accord was heard by the immortal Father. Indeed, the many fervent souls, by singing night and day for three or four months, managed to redress the blasphemy accrued during sixty years of French Protestantism.[87]

Even with an imperfect understanding of the Latin words, devout Catholics clearly knew the Te Deum by heart. Long before the weary and ecstatic pilgrims sang the Te Deum at the end of their processions in 1583, the hymn had been sung

83. See Ricquier, *Recueil mémorable,* 13. Also see the detailed descriptions of the singing in Alexandre-Eusèbe Poquet, *Histoire de Château-Thierry,* 2 vols. (Château-Thierry: A. Laurent, 1839–1840), 1:354–359; Pussot, *Journalier,* 18; and the pamphlet *Le vray discours des grandes processions qui se font depuis les frontires* [*sic*] *d'Allemagne jusques à la France, dont jamais n'en fut faicte de semblable, & comme plus amplement vous sera monstré dans le discours* (Paris, 1583).

84. Meurier, *Traicté,* fols. 25r–29r; appendix, "Ensuivent les litanies, hymnes."

85. Ibid., 43r. "Car non seulement les gens d'Eglise & de Religion l'ont adoré (the host) teste nue & pieds nuds, & ont chanté infinies louanges & de jour & de nuit, & à toute heure: mais aussi les personnes seculieres de toute qualité, jusques aux femmes, & petits enfans, qui sçavent la plus part de l'office du Sacrement, comme s'ils avoient esté tousjours nourris entre les Ecclesiastiques."

86. Ibid., fol. 9v.

87. Ibid., fols. 4v–5r. On the children singing and spiritual songs, see fols. 28r–v and 41v.

during processions for Corpus Christi, the feast on which Catholics displayed the consecrated host in jeweled monstrances, carrying it through city streets decked with tapestries.[88] Of the feasts occasioning general processions—the Purification, Palm Sunday, Easter, Corpus Christi, Saint Mark, and Rogations—Corpus Christi was the apotheosis of the processional calendar, with by far the longest and most lavish processions.[89] In Paris, celebrations for Corpus Christi—known as the *Fête-Dieu*—began with parish processions for which dozens of garlands of red roses were ordered each year to adorn the cross, the monstrance, and the heads of the laymen and clerics participating in the procession.[90] If the ceremonial of Saint-Jacques-de-la-Boucherie is any indication, parochial processions circumscribed the entire parish, with bells ringing from the time the procession left the church until it returned.[91] The whole city then celebrated the *Fête-Dieu* with a general procession to Notre-Dame in which, if the king was present, he walked behind the host. Here, then, we have found our model for the procession against heresy ordered by Francis I in 1535, which seems to have been the first time the Blessed Sacrament was carried in something other than a Corpus Christi procession.[92] As the religious conflict wore on, the triangulation of the Corpus Christi, the Te Deum hymn, and processions against heresy only became more complete with the general escalation in the number of religious processions: in 1553 Henry II ordered the processions, Masses, sermons, and other solemnities that marked Corpus Christi henceforth to be repeated a week later, thereby redoubling the public veneration of the Host; under Henry III Te Deum processions became the standard way to announce Catholic victories over the heretics, and—despite some concerns about its propriety—consecrated hosts were transported across the countryside in the *processions blanches* that ended with the Te Deums at their destination.[93] The feast that had originally allowed the laity to see the body of Christ and to actively participate in celebrating

88. See, for example, *Processionale usibus ac ritibus sanctae romanae ecclesiae accommodatum* (Paris: Jean Henault, 1667), which contains nine hymns for Corpus Christi, of which *Te Deum laudamus* is ninth and the last (165). I thank Elisabeth Le Guin for consulting this source for me. Normally one would not expect to find the Te Deum in processionals, as it was sung just upon entering the church and was therefore technically not part of the procession. For the description of a procession of the Holy Sacrament in sixteenth-century Rome, see Michel Brenet [Marie Bobillier], "La musique dans les processions," *La Tribune de Saint-Gervais* 2 (1896): 65–68, 81–84, 99–103, at 102.

89. Meurier, *Traicté,* fols. 8r–v, 13v.

90. Diefendorf, *Beneath the Cross,* 39.

91. Portions of the ceremonial of that church are reproduced in Laurence Fritsch-Pinaud, "La vie paroissiale à Saint-Jacques-de-la-Boucherie au XVe siècle," *Paris et Île-de-France* 33 (1982): 7–97; see 79, 84–85, 88, and passim.

92. Diefendorf, *Beneath the Cross,* 46.

93. On Henry II, see ibid., 33; on the *processions blanches,* see Meurier, *Traicté,* esp. fol. 42r–v.

its corporate symbolism thus spun off into other processions in defense of the (Catholic) body politic, making the Te Deum a battle hymn of eucharistic devotion and, ultimately, of Catholic kingship in France.

Indeed, the Te Deum hymn proved so emblematic of Catholicism that it became a script for blasphemous processions as well. Claude de Sainctes relates how those who sacked Catholic churches in 1562 took pleasure in dressing one of their ranks as a priest and making a triumphal procession through town, "singing the *Te Deum laudamus* or *Requiem* in derision."[94] Here a heretic-priest presided in the topsy-turvy Kingdom of Misrule and overturned the Catholic order of society with the rituals of Carnival-time justice. But the sacred parodies tolerated by the church in the name of "Paschal laughter" and the "feast of fools" (during which the "faithful" had once held drunken orgies on church altars) here bore terrifying results, showing how quickly Carnival festivities might turn to iconoclasm and how easily games and farces could end in murder.[95]

From the outset of the Wars of Religion, singing "seditious songs" in public could be punished with hanging, so disruptive did they prove to civil peace.[96] The crown banned the singing of psalms in assemblies, and Catholic clergy began to think of ways to stem the tide of Protestant conversions precipitated by the seductive effect of psalms. We have already read of Edmond Auger's hopes to have Pierre de Ronsard write Catholic songs in French. If a collection was planned, only the Te Deum from it ever materialized, in the form of a verse paraphrase Ronsard published a few years later. It appeared at the end of his *Discours des Miseres de ce temps* (1567), a venue that juxtaposed the hymn with the royalist politics of the pamphlet literature he collected there and certainly promoted the hymn as a battle song of French Catholics. This may explain how his paraphrase so easily made its way into

94. Claude de Sainctes, *Discours sur le saccagement des Eglises Catholiques* (Paris: Claude Fremy, 1563), reprinted in *Archives curieuses de l'histoire de France,* ed. L. Cimber and F. Danjou, ser. 1, 15 vols. (Paris: Institute Historique, 1834–1840), 4:384. See Davis, *Society and Culture,* 152–187, for a deft analysis of contemporary religious riots including this mock Te Deum (p. 180). Te Deums and Requiems often mirrored one another. For instance, L'Estoile relates that Henry IV ordered a Te Deum to be sung for the submission of Lyons in February 1594, at which the Leaguers in Paris ordered a Requiem to be sung instead (*Mémoires-journaux,* 6:156). Even Catholics parodied the Te Deum. See George Nugent, "Anti-Protestant Music for Sixteenth-Century Ferrara," *Journal of the American Musicological Society* 43 (1990): 228–291, and Bonnie J. Blackburn, "*Te Matrem Dei laudamus:* A Study in the Musical Veneration of Mary," *Musical Quarterly* 53 (1967): 53–76.

95. On the "feasts of fools" and "Paschal laughter" see Michael Bakhtin, *Rabelais and His World,* trans. Hélène Iswolsky (Bloomington: Indiana University Press, 1984); on riots developing from Carnival celebrations see Davis, *Society and Culture,* 152–187, and Emmanuel Le Roy Ladurie, *Carnival in Romans,* trans. Mary Feeney (New York: G. Braziller, 1979).

96. See van Orden, "Cheap Print and Street Song."

Example 4.2. The chant *Te Deum laudamus,* verses 1–5, in *Heures de Nostre Dame* (1583). French paraphrase, "O Seigneur Dieu nous te louons" (1565), is by Pierre de Ronsard

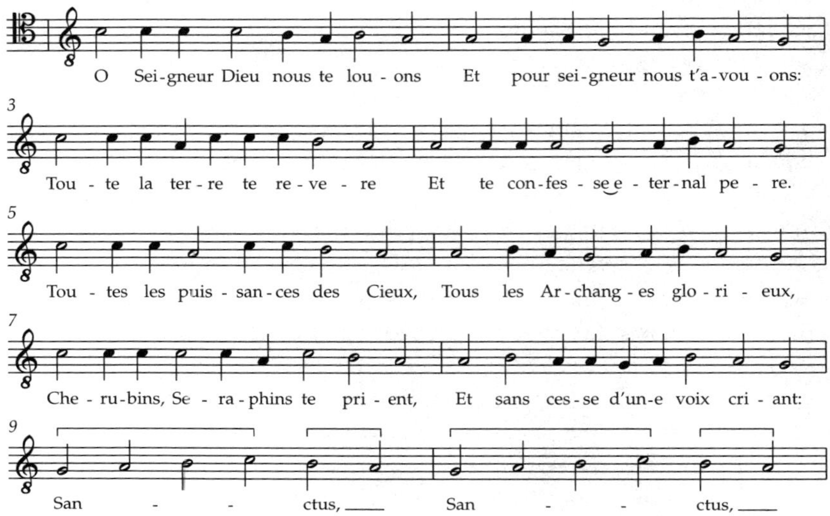

pamphlets publicizing royal victories, such as *Victoire obtenue par monsieur le Mareschal de Biron* (1580), and it certainly explains the political significance of the four-voice setting made by Guillaume de Chastillon de La Tour, which appeared in a print dedicated to the king alongside a "Chant de victoire fait au nom du Roy" and "Vive Henri, le beau Soleil de France." [97]

Ronsard dedicated his translation to Jean de Monluc, bishop of Valence, "pour chanter en son église," which raises the question of how it might have been sung.[98] The chant melody was well known, though at first glance it seems difficult to squeeze Ronsard's steady parade of eight-syllable lines and regular quatrains into a melody serving very unequal lines of Latin prose. But the chant tune is comprised of short phrases that begin on a recitation tone, which allows them to collapse and expand as necessary to accommodate the long and short lines of prose. Example 4.2 presents the version of the chant printed in the *Heures* of the king's penitential confraternity with a rendering of how Ronsard's text can be sung to it.[99]

97. Ronsard, *Oeuvres complètes,* 2:1080–1082. Guillaume de Chastillon, sieur de La Tour, *Airs* (Caen: Jacques Mangeant, 1593), fols. 6v–7v, 10v–12v, and 3v, respectively.

98. Ronsard, *Oeuvres complètes,* 2:1605.

99. *Heures de Nostre Dame a l'usage de Rome, selon la réformation de nostre S. Pere Pape Pie V* (Paris: Jamet Mettayer, 1583, with musical supplement by Le Roy & Ballard), 11–12. Note values quartered.

Of course, the practice of fitting new texts to well-known tunes was extremely common at the time. Reams of chansons printed up in small anthologies include no music but instead merely name a popular tune or *timbre* to which the poetry can be sung.[100] Occasionally plainchant melodies were used as *timbres.* This occurred only rarely, but those occurrences verify the practice of singing new texts to plainchant *timbres.* To wit, Ronsard wrote a poem on the *timbre* of the litany for rogation days, *Te rogamus audi nos,* and Christophe de Bordeaux included a song on the same *timbre* in his collection of spiritual chansons.[101] No doubt Catholics knew the litany not only from their Books of Hours, which always included the Litany of the Saints, but also from the great rogation processions during which it was sung by clergy and layfolk alike. In contrast, Leger Bontemps recommended the tune of Marot's "Estant assis aux rives aquatiques" for his translation of the Te Deum, but there is no reason not to suppose that Ronsard expected Monluc and others to sing his Te Deum to the original chant melody.[102]

Early on in the wars, Monluc had championed the idea of letting Catholic congregations sing prayers and psalms together in French in church. The French delegation to the Council of Trent in 1563 even proposed that Catholics sing vernacular canticles during Mass.[103] Ronsard's "O Seigneur Dieu" appeared during this first tentative step toward vernacular Catholic hymnody.[104] But the council rejected the idea, and it was abandoned until around 1580. In his book on processions, Meurier tells us that one blessing of those years was the invention of new spiritual songs by those who formerly had invented "chansons folles & mondaines." He also relates that the pilgrims sang these new songs during the summer of 1583.[105] Here he is

100. See van Orden, "Cheap Print and Street Song."

101. Ronsard, *Oeuvres complètes,* 2:617; Christophe de Bordeaux, "'Voulez ouyr chanson chanter' autre chanson nouvelle qui se chante a plaisir de ces apostats regniez sur le chant 'Te rogamus audi nos,'" in *Beau recueil de plusieurs belles chansons spirituelles* (Paris: for Magdeleine Berthelin [after 1563], fol. 3v.

102. Leger Bontemps, "'Nous te louons majesté souveraine' Chanson au Cantique contenant la louange de nostre Dieu, suyvant le Te Deum, sur le chant 'Estant assis aux rives aquatiques,'" in Christophe de Bordeaux, *Beau recueil de plusieurs belles chansons spirituelles,* fol. 69v.

103. On Monluc and this proposition, see Michel Jeanneret, *Poésie et tradition biblique au XVIe siècle: Recherches stylistiques et paraphrases des "Psaumes," de Marot à Malherbe* (Paris: Librairie José Corti, 1969), 201–203.

104. There were, of course, prior translations of the Te Deum. In the Middle Ages, they are found with French verse translations of the psalms, hymns, and common prayers such as the Pater Noster, Ave Maria, and Gloria. See Denise Launay, *La musique religieuse en France du Concile de Trente à 1804* (Paris: Société Française de Musicologie, 1993), 27–28. There were also "Huguenot" versions of the Te Deum in French, but not before the coronation of Henry IV. On "Dieu, nous te louons," a translation in measured verse attributed to d'Aubigné and set by Le Jeune, see His, *Claude Le Jeune,* 360–363.

105. Meurier, *Traicté,* fol. 41v.

surely speaking of spiritual songs in French, and perhaps specifically of French versions of hymns, which began to appear in print the year before the *processions blanches.* The first polyphonic collection of explicitly Catholic songs in French was written by Anthoine de Bertrand and published in 1582; it includes—notably—a four-voice setting of Ronsard's "O Seigneur Dieu" and a dozen well-known Latin hymns, most with alternate versions in French.[106] Bertrand left in Latin hymns that Catholics surely knew, such as *Ave maris stella* and *Pange lingua,* adapting the plainchant in the superius, adding rhythm, and harmonizing it in the same spirit as Dufay's fifteenth-century harmonizations. Other settings provide both Latin and French texts, but the Te Deum sets only Ronsard's French paraphrase, which makes sense given how unwieldy the Latin prose is compared to Ronsard's convenient strophes. Likewise, Bertrand abandoned the plainchant in favor of a tuneful newly composed superius.

Bertrand's collection is of particular interest because, as Jean-Michel Vaccaro showed, he composed it under the influence of the Jesuits in his hometown of Toulouse and in the wake of a spiritual conversion not unlike the sort referred to by Meurier.[107] It was the Jesuit Edmond Auger—later the confessor of Henry III and instigator of the king's *confrérie des pénitens*—who established the penitential confraternities in Toulouse following the Jubilee in 1576 that precipitated Bertrand's Catholic zeal. And although, as the print's preface tells us, Bertrand was killed "by those who detest these ecclesiastical hymns," the *Airs spirituels* came to light nonetheless, probably thanks to the Jesuits.[108] The print ends with a sonnet by Father Michel de Bonnières praising Bertrand's works, praise echoed by Father Michel Coyssard in his *Traicté du profit que toute personne tire de chanter en la Doctrine Chrestienne,* published by the Jesuit press of Lyons in 1608.[109]

Coyssard's treatise locates Bertrand's work in the midst of Jesuit projects to combat Protestant psalms with Catholic hymns, but we should be clear about where they might have been sung. Auger's 1563 letter to Rome suggested that Ronsard write spiritual songs in French, not for use in church—which is what Monluc proposed—but for Catholics to sing "at home, in shops, and while travelling."[110] It is

106. Anthoine de Bertrand, *Airs spirituels contenant plusieurs hymnes & cantiques* (Paris: Le Roy & Ballard, 1582), fols. 13v–14r.

107. Jean-Michel Vaccaro, "Le livre d'airs spirituels d'Anthoine de Bertrand," *Revue de musicologie* 56 (1970), 35–53. Vaccaro provides a transcription of the superius and bassus of "O Seigneur Dieu."

108. The preface is reprinted in Lesure and Thibault, *Bibliographie des éditions d'Adrian Le Roy et Robert Ballard,* 43–44.

109. See Vaccaro, "Livre d'airs spirituels," 43–44, and Gérald Pau, "De l'usage de la chanson spirituelle par les Jésuites au temps de la contre-réforme," in *La chanson à la Renaissance: Actes du XXe colloque d'études humanistes du Centre d'Études Supérieures de la Renaissance de l'Université de Tours, juillet 1977,* ed. Jean-Michel Vaccaro (Tours: Éditions Van de Velde, 1981), 15–34.

110. Cited in Kennedy, "Jesuits and Music," 82.

nice to imagine pilgrims and shopkeepers singing Bertrand's *airs,* but they were probably too complex. Although the polyphony is simple compared to the riches of Bertrand's secular chansons, the settings are far from strictly homophonic (so far as one can judge from the two remaining parts), and their wide range of note values and rhythms would have made them difficult to learn and fit together. These are not street songs, even given the remarkable ability of the faithful to memorize Latin hymns as long and irregular as the Te Deum. Ten years later, Coyssard himself produced a more battle-worthy set of hymns "pour chanter avec la Doctrine Chrestienne": the *Paraphrase des Hymnes et Cantiques spirituelz* (1592), which was written with detachable superius parts for those who found the four-voice polyphony too cumbersome.[111] Here was a nice vernacular compendium of favorite hymns (including the Te Deum), proses, Marian antiphons, common prayers, and commandments of the sort by which Jesuits had been propagating the catechism as far afield as Brazil.[112] The Jesuits believed that Christian doctrine would be impressed on children more easily through tuneful melodies, and it is not hard to imagine how the swinging triple time Coyssard used to open his Te Deum paraphrase could have taught the children the most joyful side of praise and obedience.[113]

Coyssard's *Paraphrase des Hymnes et Cantiques spirituelz* is only the most literal example of catechistic song. Viewed with a broad perspective, the penitential confraternities, the *processions blanches,* and even the processions of Corpus Christi all instructed Catholics by involving them in church rituals and teaching them songs that symbolized fundamental doctrines. One need only think of the origins of the Corpus Christi feast—its Office penned by Saint Thomas Aquinas—which was instituted in the thirteenth century as a vehicle of reform *avant la lettre.* Its display of the host combated popular heresies by allowing the laity to view the Eucharist and to devote themselves to this central sacrament.[114] So too, orthodoxy triumphed in the late Middle Ages with the founding of confraternities, which created new possibilities for devotion that included not only hearing Mass, but prayers, singing, and penitential flagellation. These appropriations of monastic practices expanded into the penitential confraternities of the sixteenth century, just as Corpus Christi processions inspired the *processions blanches* against heresy, and while the cuts in a pilgrim's feet might have been the most visible sign of faith, it was through singing

111. See the excellent chapter on Coyssard in Launay, *Musique religieuse,* 119–136, as well as Pau, "De l'usage de la chanson spirituelle."

112. Paulo Castagna, "The Use of Music by the Jesuits in the Conversion of the Indigenous Peoples of Brazil," in *The Jesuits: Cultures, Sciences, and the Arts, 1540–1773,* ed. John W. O'Malley et al. (Toronto: University of Toronto Press, 1999), 641–658.

113. Michel Coyssard, *Traicté du profit que toute personne tire de chanter en la Doctrine Chrestienne* (Lyons: Jean Pillehotte, 1608), 20ff.

114. Meurier, *Traicté,* fols. 13v–16v.

that Catholics trumpeted their faith to the world at large. This was a time when the *Pange lingua* defined confessional politics, when *Ave*s professed fundamental articles of religion, and when the Te Deum reinforced that old Gallican belief in "one faith, one king, one law." Henry III's Te Deum ceremonial featured the jewel in the crown of the Catholic hymnal, a song that the faithful knew by heart. With it, the Te Deum ceremonial attempted to capture for kingship the same popular sentiments that came together when the Te Deum was sung in Corpus Christi processions and the *processions blanches,* drawing citizens to the streets for huge general processions; pairing the hymn with festive fireworks, free food, and wine; and anticipating, no doubt, that the cries of "Vive le Roy!" would reaffirm the divine right of the king. Processions and hymnody had long since proven their usefulness in bringing layfolk round to orthodoxy, and with religious orthodoxy came benefits to political orthodoxy.

THE LITURGY OF ROYALTY

Although many of their outward forms were the same, the king's ceremonial differed from the other processions I have been discussing with regard to kingship. In transferring Catholic identity to the political realm, the Te Deum ceremonial used the corporate sensibilities developed through Catholic practices to reaffirm each person's sense of belonging to the body politic. Even as it celebrated the king as a Deus Sabaoth, the Te Deum turned the fete of terrestrial power toward its ultimate source by reminding French subjects of the sanctity of their relationship with the king. We have examined the ceremony's public representation of the monarch and the systems aiming it at the largest share of the population; let us now consider those elites who gathered inside Notre-Dame to hear the Te Deum: the clergy, parlementarians, and noblemen who—having participated more regularly in royal ceremonial—understood its nuances.

One important feature of the ceremonies at Notre-Dame was a psalm sung in the form of a motet. The psalm gave Te Deum ceremonies more substance and particularized their messages according to the occasion. Looking through the Te Deums in Godefroy's *Cérémonial françois,* one sees, for example, that Psalm 20 was sung for the Te Deums celebrating the royal births of 1601, 1607, and 1608, and that Psalm 19 was sung for the Te Deums celebrating the marriage of Henry IV (1601), the peace treaty that followed from it a few weeks later, the marriage of Louis XIII (1615), and the victory at Île de Ré (1627).[115] After battles led by the king himself, such as those of 1620 and 1628, Te Deums employed the final verse of Psalm 19,

115. It was also sung for the victory of Henry IV at Amiens in 1597 (Fonds Godefroy 426, fol. 69Br, Bibliothèque de l'Institut) and the victory of Louis XIII at Montpellier in 1622 (Nouvelles acquisitions françaises 7239, fol. 141r–v, Bibliothèque nationale de France). Royal births had long been celebrated with motets and *feux de joye.* See Cazaux, *Musique à la cour de François Ier,* 163–167.

"Domine salvum fac regem" (God save the king.)[116] All these psalm texts played noteworthy roles in the coronation. For ecclesiastics and those close to the court—an audience used to decoding the signs of power with intertextual readings—the "coronation psalms" reminded those who recognized them that no matter what state event the Te Deum celebrated, thanks to God always began with thanks for their sovereign. Indeed, by pairing these psalms with the Te Deum, the ceremonial intimated that praise to God was synonymous with thanksgiving for the Holy Balm received by French kings.

The portions of the Te Deum ceremonies that took place in Notre-Dame are also of prime musical interest because it was here, with the musical forces available only in cathedral and royal chapel, that the proceedings impressed the king's *royale grandeur* upon those present through sheer musical splendor. The accounts gathered by Godefroy specify that the Te Deum hymn was sung *en musique* at Te Deum ceremonies; at coronations, the Te Deum was "accompagné d'orgues, & autre Musique" or "achevé en Musique par la Chapelle du Roy" after its opening intonation.[117] *En musique* indicates polyphonic performance, for which musicians could have chosen from a number of contemporary settings. Amid the output of the king's printers, for example, we find settings of the Te Deum by Jean Maillard (1564), Jacobus de Kerle (1572), and Didier le Blanc (1584), as well as a stunning four settings by Eustache Du Caurroy published in his monumental *Preces ecclesiasticae* (1609), a print that consummated forty years in royal service.[118] Save for the version by Le Blanc, these settings include music only for the even-numbered verses (and the second part of verse 5), leaving the odd-numbered verses to be sung as plainchant or played on the organ. The long tradition of performing the Te Deum *ad alternatim* with organ—one noted by observers as early as the fourteenth century—doubtless also inspired Pierre Attaingnant's publication back in 1531 of a keyboard setting that richly embellishes the odd-numbered verses of plainchant, one of the first such intabulations ever issued in print.[119]

---

116. *Laudate Dominum omnes gentes* does not reappear after 1587 in the Te Deum ceremonies documented by Godefroy.

117. Also consider the account of the Te Deum for the victory in Montpellier (31 October 1622): "On à sonné une clochette qui servoit de signal aux Orgues, par ou on à commencé et en suitte s'est chanté Le Te Deum en musique, par les chantres et musiciens qui estoient au letrain, L'orgue faisant les Intermeses des versetz. . . . On à apres chanté L'exaudiat en musique excelente sans intermese en forme de mottet" (Nouvelles acquisitions françaises 7239, fol. 141r–v, Bibliothèque nationale de France).

118. On Du Caurroy's career see Marie-Alexis Colin, "Eustache Du Caurroy: un compositeur français aux confins du XVIe et du XVIIe siècle," *Acta Musicologica* 73 (2001): 189–258.

119. On the use of instruments in performances of the Te Deum hymn, see Winfried Kirsch, *Die Quellen der mehrstimmigen Magnificat- und Te Deum-Vertonungen bis zur Mitte des 16. Jahrhunderts* (Tutzing: Hans Schneider, 1966).

Polyphony and organs added musical luminescence to Te Deum ceremonies, referring with sonic force to the deafening musical high point of king's *sacre et couronnement,* when trumpets, clarions, drums, and cries of "Vive le roy!" gave way to the Te Deum. At no other point is organ mentioned in coronation descriptions. It is not simply that the polyphony marked Te Deum ceremonies as significant events, but, more particularly, that the organ and *alternatim* settings recalled the moment of enthroning. Radiant with God's blessing, the *Roi-Dieu* sat above the congregation, the musical excess bearing him aloft in a striking parallel to the motets that distinguished the elevation of the consecrated host as the spiritual apex of the Mass. Music functioned as a crucial part of the staging, flooding the scene intermittently with the brilliance it alone could bring to the ritual and, in polychoral works and *alternatim* settings, emphasizing the spatialized and external dimension of sovereignty elsewhere represented by the centralized geometry of court spectacle. It often sounds trite to assert that polyphony, through its very excess, signified greatness, transcendence, and divinity. But music for the coronation and for the Te Deum ceremonial was arguably noisy and grand and expansive with the intention of conveying royal majesty by pulling out all the stops (literally), writing for as many voices as one could muster, and communicating the momentousness of the occasion with as great a noise as possible, negating the aesthetic of textual audibility favored by many humanists. At a time when important public cries were announced by four trumpets instead of one and Catholic ritual cued the faithful to rapture with organs and polyphonic masses, more really did count for more.

The psalms employed in the Te Deum ceremonial also occupied theatrically prominent places in the *sacre et couronnement.* As the king entered Reims Cathedral, the canons led the procession singing Psalm 20, "Domine in virtute tua laetabitur rex" (The king shall rejoice in thy strength, O Lord) in *fauxbourdon.*[120] The psalm "made public testimony of the joyfulness that everyone felt for this solemn act" at the outset of the ceremony when the congregation caught their first glimpse of the royal candidate, who would not be seen again until his enthronement.[121] It

120. The terms "psalm," "verse," and "antiphon" are variously used to describe this processional piece; in fuller accounts only the first two verses are given. It may well be that the first two verses served as an antiphon for the psalm, which was sung during the procession and finished at the altar. See Godefroy and Godefroy le Jeune, *Cérémonial françois,* for the seventeenth-century *formulaire moderne* (1:58) and for the coronations of Francis I (1:247), Henry IV (1:358), and Louis XIII (1:408–409). Also see [Nicolas de Thou], *Ceremonies observees au sacre & coronement,* fol. 15r.

121. Some ceremonials prescribed the singing of Psalm 20 upon the king's arrival in Reims, which usually occurred the day before the coronation in the form of a triumphal entry that culminated with a Te Deum and other services at the cathedral. See [Jean de Foigny], *Le sacre et coronnement du Roy de France, avec toutes les Ceremonies, prieres et oraisons* (Reims: J. de Foigny, 1575), fol. A2v. Foigny's is not an account, but rather a vernacular ceremonial pulled together before the coronation of Henry III from the cathedral ceremonials.

is a true coronation psalm in which God sets "a crown of precious stones" upon the head of the king and lays "honor and majesty" upon him, blessings juxtaposed with a curse upon God's enemies, who will be devoured in fire, swallowed up in wrath, and their seed destroyed from among the children of men. As the king made his way to the altar, then, Psalm 20 celebrated the crown, majesty, and the sword.

Psalm 19 appeared, in part, the evening before the coronation, when the future king heard a brief service consisting of responds, a Marian antiphon, and prayers, after which he changed clothes, heard vespers, and made his confession.[122] Following the Marian antiphon, the "Domine salvum fac regem" was sung, which led Edward Lowinsky, François Lesure, and others to dub the motets setting this text "coronation motets."[123] This last verse of Psalm 19 received three polyphonic settings in the sixteenth century, all by French composers. Jean Mouton used it to open his motet of that title, while Jean Maillard and Guillaume Costeley employed the verse at the head of a text that cobbled it together with verses from Psalm 20. There is no evidence that the motets were performed during any coronations, though they could have been sung during the Coronation Mass, which virtually no sources describe in any detail. Motets were often sung at the Offertory, Elevation, or end of Mass, and a *Domine salvum fac regem* would have been a most fitting recessional. We know that the Royal Chapel sang the whole of Psalm 19 at the end of each day's *grande messe,* a tradition transformed by the early seventeenth century into the daily singing of a *Domine salvum fac regem* motet at the close of High or Low Mass for the king.[124] Perhaps the sixteenth-century *Domine salvum fac regem*

122. Godefroy and Godefroy le Jeune, *Cérémonial françois,* 1:53.

123. Edward E. Lowinsky, *The Medici Codex of 1518: A Choirbook of Motets Dedicated to Lorenzo de' Medici, Duke of Urbino,* Monuments of Renaissance Music 3–5 (Chicago: University of Chicago Press [1968]), 1:73, 157–158; François Lesure, "France (I)," in *The New Grove Dictionary of Music and Musicians,* ed. Stanley Sadie, 2nd ed., 29 vols. (London: Macmillan, 2002), 9:143. The circumstances for which Mouton may have composed his motet are evaluated in Albert Dunning, *Die Staatsmotette, 1480–1555* (Utrecht: A. Oosthoek, 1970), 104–108. Also see Cazaux, *Musique à la cour de François Ier,* on Mouton's *Domine salvum fac regem* (168–169) and coronation music from the time of Francis I (167–175).

124. On the singing of the *Exaudiat* in the chapel of Henry III see [*Réglements* for Royal Households], Fonds Dupuy 489, fols. 5r, 6v–7r, Bibliothèque nationale de France. I thank Marie-Alexis Colin for generously sharing her transcription of this document with me. On the seventeenth-century practice, see Anthony, *French Baroque Music,* 418n. 2, and Irving Godt, "Guillaume Costeley: Life and Works" (Ph.D. diss., New York University, 1969), 124–129. The *Exaudiat/Dominum salvum fac regem* settings contained in the Deslauriers Manuscript (Rés Vma ms 571, Département de la Musique, Bibliothèque nationale de France), a compilation of sacred repertory from the first half of the seventeenth century, give some idea of how the psalm was performed: they include an anonymous *Exaudiat* among the "faux bourdons pour les terminaisons des psaumes" (fols. 127v–128r), an indication that the final verses often received some sort of special polyphonic treatment; an anonymous *Exaudiat* for four voices that employs a six-voice setting of verse 10 ("Domine salvum fac regem") as a refrain af-

motets stood in for or amplified the singing of Psalm 19. Whether or not they were heard at coronations, they captured the spirit and the letter of the *sacre,* making them apt for state occasions that—like the Te Deum ceremony—cultivated the coronation's tone. No doubt such a motet followed up the threefold cry of "God save the king" at the Te Deum in Notre-Dame for Louis XIII's victory in 1628.

Psalm 19, "Exaudiat te Dominus," was known commonly as the *prière pour le roy,* and it crops up not just in masses for the king but in chansonniers and civic ceremonies such as Henry IV's entry into Rouen in 1596.[125] The psalm's assurance that the Lord will save his anointed, coupled with the closing verse—"God save the king and hear us on the day in which we shall call upon You"—made it the ideal prayer for faithful subjects. Small wonder that the *Heures de Nostre Dame . . . pour la congregation roiale des penitens* (1583) instructs confreres in the royal order to kneel and sing the psalm each morning after the common mass.[126] In this respect, the royal confraternity mirrored the practices of the royal chapel, though it did so by enlisting singers from the *musique de la chambre* and courtiers to pray for the king. The confraternity included ten singers from the *chambre* (eight men and two boys), who were treated as full-fledged members and expected to sing at Wednesday services and in all the feasts and processions of the order; six more singers (four men and two boys) were engaged to sing *fauxbourdon* and polyphony at all the services.[127] Distinguished singers from the *chambre* counted among the founding members of the confraternity, most notably the famous castrato Etienne Le Roy and Girard de Beaulieu, both of whom had starred in the *Balet comique de la Royne* a few years before.[128]

Clearly the penitents indulged in polyphony, but like the pilgrims in the *processions blanches* that would traverse northern France that summer, the most striking aspect of the new order's music-making was congregational singing. The founding of the confraternity occasioned the publication of a book of statutes written by Auger, which provides details of the musical performances expected at its services. Statute 28 states that "everyone will learn to sing and to perform the ceremonies well." Further along the statutes specify that the service music will be performed by

---

ter each verse (fols. 67v–69r); an anonymous setting of verses 10, 7, 4, and 1 of Psalm 19 (fols. 90r–91r); and a *Domine salvum fac regem* for five voices and basso continuo by Antoine Boësset (fol. 175r–v).

125. For the entry see Godefroy and Godefroy le Jeune, *Cérémonial françois,* 1:945. For a charming *ad alternatim* setting of the *prière pour le roy* in the homophonic style of the *air de cour* with the solo sections set in a lilting triple time, see Nicolas Le Vavasseur, *Airs à III. IIII. et V parties par Nicol. Le Vavasseur* (Paris: Pierre Ballard, 1626). The inclusion of this simple motet at the end of a book of *airs de cour* suggests its frequent and nonliturgical use.

126. *Heures de Nostre Dame a l'usage de Rome,* 351.

127. Edmond Auger, *Statuts de la Congregation des pénitens* (Paris: Jamet Mettayer, 1583), 59–61.

128. See Yates, *Astraea,* 178–179.

two choirs led by choristers drawn from the ranks of the professional singers; either the congregation and the professional singers, taken together, will be divided into two choirs, or the congregation will sing *ad alternatim* with the professionals.[129] We know that for chapter meetings the Te Deum was sung by two choirs, perhaps with the confreres taking the odd-numbered verses and the "Musique" performing the even-numbered verses in polyphony; for the same meetings, the *Veni creator spiritus* was begun by the rector and continued by the confreres. Perhaps this was performed *ad alternatim* as well; I suggest this on the basis of the two *ad alternatim* settings of the hymn flanking the pair of Te Deum settings in Du Caurroy's *Preces ecclesiasticae II,* a book that was dedicated to Henry III's sister Marguerite by a man who served the last Valois king throughout his reign.[130] The group of *Veni creator spiritus* and Te Deum settings, all in the *ad alternatim* form that implied liturgical origins, fittingly memorialized a king for whose Order of the Holy Spirit the Pentecostal hymn must have been something of an anthem. Moreover, the pairing of the *Veni creator spiritus* and the *Te Deum laudamus* drew attention to the mystic body of the king by recalling the ceremonial for the exposition of relics, which began with the *Veni creator spiritus* and ended with the *Te Deum laudamus.*

The polyphonic verses of Du Caurroy's expansive settings were for experts only, but the confreres could have managed the plainchant verses in between. And like any Catholic congregation making its way through the day's hymns, with the help of the choir they surely could have negotiated as basic a chant harmonization as this anonymous *Exaudiat te Dominus* published by the Ballards in 1599 (ex. 4.3).[131] Barely more than root-position chords harmonizing the chant tune, this was *fauxbourdon* for beginners and hardly something to put one's name to, even given the evidence of a French predilection for extremely simple *fauxbourdon* settings exemplified by the Magnificats of Claudin de Sermisy. It made sense to publish the straightforward *Exaudiat te Dominus* for its liturgical and paraliturgical value as a *prière pour le roy,* even if it lacked polyphonic artistry.[132]

Whatever its repertory and performance practices, the Congregation of Penitents insisted upon communal singing, an insistence that surely witnesses Auger's hand in the organization of the order, for as we have already seen, Auger credited sing-

129. Auger, *Statuts,* 26, 66, and 68–69. Statute 28 reads: "Chacun apprendra à Psalmodier & faire les ceremonies bien & devotement selon qu'elles seront instituees, ordonnees & enseignees." On Auger and the king's penitents, see Jacques-Auguste de Thou, *Histoire universelle,* 9:67–69.

130. On the dedication and organization of the print see Eustache Du Caurroy, *Preces ecclesiasticae* (Paris, 1609), ed. Marie-Alexis Colin (Paris: Klincksieck, 2000), xv–xx .

131. *Octo cantica divae Mariae virginis . . . quaternis vocibus* (Paris: Veuve Robert Ballard and Pierre Ballard, 1599), fols. 14v–15r. Transcription my own; note values halved.

132. Here we should note the Jesuits' affinity for *fauxbourdon.* See Kennedy, "Jesuits and Music," 79.

EXAMPLE 4.3. *Exaudiat te Dominus,* verse 1 (1599). Bassus reconstructed

ing with the power to bring about Catholic victory in France. In his plan, the congregation deployed music on two fronts, one internal and the other external. In the first place, the statutes aimed to transform the confreres through the self-surveillance for which the Jesuits were so well known. Members had to confess twice a month—once a year was the norm—and perform a statutory review of their conscience each night before retiring, noting their sins, repenting, and asking God for forgiveness. This continual assessment of one's personal behavior was one mechanism by which the civilizing process came to instill obedience in courtiers, and in this respect we might see the congregation as a test case for the social regulations that later came to dominate court life. But we must also remember that *civilité* taught its lessons in a second way—by controlling the outward signs by which one displayed moral uprightness to the outer world. And not only did good behavior confirm inner goodness, but learning to perform well mechanically—to control one's body in order to write well, walk, sing, dance, and fence—taught the basics of honor and morality and, by standardizing behavior, refined the gearing of social interactions. Auger's insistence that the confreres learn to sing in order to perform the ceremonies of the order "well and devoutly" obliged the confreres to participate fully in them. By singing the liturgy, wearing the robes, and marching in the processions of the congregation, their words and steps were scripted, and passivity was made impossible by requiring all to follow the liturgy and the patterns of *ad alternatim* performance. Whether their heart was in it or not, they sang "God save the king" together at the end of mass, and—by the logic of Coyssard's *Traicté du profit*—singing the "Exaudiat te Dominus" and "Domine salvum fac regem" impressed sentiments of adoration upon their souls. Like a musical catechism, Psalm 19 expressed a basic doctrine of faith at the Catholic court, and though it appears but fleetingly in the Catholic liturgy—significantly, at matins for the Feast of Corpus Christi—those faithful to the king would have known it by heart and would have felt the need to genuflect when they heard it sung at a Te Deum ceremony.

In sum, the "liturgical" portion of the Te Deum ceremonial condensed the most dramatic moments of the *sacre et couronnement* into a festive précis of divine right. It must have been hoped that the same music that punctuated the most spectacular moments of the king's investiture would resuscitate the coronation's feverish acclaim with each subsequent performance. Royalty demanded its own liturgy, its psalms and prayers for the king, its Pentecostal emblems, and this became the stuff of Counter-Reformation civility and royal ceremonial, bits and pieces drawn from Catholic ritual, the Bible, and chivalric oaths, the whole assembled in expression of the political theology of kingship.

MAJESTY TRIUMPHANT, HENRY IV

With the Order of the Holy Spirit and the Te Deum ceremonial, Henry III established two institutions that would persist relatively unchanged throughout the ancien régime. French rulers up through Charles X took up leadership of the Order of the Holy Spirit upon their coronations, and even after the fall of the Bourbons, Berlioz penned a colossal Te Deum—replete with organ and choir in traditional antiphony—in a Republican era busy reenlisting the ceremonies of kingship in state service. But not all of Henry III's creations projected royalty in ways that his successors found useful. Henry IV abandoned the Penitents of the Annunciation and suppressed penitential confraternities more generally owing to their seditious activities under the League, in whose processions penitents marched, still barefoot but sporting helmets and guns.[133] Even had some Leaguers not worn white robes, the guise of a penitent king ill suited Henry IV, who no doubt learned from the poor reception that greeted Henry III's excessive demonstrations of piety.

133. See Yates, *Astraea,* 196.

Rather, military victories shaped the public image of Henry IV, winning him the epithet of "le Grand" as early as 1594. Henry IV was just as shrewd a negotiator as Henry III and just as desirous of diplomatic resolves, but he was moreover a brilliant commander with decades of experience in battle. His legendary cavalry victories at Coutras (1587) and Ivry (1590) made equestrian images of him indispensable, and one need only consider the statues of the king at La Flèche and on the Pont Neuf, or any of the innumerable paintings and engravings showing him on horseback, to see how the older, static genre of the royal chevalier was revived around his person in ways that would shape portrayals of Louis XIII and Louis XIV. After decades of boy-kings and royal impotence, Henry's vigorous health radiated command. Onlookers in Lyons, for example, praised his ability to battle the inclement heat the day he made his entry there in 1595, noting his "lively and active complexion, contrary to those soft and delicate natures who can only live in the shade in summer and in winter only near a fire."[134] A king of action and heroism, a king born into war, a general for twelve years, who had only charges and assaults for exercise, a tent and the open heavens for a palace, a lance for his scepter, and a breastplate for royal insignia, this was a king whose deeds easily became the stuff of myth.[135]

Nonetheless, upon the death of Henry III in 1589, Navarre faced enormous resistance. Only after many battles and abjuring Protestantism was he finally crowned at Chartres Cathedral on 27 February 1594 at the advanced age of forty. Yet the coronation was full of anomalies, from questions over his right to the throne and the strange five-year interregnum to the substitution of holy oil from Tours for that from the Holy Ampulla and of Chartres for Reims. Moreover, at the time of the *sacre* Henry IV controlled only a fraction of the kingdom. The League held most of the north, including the provincial capitals of Burgundy (Dijon), Normandy (Rouen), Picardy (Amiens), and of course, Paris and Reims, which is why the coronation took place in Chartres. They also controlled Toulouse and Marseille in the south. Paris was key, and Henry quickly set about negotiating a settlement with the moderate Leaguers in the capital, an accord helped along greatly by popular sup-

134. [Pierre Matthieu], *L'entree de tres-grand, tres-chrestien, tres-magnanime, et victorieux Prince Henry IIII. Roy de France & de Navarre, en sa bonne ville de Lyon, Le IIII. Septembre l'an M.D.XCV* (Lyons: Pierre Michel, n.d.), 6. "Le Roy estant descendu du batteau quoy que la chaleur de la saison fut violente ne se voulut enfermer sans exercice, mais contentant sa vive & active complexion contraire à ces molles & delicates natures qui ne peuvent vivre en esté qu'à l'ombre, ny en hyver qu'aupres du feu, voulut revoir toutes les beautez dont la Nature & l'art ont enrichi la Clare en attendant l'heure de son disné."

135. I gloss [Pierre Matthieu], *Les deux plus grandes, plus celebres et memorables resjouissances de la ville de Lyon. La premiere, Pour l'entrée de . . . Henry IIII . . . La seconde, Pour l'heureuse publication de la paix,* (Lyons: Thibaud Ancelin, 1598), 65.

port for the country's newly Catholic king, a widespread desire to end the wars, and the submission of Lyons to Henry just before his coronation.[136] He took Paris —peaceably—the following month, celebrating his victory by marching in a Te Deum procession to Notre-Dame.[137] In September 1594 he made his first official entry into the capital, again in the form of a general procession and Te Deum.[138] By choosing to enter Paris with a Te Deum procession, Henry personally capped the series of six Te Deums that Parisians had witnessed between March and September of that year announcing the submission of city after city to the king.[139] Henry's Te Deums seem calculated to affirm the sacredness of his kingship in a Paris still uneasy over the Catholicity of its king, who received papal absolution only the following year.

Henry's choice of the Te Deum as the most appropriate ceremonial to seal his coming to power is nonetheless surprising. Kings traditionally made royal entries at the beginning of their reigns, when Paris and provincial capitals rolled out the red carpet for the new sovereign, hoping through their tributes to gently instruct young monarchs on their obligations even while securing future favor. Charles IX embarked on a royal tour in 1564 to celebrate his majority, a progress of almost two years that included entries into Troyes, Lyons, Valence, Avignon, Narbonne, Toulouse, Bordeaux, Bayonne, Angers, and Tours.[140] Such ceremonial campaigns were not far from military ones, for they allowed kings to take possession of their realm in a process Geertz has described as ritual marking of the land: like a wolf or tiger spreading his scent through his territory, the king's visits, conferences, banquets, and jousts marked his kingdom as almost physically part of himself.[141] The fact that even normal times left the court always on the move—Henry III resided as little as two or three months in Paris some years—attests to the age-old form of exerting domination by constant prowling across the land.

EARLIER in this chapter, I suggested that we might consider the Te Deum ceremonial a Catholic version of the Roman triumph. With Henry IV we can observe the inverse—his royal entries often projected themes of the Catholic *sacre* onto those of a conquering king. Of course, the recovery of imperial ceremonial in classically inspired entries had a natural tendency to hint at its continuity in Catholic rituals

136. On Henry IV's coronation and coming to power, see Holt, *French Wars of Religion,* 158–161.

137. The taking of Paris is described in L'Estoile, *Mémoires-journaux,* 6:185–200, and d'Aubigné, *Histoire universelle,* 9:19–24.

138. L'Estoile, *Mémoires-journaux,* 6:226–227.

139. Ibid., 6:210–227.

140. See Graham and Johnson, *Royal Tour of France.*

141. Geertz, "Centers, Kings, and Charisma," 125.

of investiture. We have only to note how the Herculean arches of entries gave way to Te Deums at their end. But a more pointed effort to turn entries into public *sacres* seems to be at work in the first difficult years of Henry's reign. At Lyons (1595) and Rouen (1596), elements of the coronation were restaged in order to establish the pacts between sovereign and subjects that should have been sealed with the coronation oaths. It is as though the violence of Henry's fight for the crown was too great to have been legitimized at the coronation in Chartres. Moreover, because his *sacre* occurred before the country was truly his, the ceremony itself lacked legitimacy.

My attention was initially drawn to the entries at Lyons and Rouen because they were unusually full of music. At Rouen there were musical performances at three stations and a musical theme at a fourth, while the entry at Lyons likewise had three stations with music, including a great pavilion holding three choirs of instruments and singers.[142] Music had not been used on this scale before.

Moreover, in Lyons and Rouen, music was used charismatically. In Lyons Orpheus awakes and sings at the sight of the king, his magical harmonies materializing the marvels awaiting Henry inside the city and setting off the bells and noisy cannonades that began as Henry passed through the principal gate. In Rouen it was Amphion who sang the entry into being at the first station, a broken gate to one of the city's bridges. There Amphion's music reassembles the fallen stones of the city's walls, towers, and fortresses, just—the description tells us—as Henry's virtues will restore the ruins of France.[143] A marble plaque engraved with the words "Gallicarum urbium restitutori" spelled out the theme of urban renewal that the city fathers doubtless planned to discuss during the king's four-month stay in the city he had so brutally besieged only a few years before. On the next arch, a devastated Rouen, figured as a woman, reclined on the wreckage of the city and offered her heart to a life-sized effigy of Henry, who reached out to help her to her feet. Together, the arches set the tone for the entry, making princely virtues of coalition and reconstruction and casting Henry as the Amphion whose soft touch commanded the immediate beautification of the city made in preparation for the entry as well as the long-term renovation of this crucial port joining Paris to the sea. As in Lyons, music materialized the marvels that followed, but the choice of Amphion spoke to the specific plight of Rouen.

---

142. [Matthieu], *L'entree de tres-grand . . . Prince Henry IIII . . . en sa bonne ville de Lyon,* and *Discours de la joyeuse et triomphante entrée de tres-haut, tres-puissant et tres-magnanime Prince Henry IIII de ce nom . . . faicte en sa ville de Rouën, capital de la province & duché de Normandie, le Mercredy saiziéme jour d'Octobre, 1596* (Rouen: Jean Crevel, 1599).

143. *Discours de la joyeuse et triomphante entrée,* 34.

Both entries featured a singing Apollo, who, like Amphion and Orpheus, possessed a king's ability to reorder the world. In Lyons Apollo stood at the center of a Temple of the Houses of Bourbon and Navarre, flanked by statues of French kings from Pharamond to Henry III and representing a God-king or "Apollon Gaulois."[144] Like the lute-strumming "Monarque melodieux" who had appeared in the 1564 entry of Charles IX in Lyons, this musical Apollo signified Henry's godly ability to achieve concord among his subjects by referencing the divine origin of the lyre (a gift from Mercury to Apollo) and the fable of Apollo raising the walls of Troy with his songs. Given that the French understood themselves to be descended from the Trojans via Pharamond, Apollo belonged to the mythology of their classical past, his lyre and the solar imagery that would become so pervasive in the time of Louis XIV already signifying kingship in sixteenth-century France. In short, music gave visceral form to Henry's divinity and presented his rule as a force of the universe. Although Henry III's reputation as a musical connoisseur justly outshines that of his successor, only under Henry le Grand did entries and public ceremonies use music to thematize social harmony in truly significant ways.

With its early support of Navarre, classically minded Lyons had little to prove politically and was most keen to uphold its tradition of welcoming kings with allegorical references drawn from a secular past. Rather different was the tenor of Henry's entry in Rouen, a city still coming to terms with the king whose coronation they had rejected and whose authority they resisted under arms.[145] Henry had besieged the Norman capital for five months in 1592, and it remained a League stronghold up until April 1594, a fact that no doubt occasioned Henry's subsequent visit to the rebellious city. There the entry thematized not only the civic renewal they hoped would come of religious unity under their newly Catholic king, but the sacredness of kingship itself.

In Rouen the *sacre et couronnement* became the organizing principle of Henry's entry. And as in the Te Deum ceremonial, it was music that keyed the entry's symbolism to the coronation. The entry began outside the city, where Henry was greeted by the city religious singing *Exaudiat te Dominus.* From there the proces-

---

144. For contemporary references to an "Apollon Gaulois," see the frontispiece of Jean Dorat, *Magnificentissimi spectaculi, a Regina Regem Matre in hortis suburbanis editi* (Paris: Frederic Morel, 1573); the fold-out engraving in [Matthieu], *L'entree de tres-grand . . . Prince Henry IIII . . . en sa bonne ville de Lyon,* Figure M; and Ronsard, "La lyre," in *Oeuvres complètes,* 2:695. On French appropriations of Hercules and Apollo in royal contexts see Giesey, "Models of Rulership."

145. On the religious culture of Rouen under the League, the economic crisis precipitated by League policies, and the negotiations with Henry IV that brought the standoff to a close, see Benedict, *Rouen during the Wars of Religion,* 167–232.

sion made its way past the arches with Amphion and Lady Rouen and stopped just inside the city, where a large spherical machine covered with clouds hovered over the street (see fig. 4.4). As Henry passed beneath it, a great harmony of voices and instruments mixed together emanated from above, and among them one could hear a voice, "rather deep like that of God," issuing commands:

> Esprits des Cieux mouvans, stables Intelligences,
> Fendez soudain la voye à l'Esprit qui descent,
> Pour faire dans un Ciel gros de mille influences,
> A un Monarque unique un unique present.
> Ce Monarque est mon Oinct; il me craint, & je l'aime,
> Rendez luy de l'honneur, & l'honneur sera mien.
> Tout Roy d'homme est faict Dieu: c'est un autre moy-mesme,
> Qui sur terre dispose & du mal & du bien.[146]

> [Spirits of the moving Heavens, stable Intellects, quickly cut a path for the Spirit who descends to give a unique present in a Heaven full of a thousand influences to a unique Monarch. This Monarch is my Anointed one; he fears me, and I love him, render honor to him and the honor will be mine. Each King of men is made God: he is another of myself who dispenses on earth both bad and good.]

At the same instant, the clouds above Henry receded to reveal a celestial sphere full of an infinity of flaming stars and dazzling planets. The sphere turned smoothly and quickly in contrary motion to the vault around it, and from it descended an angel, who presented the king with a crown of gold and a sword gilded with fleurs-de-lis. The angel spoke by virtue of a long tube, into the other end of which a child declaimed two poems announcing God's transformation of Henry from *chef de guerre* into *chef de paix.* As in the coronation ceremony, the sword symbolized not war but an inviolate peace, for just as God gave the king power to defend the crown and van-

---

146. The station is described in *Discours de la joyeuse et triomphante entrée,* 41–45; this poem appears on p. 41. "Il oüit une grande & fort plaisante harmonie de Musique de voix & sons d'instrumens meslez ensemble, procedant d'enhaut; & parmi le tout estoit entendue une voix assez grave comme de Dieu, parlant & commandant aux esprits & intelligences Coelestes ainsi que s'ensuit. *Esprits des Cieux mouvans.* . . . Et au mesme instant s'ouvrit sur la teste de sa Majesté une grande nuee descouvrant un Ciel sphaerique fort artificiel, & plein d'une infinité d'estoilles & astres flamboyans, avec tous les ordres des corps Coelestes y figurez, & donnans telle clarté, qu'ils esbloüissoient les yeux des regardans. La Sphaere de ce Ciel se tournoit sur son pivot, de mouvement contraire à la voulte qui l'embrassoit; & toutes deux alloient d'une admirable roideur & vistesse. Du milieu d'icelle Sphaere sortit un Ange si industrieusement & artistement faict, & ayant ses mouvemens si faciles, si promptz & si libres, qu'il sembloit estre vif: Et descendant il presentoit au Roy d'une main une couronne d'or, & de l'autre une espee dedans son fourreau couvert de fleurs de lys d'or, & ensemble luy prononçoit ces vers qui ensuivent, par le moyen & industrie d'un enfant, resonnant sa voix dedans un canal qui la venoit redonner bien intelligible dedans le corps dudict Ange. *Monarque arreste toy.*"

quish his enemies, by the sword the king struck dread into enemies of the *patria* and fended off war. The treaty with Spain and the Edict of Nantes were still over a year off, but even at this early date it was evident that Henry could make the peace a binding one when it came.

Clearly, this part of the entry reenacted Henry's *sacre et couronnement,* transfer-

FIGURE 4.4. Heavenly vault. In *Discours de la joyeuse et triomphante entrée de tres-haut, tres-puissant et tres-magnanime Prince Henry IIII de ce nom . . . faicte en sa ville de Rouën* (1599)

ring the ceremony to the street and, in the process of condensing and simplifying its ritual investment of divine right, rendering a number of its mysteries quite spectacular. The *sacre* strove to manifest God's sanction through the liturgical emphasis on the Holy Oil, but how much simpler to have God proclaim from offstage, "This Monarch is my Anointed one; I love him; He is a God; Honor him." And theater could show what the *couronnement* could not: that the crown and sword—both believed to have belonged to Charlemagne and to place French kings in a lineage originating with the saintly emperor—were fashioned in heaven and delivered by angels. Stage machinery permitted spectators to marvel at the inner workings of the heavens even as a chorus and orchestra made audible the celestial music of the spheres as the "Esprits des Cieux mouvans" whirled them and the "stables Intelligences" held the whole fast. Pivoting with "admirable roideur & vistesse," the sphere's contrary motion against the vault mapped the fortunes of the king with the conjunctions and oppositions among its heavenly bodies and illustrated the mutable proportions among sounding intervals that created the dissonance and resolution of harmony. The machine appeared as a gigantic armillary sphere with music propelling it, but the sum of these parts was greater, for it materialized the éclat of majesty. Though not terribly subtle, the restaged *sacre et couronnement* probably seemed warranted in a city that so vigorously denied Henry's right to the throne. And though shocking, its directness presents a valuable reading of the *sacre et couronnement* as a ceremonial hinging on oil, a crown, a sword, and lots of singing, one best represented to the public in the modes of musical theater well supplied with fancy machines, magic voices, good lighting, and sumptuous props. We might even see in the pairing of this apparition with the *Exaudiat te Dominus* at the outset and the Te Deum at the entry's end a conscious desire to thematize the coronation across the whole event.

Henry passed by columns supporting Justice, Victory, Saint Louis, and Royal Renown, an obelisk depicting his virtues through "hieroglyphs" showing the labors of Hercules, and an arch celebrating his universal majesty, where, holding scepter, *main de justice,* and sword, his crowned likeness received a laurel wreath and a garland of stars which, "surpassing the royal crown, denoted that his clemency exceeds his force, in this way imitating the goodness of God, of whom he is the image."[147] A band of cornets played a motet on the text "Quis novus Euagoras regnum" from their gallery in one arch as he received the keys to the city, and Henry came upon one last musical scene: a beautiful garden full of grapes, rabbits, and singing birds. The garden represented the abundance of a peaceful countryside, but through it slithered a large python. As Henry passed, Apollo appeared and slew the python of rebellion with his bow and arrow.

147. Ibid, 65. "Une guirlande d'estoilles surpassant la couronne Royalle, denotoit que sa clemence excede sa force, imitant en ce la bonte de Dieu, duquel il est image."

The grisly rite of Apollonian purification rescued musical bounty and repose from the monster that threatened it. Apollo might soon return to his lyre, but innocence could not be regained, and this held true for the portrayals of Apollo in both Rouen and Lyons. Absent are the bevy of Muses and Cupids usually surrounding Apollo. War banished the Shepherd-God who strummed his lyre in Parnassian idylls, replacing him with more muscular depictions of Apollo in which the eagle and sun are far more prominent, pythons lurk underfoot, and snakes entwine staves in ornamental borders (see plate 2).[148] The bucolic view of life so pronounced in fete, chanson, and theater under Charles IX promoted a relaxed vision of Apollo that ill fit the monarchy in an era of pervasive war and regicide, and while *bergeries* did once again return to a society able to imagine itself safe from dissolution, the horrendous specter of civil war remained.

Henry's power as a man of arms ultimately brought a strong peace to France. His first foreign war, declared on Spain in 1595, united the country against a common enemy in such a way that his rout of the Spanish at Amiens in 1597 became an emblem of peace. Not usually given to superlatives (and certainly not when speaking of Henry III), Pierre de L'Estoile wrote in his journal, "Thursday the 25th of this month, Amiens was rendered to the King, who retook it not by craftiness, but by the most memorable effort and by the greatest glory of arms in the world."[149]

Both Henry's spirit and that lasting peace were conceived in musical terms. One commentator described the Peace of Vervins as "a work and extraordinary effect of [Henry's] holy and sublime spirit, which conducts the harmony of the Angels and the Heavens, which by discordant accords maintains the elements, and which for our good has reconciled these two great Monarchies."[150] The celebrations at Notre-Dame de Paris themselves burst with musical symbolism: on the king's side of the altar was the *musique de la chambre* with singers, lutes, viols, and other soft instruments, while the Spanish delegates were treated to the louder Royal Chapel, cornetti, and sackbuts. These combined forces performed a Mass of the Holy Spirit for double choir, and at the Elevation the *musique de la chambre* performed "an air so sweet and so harmonious that it seems as if the Angels of the Heavens are coming

148. Also see Peter Paul Rubens, *La proclamation de la régence.* According to one legend, it was Apollo who gave the caduceus to Hermes. The caduceus represents peace with its reconciliation of opposites, but it also speaks directly to a peace forthcoming from Apollo's slaying of the python.

149. L'Estoile, *Mémoires-journaux,* 7:106. "Le jeudi 25e de ce mois, Amiens fust rendu au Roy, qui la reprist, non par ruse, mais le plus mémorable effort, et par la plus grande gloire des armes du monde."

150. [Matthieu], *Les deux plus grandes . . . resjouissances,* 65–66. "Nous serions bien aveugles si nous ne recognoissions la lumiere qui dissipe nos tenebres, nous n'aurions point de jugement si nous ne remarquions que c'est un ouvrage & un effect extraordinaire de ce sainct & sublime esprit qui conduit l'harmonie des Anges & des Cieux, qui par discordans accords entretient les elemens, & qui pour nostre bien a reconcilié ces deux grands Monarchies."

by their celestial song to ravish the assistants in devotion and excite them to the holy and celestial meditation of the joy of the Heavens and the glory of the Blessed."[151] Here Henry rules over the softer and more ravishing *musique de la chambre* in a battle of the bands in which *alternatim* performance cedes to a consort of unlike forces, and harmony is conceived as the coming together of dissimilar things in a concord only calculable through the sublimity of musical proportions.

The king's lyre represented a hard-won peace, the making of concord from discord, the resolution of dissonance, and a controlled way to unite unlike matters by placing them in sympathetic relationship to one another. Politically, the lyre signified Bodin's "harmonicall governance," which emanated from an absolutely powerful sovereign. In religious terms, it signified the goals of the Edict of Nantes. Though famously misunderstood as an edict of tolerance, the Edict of Nantes was actually intended to create religious concord through unity, not religious coexistence. Through peaceable means, it aimed to ease the transition to a Catholic consensus it never achieved; its failure forced Louis XIII to set out on military campaigns against Huguenot strongholds in 1620–1622 and to besiege La Rochelle in 1627–1628, and led Louis XIV to revoke it entirely in 1685. "Harmony" glossed centralization, absolute authority, and unity. Apollo represented a heliocentric political philosophy with the king at its center and his lyre the web of authority generated by its God-given, immutable chords.

LOUIS XIII AND FINAL VICTORY

The fall of La Rochelle in October of 1628 cast the ultimate blow to Huguenot resistance. Though heavily besieged at various times after its capture in 1568 by the Protestants, La Rochelle had always withstood assault. The citadel's thick walls and the proud towers of this fortified port on the Gulf of Gascogny stood for the Protestant cause, and with an avenue to the sea and England, it remained well defended. But the decisive battles of 1622, during which Louis XIII took the region of Poitou in a brilliant victory at the Île de Ré and then swept south to Montpellier, spelled the end for La Rochelle, which stood alone after the Peace of Montpellier. It was only a matter of time before Louis turned his attention to this thorn in the side of French Catholics and routed the Huguenots once and for all.

The final siege began in August 1627 and ended over a year later. Nearly half of

151. *Les Ceremonies observees a la solennisation de la Paix, en l'eglise nostre Dame de Paris, le 21 Juin 1598* (Paris: Denis Binet, 1598), 5, 10–14. "La preface se chante, les Chantres alternativement disent le *Sanctus* a deux choeurs, la consecration se faict, l'eslevation, l'adoration, pendant laquelle la musicque de chambre seule chante un air si doux & si harmonieux qu'il semble que ce soyent Anges du Ciel qui viennent par leur celeste chant ravir les assistants en devotion, & les exciter à saincte & celeste meditation de la joye des Cieux & gloire des Bien-heureux" (13–14).

the population of 25,000 perished, and another 5,000 fled, so that Louis and his conquering army were greeted only by phantoms when they took the keys to the city on 30 October 1628.[152] Bread was distributed to the starving inhabitants, royal troops and artillery moved in, and courtiers in the entourage of the king peopled the city so that it "appeared as another Paris." With the return of the Catholic orders, Mass was once again heard at the Church of Sainte Marguerite on All Saints' Day: "It was a great joy to see the Crosses of the Capucins & Recollects that they carried in Procession through the streets, singing the Te Deum; also sixty or eighty of the Brothers of Charity came from their Salles quarter with their Cross to the same Church singing antiphonally."[153] The impromptu Te Deum procession launched a series of ceremonies carefully staged to cast the king's victory as a victory for God and to emphasize his clemency toward the conquered.[154] As the friars made their way to the church, three hundred Rochelois awaited Louis's formal entry into the city. By commandment, they threw themselves down in the muck, heads bare, crying in high and trembling voices, "Vive le Roy qui nous a faict misericorde" when he appeared. Likewise, within the city, the members of the law courts were made to kneel on the cobblestones, forbidden to make speeches, and instructed to cry out "Vive le Roy qui nous a fait grace" upon the king's arrival. Louis distributed some bread to thankful women and then proceeded to the church, where he was received by the archbishop of Bordeaux.

During the service, the archbishop publicly remarked the "devotion, douceur & clemence" of the king; in the sermon, the king's preacher, Father Souffran, likened him to God in the manner of the phrase "Rex regum et Dominus dominatium" by calling him "le grand Roy des Roys." But it was the king himself who put the most effective spin on the ceremonies when he sang the Te Deum from beginning to end along with the celebrants: "It was in this way that the King gave true testimony to great piety, singing the Te Deum himself . . . from the beginning to the end, & showing that he rendered great glory to God for such a glorious victory."[155] Louis's chanting of the Te Deum was certainly an extraordinary public performance for a king, one remarked with some precision in an account from the *Mercure François.*

152. On the battles of 1620–1629 see Holt, *French Wars of Religion,* 177–187; the description of Louis XIII's entry into the city is drawn from *Mercure François* 14 (1629): 703–717.

153. *Mercure François* 14 (1629): 708. "Ce fut une grande joye de voir les Croix des Capucins & Recolects qu'ils portoient en Procession par les ruës chantans le Te Deum, comme aussi les Freres Religieux de la Charité vindrent au nombre de 60. ou 80. de leur quartier de Salles avec leur Croix jusques dans ladite Eglise Psalmodians."

154. Jouhaud, "Printing the Event."

155. *Mercure François* 14 (1629): 712. "Ce fut alors que le Roy rendit vrayement témoignage d'une grande devotion; chantant luy mesme le Te Deum . . . depuis le commencement jusques à la fin, & faisant paroitre qu'il rendoit à Dieu une grande gloire pour une si glorieuse victoire."

Under normal circumstances Louis would have been one object of the hymn, conflated with the Lord as the "te" of "Te Deum laudamus, te Dominum confitemur, te aeternum Patrem omnis terra veneratur." Had he remained silent on his throne before the altar in La Rochelle, the congregation's cries of "Tu Rex gloriae" would have blessed him as well, but by singing along "from the beginning all the way to the end," he channeled the praise heavenward, amplifying it with his own. For the duration of the hymn, Louis stepped in as high priest, leading his subjects in thanksgiving for the final defeat of heresy in a France that had been ravaged by religious war for over sixty-five years.

Louis's thanks to God may have welled up uncontrollably in song, or it may have been an official act of the sort that remained mindful of a past when the king's piety and religious orientation came under greater scrutiny. Either way, there is no doubt that he often employed his considerable musical abilities in acts of devotion. Indeed, Louis had filled the tedious hours of the siege by writing motets. When Pentecost approached, having left his singers and musicians at home, he showed the clerics how to notate the psalms for vespers, and on the day of the feast "one saw this pious monarch like David amid his singers, animating them with his voice, directing them with his movements in singing the Psalms from the notes that he had marked, and the whole with accords so just and sounds so measured that one was charmed by their harmony."[156] So, too, Louis took it upon himself to lead the Te Deum when he finally took the city, blessing it with particular efficacy on that historic day.

The simple service at the front combined royal entry with Te Deum in a gritty ceremony that shows both at their most elemental. The faux triumphal procession staged for official entries is here a very real one shorn of arches and theaters and festive uniforms. Even the artillery fire that usually accompanied Te Deums is more genuine, emanating as it did from cannons just drawn into the city from the battlefield. The procession of the religious orders, too, is a stripped-down version of the bell-ringing civic processions to the cathedral or church and bonfires that comprised standard Te Deum ceremonies. Only its essence remained as the monks made their way to the church, singing the Te Deum and retaking the city for Catholicism. By singing and carrying the cross through the city, Catholics won a war waged not just in battles and sieges, but in processions that turned to riots, in

156. On Louis XIII's musical abilities and the motets he composed at La Rochelle, see Durosoir, *Air de cour,* 178–180, this quotation at 179: "on vit à ce jour ce pieux monarque, comme David au milieu de ses chantres, les animer de sa voix, les diriger par ses mouvements en chantant les Pseaumes selon les notes qu'il avoit marquées; & tout cela avec des accords si justes & des sons si mesurés qu'on étoit charmé de leur harmonie."

the penitential *processions blanches,* and in the purges by Holy Leaguers and Confreres of the Holy Spirit that rendered the proximity of ritual and violence so horribly clear. Four days later the victory of eucharistic devotion was sealed with a Corpus Christi procession in which Louis walked behind the Holy Sacrament.

No garlands of red roses marked the festivities, nor colorful banners, but there was song. There was a Te Deum procession with singing clerics, a triumphal entry with a *Te Deum laudamus* sung by the king at its end, and a Corpus Christi procession with its *Pange lingua* and Te Deum hymns. These battlefield rituals are not far from those of the knight who drives his sword into the earth, kneels down, and prays before his makeshift cross, or of Charlemagne's companions, who, in the *Chanson de Roland,* sang a Te Deum after his victorious combat.[157] They invoke a wrathful, judgmental God, an Old Testament Deus Sabaoth whose bloodlust stirs an inhuman strength within the warrior. Like the lions and leopards associated with his powers, the warrior-king kills with the ferocity of an untamed nature, taking the country in campaigns of slaughter. Only the coronation legitimizes his violent, inhuman power and transforms it into the productive power of the governor. As Valerio Valeri characterized the process, the coronation unites the king with the people, both directly and through the gods that represent the moral principles of the community.[158] The king does not have to be a true invader: the rite of coronation takes on the task of representing the king's passage from the human to the divine and his triumphal return (symbolic "invasion") to the community he will rule. But when the king *is* a true invader, how vivid the function of ritual and, in these rituals that have been stripped down to little more than ritual singing, how essential the function of music. The rites in La Rochelle transformed Louis from the source of a horrific siege into a beneficent monarch, his clemency, pity, and healing touch represented by ceremonial pardons, the symbolic distribution of bread, and plainchant. From the monstrous inferno of siege machines and cannons, the king was borne into the city by the even harmony of *fauxbourdon* and hymns, his brutality transformed into Christian divinity by a ritual in which he shed his armor for the priest's chasuble and, leaving the mundane world of speech behind, showed his extraordinary piety by leading his subjects in the enchanting strains of the *Te Deum laudaumus.*

---

157. See Žak, "Das Tedum als Huldigungsgesang," 26.
158. Valeri, "Regalità," 748–749.

FIGURE 5.1. "Ballet of the Provinces." Engraving by Jean Cousin in Jean Dorat, *Magnificentissimi spectaculi* (1573). By permission of the British Library, shelfmark 837.e.41

5
PYRRHIC DANCE AND THE ART OF WAR

IN A BALLET libretto written near the end of his life, Descartes called war a "ballet for the birth of peace."[1] He was hardly alone in likening the soldiers who lined up for pitched battles to troupes of dancers. Indeed, the orderly battalions of seventeenth-century armies—clean, rested, well equipped, and dressed in uniforms—sooner resembled a corps de ballet than the ragtag bunches of conscripts that constituted the French armies of his youth.[2] Agrippa d'Aubigné said as much when describing the *Ballet des provinces* performed for the visit of the Polish ambassadors in 1573: "The Polish admired the disentangled confusions, the well-formed ciphers of the ballet, the different musics, and said that no King on earth could imitate French dance, but I would prefer that they had said this of our armies."[3] Brantôme, in more flattering tones, described the same ballet as a model battle in which the sixteen ladies representing the provinces of France lined up like a small battalion and marched to a pleasant war tune (see fig. 5.1).[4] Catherine de' Medici's ladies-in-waiting impressed the foreigners with their training, solid judgment, and excellent memories, which allowed each to find her proper position within the dizzying sequence of turns, interweavings, mélanges, and halts that made up the figures of the ballet. In 1581 Beaujoyeulx too described his choreography as a *bataille rangée* owing to the order, studiousness, and attention to position it required.[5]

1. Descartes, *Oeuvres,* 5:616–627.

2. On equipment and uniforms see Wood, *King's Army,* esp. 165–166; and Lynn, *Giant of the* Grand Siècle, 169–180.

3. D'Aubigné, *Histoire universelle,* 4:156. "Les Polonnois admirerent les confusions bien desmeslees, les chiffres bien formez du ballet, les musiques differentes, et dirent que le bal de France estoit chose impossible à contrefaire à tous les Rois de la terre: J'eusse mieux aimé qu'ils eussent dit cela de nos armies."

4. Brantôme, *Recueil des dames,* 54. Also see Dorat, *Magnificentissimi spectaculi,* fol. c1r–v, and Franko, *Dance as Text,* 21–23.

5. Beaujoyeulx, *Balet comique de la Royne,* fol. 56r. "A la moitié de ce Balet se feit une chaine, composee de quatre entrelacemens differents l'un de l'autre, tellement qu'à les voir on eust dit que c'estoit une bataille rangee, si bien l'ordre y estoit gardé."

Though performed by beautiful women, the discipline of geometrical ballets was martial.

By likening war to ballet (rather than ballet to war), Descartes reversed the analogies of the sixteenth century. This reversal draws attention to a transformation that occurred between those first *ballets de cour* and the mid-seventeenth century: warfare became more balletic. Any who turned out to watch pitched battles—gladiatorial spectacles par excellence, and ones that drew significant crowds of onlookers—might have observed a new precision in the open field, where foot soldiers could fall into a variety of formations, hold together in battle, and execute complex maneuvers in synch. The discipline observed in ballet was eventually implemented in battle.

By the eighteenth century, the mechanism of the disciplined mass or docile troop cast military discipline over the entire social body. This "military dream of society," as Michel Foucault called it, took the disciplinary techniques of the modern army as the principle for maintaining the absence of warfare in society more generally. "In the great eighteenth-century states, the army guaranteed civil peace no doubt because it was a real force, an ever-threatening sword, but also because it was a technique and a body of knowledge that could project their schema over the social body."[6] Discipline was made national and brought to state administration; it trained individual bodies, located them in ranks, and combined their discrete activities in the meticulously coded cogs of a vast machine. In Foucault's analysis, the political ramifications of military discipline were extended from those of a larger and stronger army to those of economic and demographic forces finally marshaled.

Foucault's study moved forward from the seventeenth century, when military discipline might be observed in selected armies. But before military discipline was applied to soldiers, its technologies were discovered by disciplinarians working in a variety of social spheres, something Foucault alludes to with his references to Jesuit education. As the *Ballet des provinces* suggests, the "military dream" was more easily nurtured in the theatrical hothouse of court spectacle than out in the open field of battle. My intention in this chapter is not to revisit the *ballet de cour,* but, rather, to follow up the lead suggested by those early court ballets and consider civility from the broader perspective framed by military discipline. Catherine de' Medici's ladies-in-waiting were at once refined and martial, civilized and military, and they elicited admiration not simply for the art of *ballet de cour,* but for a kind of musical and muscular regimentation that we find being researched and implemented elsewhere in society at large. We discover similar discipline in the performances of urban youth groups, Jesuit collegians, city militias, and the drills of forward-thinking military strategists. Perhaps more important, in the military ethos

6. Michel Foucault, *Discipline and Punish: The Birth of the Prison,* trans. Alan Sheridan, 2nd ed. (New York: Vintage Books, 1995), 168.

of these social organizations, we discover sources of discipline that would only subsequently be channeled into armies.

Between the regulated gestures of court ballet and the *civilité* of seventeenth-century battalions stood a category of dance that Plato called "dances of war" or pyrrhic (*Laws* 7.815a):

> It represents modes of eluding all kinds of blows and shots by swervings and duckings and side-leaps upward or crouching; and also the opposite kinds of motion, which lead to active postures of offence, when it strives to represent the movements involved in shooting with bows or darts and blows of every description.[7]

Revived in the sixteenth century, the idea of dancing into battle placed music at the center of projects to order not just individual bodies, but whole battalions. The musical technology at work in dance could be applied to ever greater numbers.

Moreover, pyrrhics preserved the violence of armed combat, containing brute force in choreographies that seemed too driven by blind rage to have been rehearsed in advance. The collisions of swords and shields caused a disarray that seemed impossible to overcome, yet the dancers fell back in orderly ranks and renewed their attacks again and again in time to the music. While the cool-headed geometry of *ballets de cour* suggested the futuristic benefits of fully "automated" battalions, pyrrhics presented the more immediate possibility of marshaling the bloodlust of individual soldiers and deploying it in rough synchronization. Eventually the musical precision, attention to gestural detail, and coordination of individual bodies attained in the ballet-machines at court would be extended to the army. Fighting would become more precise, but only when the institutional structures of armies allowed for the training required to achieve it.

The pyrrhic forms an obvious hinge between the practices of military and civil society (drill and dance, infantry and corps de ballet), introducing themes that will be developed further in the following chapter (cavalry and equestrian ballet). The classical history of the pyrrhic, however, sets this chapter apart, for it adds a uniquely humanistic element to our history of music, discipline, and arms. Military discipline was revived in imitation of the ancient Spartans and Greeks, making it a truly Renaissance project. Beginning with Machiavelli, writings about the infantry mined classical sources such as Vegetius, Aelianus, and Xenophon for information on the training of foot soldiers, while pedagogues fascinated by the classical *gymnasium* pulled from Plato and others descriptions of pyrrhics that developed strength in youths. These two strains of what we might think of as "military humanism" intertwined, and in them—taken together—we can rediscover the disciplinary ideal that

7. Plato, *Laws,* trans. R. G. Bury, 2 vols. Loeb Classical Library (Cambridge, Mass.: Harvard University Press, 1967–1968), 2:91.

motivated not just the physical practices of the Renaissance, but the political practices they sustained.

CLASSICAL PYRRHIC DANCE

Let us begin with the classical sources that first brought the mutual relevance of dance and military discipline to the attention of Renaissance humanists. Most interested sixteenth-century readers would have known Plato's accounts of the pyrrhic in the *Laws,* which prescribed it for the education of both boys and girls (*Laws* 7.796). Subsequent writers likewise affirmed the usefulness of the pyrrhic for training warriors, to the extent that one could fairly rank it as part of a classical education, or at least a Platonic one. In Crete, apparently, everyone studied the pyrrhic, "not only private persons and those of low estate, but also Princes and Magistrates," and in Athens it was a regular part of military celebrations, religious rites, and the Panathenaea festival.[8] It developed fitness, agility, and grace in children, even while teaching the inner virtues of resilience and discipline. Seen in this light, pyrrhics fit well into programs of moral education and, in the Renaissance, inspired depictions such as that in the lastingly popular *Artis gymnasticae . . . libri sex* (1569) of Hieronymus Mercurialis, in which the great educators Plato, Socrates, and Aristotle oversee a performance of the dance (see fig. 5.2).

By the time of Plato (c. 427–347 BCE), the pyrrhic had already enjoyed a long history in Crete, mainland Greece, and Asia Minor.[9] As early as the seventh century BCE the Spartans were using it for military training, a practice that persisted into the third century CE, and one to which we will turn momentarily. The Athenians seem to have reserved it for youths and festivities, though warriors did dance pyrrhics as well. Plato's Athenian contemporary Xenophon (c. 420–350 BCE) beautifully described a *convives militares* that took place during Cyrus's retreat from Persia, an expedition in which Xenophon fought as a Greek mercenary. As the army marched into Paphlagonia, local ambassadors greeted the soldiers with the hope of minimizing any plunder, occasioning the Greeks to receive them at a banquet with entertainment provided by the soldiers (*Anabasis* 6.1.1–13).[10] Two Thracians danced in full armor to the flute and then parried with their swords until one struck the

8. Lucian of Samosata, *Les oeuvres de Lucian de Samosate,* trans. Filbert Bretin (Paris: Abel l'Angelier, 1583), 359. "Non seulement les personnes privees & de bas estat, mais aussi les Princes & Magistrats."

9. See Barbara Palfy, "Pyrrhic," in *International Encyclopedia of Dance,* founding ed. Selma Jeanne Cohen (New York: Oxford University Press, 1998), 282–283; and Margaret M. McGowan, "A Renaissance War Dance: The Pyrrhic," *Dance Research* 3 (1984): 29–38.

10. Xenophon, *Anabasis,* trans. Carleton L. Brownson, rev. John Dillery, Loeb Classical Library (Cambridge, Mass.: Harvard University Press, 1998), 466–471.

FIGURE 5.2. "Pyrrhichia saltatio." In Hieronymus Mercurialis, *Artis gymnasticae . . . libri sex* (1569). By permission of the British Library, shelfmark 557*.e.26

other with such apparent violence that the Paphlagonians thought him dead. The next "entrée" (as the French would have understood it) consisted of a fight between farmers over a yoke of oxen. Then Mysus danced the "Persian" dance with a shield on each arm as if engaged with opponents, spinning on toe, turning somersaults, falling to his knees and leaping up again, and knocking his shields together like cymbals, the whole "in exact time to the flute." Finally came a series of songs and dances in arms "usually exhibited on religious processions." To further astonish the ambassadors, an armed woman closed the show by joining in the pyrrhics with the men.

Xenophon names the pyrrhic among the dances associated with religious festivities, but clearly the Greeks knew a number of armed dances and made something of a specialty of them. Most important, the Greeks did not find it unseemly that professional warriors danced. Indeed, the dances told of by Xenophon made the military prowess of the Greeks excruciatingly clear to the foreign diplomats. Lightly robed for postprandial amusement, the war dances showed off their strength in a cautionary demonstration of Greek supremacy. Xenophon's contemporaries realized the potential of the pyrrhic to conduct war through divertissement.

What Plato and Xenophon only allude to—and this is key to the reception of the pyrrhic in late-Renaissance France—is that the pyrrhic proved a useful weapon in battle. This application of the dance is clearest in the more encyclopedic sources of the later period, particularly those that describe Spartan practices, such as the *Deipnosophistae* of Athenaeus (fl. c. 200 CE) and *De Saltatione,* a text attributed in the sixteenth century to Lucian of Samosata (c. 120–200 CE). *De Saltatione* presented the most thorough classical treatment of the pyrrhic widely available to sixteenth-century readers: it circulated in numerous Latin and Greek editions, and in 1583 Filbert Bretin translated it into French, publishing it as part of Lucian's *Oeuvres.*[11] Arbeau and Tuccaro both mined *De Saltatione* for material (probably in this French translation), and the opening gambit of Arbeau's *Orchésographie* (1589) owes much to the argument and content of Lucian's chapter on dance in general, and his treatment of the pyrrhic in particular, making it a fundamental point of departure for revivals that sought to understand the dance in terms of infantry training.

Like so many of the sixteenth-century apologists for dance who read and cited him, Lucian's discussion begins in a defensive mode, for his interlocutor, Craton, has just condemned dance as soft, vain, effeminate, and inappropriate for men of liberal education. Lucian objects that the greatest heroes of Greece and Lacedae-

11. The other lengthy description of the pyrrhic available at mid-century was that in Hieronymus Mercurialis, *Artis gymnasticae . . . libri sex* (Venice: Giunti, 1569), in which classical references abound. In the seventeenth century Johannes Meursius, *De ludis graecorum liber singularis* (Leiden: Elzevier, 1625), assembled a complete collection of classical references to the pyrrhic (63–75).

monia used dance to great advantage. Achilles' son, he reminds Craton, was named Pyrrhus for his expertise in dance, and with the pyrrhic he was able to defeat Troy.[12] The Lacedaemonians, too, commanded their troops through musical means, fighting "to the sound of flutes, with measure and steps rigidly controlled by their rhythm and even giving the first signal for entering into combat with the flute. This is also why they were always victorious over all other nations, having been led by music and good rhythm." [13] Lucian leaves no doubt that fighting with rhythmic precision—*par la mesure, par la musique,* and *par la bonne cadence*—brought victory. But he never describes it in detail. Did the soldiers march together in rhythm as a means of holding rank, which would have mobilized a united front, or did the music coordinate other details of the battle, such as the discharging of javelins, as Plato's description seemed to suggest?

These lacunae stymied Renaissance readers, and as was so often the case with efforts to recreate ancient performances, historians, choreographers, and musicians filled in the blanks with material closer to hand. This sort of borrowing is evident in the most famous pyrrhic choreography from the Renaissance, the one Arbeau includes in *Orchésographie.* It consists of six *passages,* precise instructions for the swordplay, and directions for costumes and staging, by far the most extensive choreography in the treatise and the most detailed pyrrhic from the period. It closes the treatise with a bellicose flair, quitting the "dances of recreation," as Arbeau calls the social dances, in favor of something more strictly masculine.[14] In fact, the pyrrhic brings the book back to where it began, for *Orchésographie* opens with a lengthy exposition of marching, which Arbeau's interlocutor, Capriole, calls "dances of war." With marches at the beginning and the pyrrhic at the end, *Orchésographie* is framed by military practices.

The marches and the pyrrhic are not just bookends, but share a common musical and choreographic vocabulary that, as we shall see, suggests the real relevance of the pyrrhic to warfare. The basic drum pattern for marches—five minim beats followed by three minims of rest—corresponds to the rhythm for the pyrrhic, and

12. Lucian of Samosata, *Oeuvres,* 360.

13. Ibid. "Les Lacedemoniens, qui estoient estimez les plus braves hommes de la Grece, ayans appris de Pollux & Castor la maniere de mener cherelie: & que c'estoit l'effect propre de la danse, en laquelle avoient estez instruits les Cariens, peuples de Laçonie: ils meslerent le tout avec la musique: jusques à combatre en guerre au son des flustes par mesure & avec un marcher compassé à la cadence & mesmement faisoient donner le premier signal pour entrer au combat, avec la fluste. C'est aussi pourquoy ils ont tousjours esté vaincueurs de toute autre nation: estans conduits par la musique & la bonne cadence."

14. Arbeau, *Orchésographie.* See the facsimile of the Paris 1888 reprint edition, ed. Laure Fonta. Page references to the English translation of Mary Stewart Evans follow references to the facsimile in my notes.

both the march and the dance alternate left and right steps with each reiteration of the rhythmic pattern (see exx. 5.1 and 5.2). Arbeau instructs the soldier to march beginning with the left foot on the first minim (beat 1) and the right on the fifth minim (beat 2), warning the drummer to leave rests at the end of this pattern "so that if confusion should occur, through a change in step, the soldiers can mend matters and easily get back on the left foot when they hear the three rests." [15] The pyrrhic, too, begins on the left foot, with kicks (*grèves*) each semibreve in alternation.

Example 5.1. Marches. In Thoinot Arbeau, *Orchésographie* (1589)

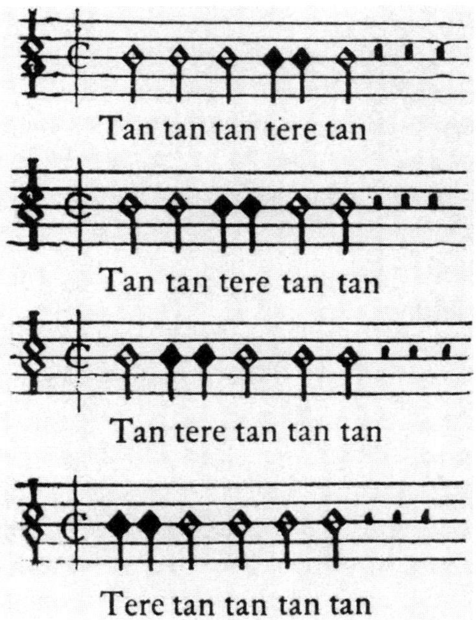

In tempo, as well, parts of the pyrrhic corresponded to the marches. Each *passage* of the dance was preceded by a circling of the hall that provided a much-needed break from the vigorous dancing. Dancers parried with sword and shield while jumping from one foot to the other at a quick pace and turning from right to left to confront new opponents. An impressive but exhausting display, the pyrrhic needed its slower "rounds" of marching between up-tempo bouts of hand-to-hand combat.

15. Arbeau, *Orchésographie,* fol. 16v; 37. "C'est pourquoy le tambour faict aucunes fois une continuation de plusieurs battements joincts ensemble, affin que sil y a de la confusion par transmutation de marches, les soldats la puissent reparer, & qu'ils se remettent tous aisement sur lassiette gauche après qu'ils ont ouy le repos du souspir ou de trois souspirs: Et cela sert grandement à faire les evolutions."

Plate 1. Giovanni Battista Rosso, *The Education of Achilles* (c. 1534–1539). Gallery of Francis I, Chateau, Fontainebleau. Réunion des Musées Nationaux / Art Resource, NY

PLATE 2. Tapestry of Henry IV as Apollo (17th century). Écouen, Musée de la Renaissance.
Photo by Gérard Blot. Réunion des Musées Nationaux / Art Resource, NY

Plate 3. Peter Paul Rubens, *Henry IV at the Battle of Ivry* (c. 1630). Uffizi Gallery, Florence. Scala / Art Resource, NY

PLATE 4. *Fete at Bayonne*, Valois Tapestries (1573). Uffizi Gallery, Florence. Scala / Art Resource, NY

EXAMPLE 5.2. Pyrrhic Dance. In Thoinot Arbeau, *Orchésographie* (1589)

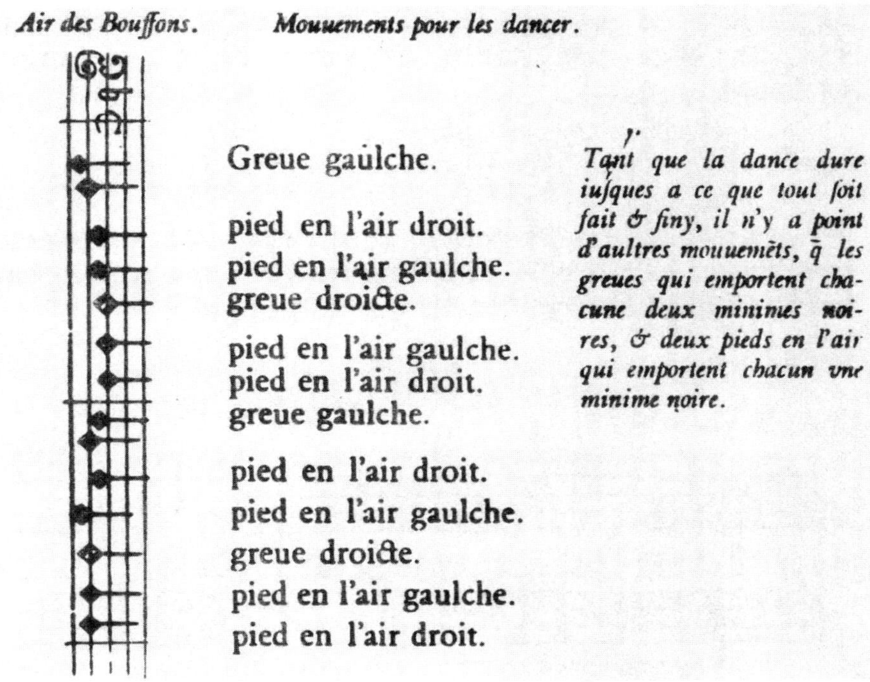
Air des Bouffons. *Mouuements pour les dancer.*

Greue gaulche.

pied en l'air droit.
pied en l'air gaulche.
greue droicte.

pied en l'air gaulche.
pied en l'air droit.
greue gaulche.

pied en l'air droit.
pied en l'air gaulche.
greue droicte.
pied en l'air gaulche.
pied en l'air droit.

Tant que la dance dure iusques a ce que tout soit fait & finy, il n'y a point d'aultres mouuemēts, q̄ les greues qui emportent chacune deux minimes noires, & deux pieds en l'air qui emportent chacun vne minime noire.

The parallels between Arbeau's pyrrhic and the marches reinforced—as he thrice repeats—that it was used in war. Indeed, it constituted not just a dance of war, but musical warfare, as Lucian made clear and others reiterated. On the authority of Plutarch, Tuccaro maintained that the Spartans always went into battle "swaying and dancing, imitating the consonance and measure of the flutes," and that the Heraclites marched into the fray to the measured proportions of music—"this good disposition gave such fright to the enemies that [the soldiers] always returned victorious."[16] In a country turned upside-down by civil strife, pillaging mercenaries,

16. Tuccaro, *Trois dialogues,* fol. 40r. "Les Laconiens & Spartiates, suyvant l'authorité de Plutarque, ne commençoyent jamais la bataille, sinon qu'en ballant & dançant, imitants la consonance & mesure des flustes." "Polygene, homme de grande authorité, raconte en son second livre des exercices, que l'experience reïteree avoit enseigné aux vaillants capitaines de la race des Eraclites, que la victoire estoit tousjours de leur costé lors que suyvant la proportion mesuree du bal & de la dance, les soldats à la cadence du son se preparoyent avec un tel ordre à donner la bataille; & ceste belle disposition donnoit tellement l'espouvante aux ennemis, qu'ils s'en retournoyent victorieux."

siege, and riot, Arbeau's pyrrhic must have suggested itself as a secret weapon. It was an advanced sort of marching and fighting to which students could graduate after they had mastered all of the other techniques in the treatise.

In Arbeau's own words, the pyrrhic helped men march as one against the enemy, which made his *passages* not just a choreography for dance, but a choreography for war. Unlike traditional sword dances such as the *matachins* and *moresca,* which referred most directly to the duel and *combat en camp clos,* the pyrrhic descended from ancient infantry maneuvers. Its significance lay in its usefulness to warfare, as a battle plan that might be used to mobilize troops with speed and cohesion. Only in the course of the sixteenth century did European infantries become tactically important in battle, and as they did so, theorists became concerned to solve the problems they presented, most of which had to do with coordinating soldiers on foot with each other and with mounted knights. In this light we should see Arbeau's treatise not as an isolated excursus on dance, but as part of the larger discourse on the art of war, one that addressed issues of concern to the foremost military strategists.

RHYTHMIC DRILL

By 1600 foot soldiers regularly clinched the outcome of battles. As early as the 1510s, Swiss foot soldiers and their German counterparts, the *lansquenets,* had become regular components of European armies, dramatically proving their worth at Ravenna (1512), Novara (1513), and Marignano (1515). Novara, for example, had been decided almost entirely by the German and Swiss despite the participation of Italian and French knights in the battle. The French forces consisted of some 10,000 foot soldiers made up of French infantry and *lansquenets* along with 1,100 heavily armored knights and 500 light horsemen. Having given up their siege of Novara when a Swiss army arrived to relieve Duke Maximilian Sforza, the French retreated a couple of miles from the city and pitched camp for the night, only to be surprised by an early morning attack by the hardy Swiss, who, having just finished a forced march the day before, nonetheless roused themselves and streamed out of Novara in search of their enemies like a hoard of "irascible bees."[17] They formed up in squadrons and fell upon the French from three sides, concentrating their attack on the backbone of the French army—the seasoned unit of German *lansquenets.* In just an hour or so, the two battalions of French infantry had abandoned the fight

17. On Novara, see Hans Delbrück, *The Dawn of Modern Warfare,* trans. Walter J. Renfroe Jr., vol. 4 of *History of the Art of War* (Lincoln: University of Nebraska Press, 1990), 79, from which this citation is taken. Delbrück's insightful study of warfare and society helped frame my research for this chapter in particular. Equally indispensable for its overview was Geoffrey Parker, *The Military Revolution: Military Innovation and the Rise of the West, 1500–1800,* 2nd ed. (Cambridge: Cambridge University Press, 1996).

in panic, leaving the *lansquenets* to face the Swiss forces alone. The *lansquenets* probably could have held on with help from the knights, but the knights were used to fighting independently, not as a tactical unit, and failed to mount a successful attack. The battle ended in a bloodbath for the *lansquenets* that brought shame to the French *gens d'armes.* Only forty knights perished in the battle, which led Italian commentators to accuse the French horsemen of cowardice.

What Novara illustrates is not just the importance of soldiers-for-hire to Renaissance warfare, but the cohesiveness upon which the strength of infantry squares depended. The foot soldiers could vanquish heavy horsemen thanks to corporate discipline, which they possessed and the horsemen did not. Indeed, of far greater importance to the military revolution than the advent of gunpowder warfare was the discipline that enabled the Swiss to mobilize large numbers of men in a coordinated fashion, for although the French and Italians amassed huge arsenals of cannon and harquebuses, no amount of fire power could compete with the tenacious Swiss squares.

The Swiss first began fighting with long spears in the fifteenth century as a defense against mounted knights. Formed up in deep squares that compressed into a bristling mass as they pushed forward into battle, horsemen found no easy way to engage these pikemen: the horses were impaled long before the knights could defend themselves (see fig. 5.3). But should the spear unit break formation, all was lost, for its advantage depended upon the force of numbers. Niccolò Machiavelli reports in his *Arte della guerra* that the Swiss killed conscripts who broke rank in battle out of fear (97). Fear itself was, indeed, their worst enemy.[18]

The courage of the Swiss solidified their tactical advantage against the horsemen and, later, against other infantry battalions. Yet their mutual trust and confidence, which emerged from years of fighting together, was hard to duplicate. Swiss conscripts surely drilled enough to be able to march in time to a drum, hold their place in formation, close ranks for battle, and turn to the right and left, but this is little more than might be expected of the city militiamen whom Arbeau likely took to be one audience for his marches. Emperor Maximilian I managed to bring together troops, arm them with spears, and drill them in close formation, creating the *lansquenets* that ultimately rivaled the Swiss. But Machiavelli, another enthusiast for foot soldiers, was never able to create a successful militia, and French commanders failed to discipline native infantry forces. The French continued to levy local troops as the century wore on, but the heart of the crown's infantry remained the Swiss

18. Niccolò Machiavelli, *Arte della guerra* (Florence: Giunta, 1521). I use the English translation, *The Art of War,* trans. Ellis Farneworth, rev. and introd. Neal Wood (New York: Da Capo Press, 1990). Where convenient, page references to this edition are given in parentheses in the body of my text.

FIGURE 5.3. Pike squadrons as seen at the Battle of Moncontour, 1569. Bibliothèque nationale de France, Cabinet des estampes, Qb1-1569

mercenaries that Francis I began to hire after Marignano, which included the halberd-toting guards that accompanied the royal family. Generally, French infantry companies were the last ones formed in a national crisis and the first to be disbanded. Incoherent and ineffectual, they filled the slots vacated by the peacetime garrisons of towns and other secondary commissions.

Infantries became fully effective only with the advent of standing armies and extensive training by drill. This development first occurred in the Low Countries under Maurice of Nassau and his cousin, William Louis.[19] Ruling together over an economically stabilized Netherlands, the princes were able to establish what was effectively the first standing army in early modern Europe.[20] Money was key to the

19. On the Princes of Orange, see Delbrück, *Dawn of Modern Warfare,* 155–171.

20. The *compagnies d'ordonnance* established by Charles VII in 1445 were permanent units that theoretically gave France a standing peacetime force of six thousand men (not including garrisons, which brought the total to close to fourteen thousand). Lynn estimates that these numbers did not in-

enterprise, for regular payment of the troops professionalized the military to a degree that had not been possible within the system of temporary levies and foreign mercenaries; but of even greater importance was the heightened discipline that could be achieved in an army with a uniform chain of command, continuity within units, a standard code of behavior, and reliable pay. Mercenaries often refused to fight when the money ran out, and, in any case, they disliked protracted sieges and campaigns that kept them away from home for long. Moreover, their preference to strike quickly and leave, coupled with the inability of the French and the Habsburgs to sustain their employment over long stretches of time, produced the feeble sort of victories that characterized Renaissance warfare, victories in which the winner found himself unable to pursue the war to its most decisive outcome precisely because the hired companies left and what gains had been made could not be capitalized upon. Such was true of the French Wars of Religion, for example, which proved so expensive that the crown could afford to keep the royal army in the field for only a few months at a time, which barely allowed for complete mobilization and produced the cycle of ineffective campaigns, ruinous expenditures, and unfavorable negotiations that underlay the series of eight civil wars waged between 1562 and 1584.[21] After each war, disbanding the army broke up most companies—including any in which a healthy camaraderie might have developed—while even at its peak numerical strength, the rabble levied to fight for the crown was often plagued by disorder and desertion brought on by severe material needs that could turn whole companies into dangerous mobs desperate for food and clothing.

Maurice built an army along wholly new lines. Companies were smaller, but the number of officers remained the same: each company of one hundred was commanded by a captain, a lieutenant, an ensign, two or three sergeants, three corporals, three runners, a captain of arms, a corporal of noble youths or privates first class, a scribe, a provost, ten privates first class, and at least two drummers. Johann Jacobi von Wallhausen reports these numbers in his *L'art militaire pour l'infanterie* (1615), a prime source of information on Nassau's techniques.[22] These smaller com-

---

crease significantly before 1659 (*Giant of the* Grand Siècle, chap. 2). If we consider that contemporaries estimated the peacetime strength of the States-General's army at something approaching forty thousand men (Delbrück, *Dawn of Modern Warfare,* 163), and that the area occupied by this army was only a fraction the size of France, we get a relative sense of how imposing Nassau's army was.

21. On the problems of the royal army see Wood, *King's Army.*

22. Johann Jacobi von Wallhausen, *L'art militaire pour l'infanterie,* trans. Theodore de Bry (Oppenheim: Uldrick Balck, 1615), 102. I use the French translation of Wallhausen's *Kriegskunst zu Fuss* (Oppenheim: Uldrick Balck, 1615) in order to confine my remarks to the texts of most influence on French speakers. Wallhausen was captain of the guards in Danzig and later ran a military school for nobles established by John of Nassau-Dillemburg in Siegen. For an assessment of his work's importance see Anglo, *Martial Arts of Renaissance Europe,* 285–290.

panies were superior in strength, Wallhausen explains, "for the fewer soldiers and the more officers one has, the better they are trained; . . . a captain with a company of 200 or 150 men can do as well as with 300 if he trains his soldiers."[23] Nassau's emphasis on leadership induced the beginnings of a professional officer corps and better command; it also meant that captains could prepare their companies for combat with a thoroughness theretofore impossible to achieve.

At the heart of the training was drill. The companies practiced forming up as quickly as possible; handling pikes, muskets, and halberds; marching in close and loose formations, marching at different speeds; turning; firing by ranks; and charging. They drilled every day, whether in garrison, camp, or the field. Unbelievers found the thought of "making exercises before the enemy" risible, but Wallhausen insisted that exercise represented the newest science.[24] From the time of their inception in the 1590s, Nassau's drills heralded a military revolution.[25] In the seventeenth century, French military academies employed masters to teach the handling of the pike and musket, making Nassau's military exercises a foundational element in noble education.[26]

Earlier theorists had recognized the value of drill, though emergency conscription permitted little more than getting troops to march in time. Machiavelli recommended that soldiers drill enough to be able to recognize drum signals and march to the beat, a suggestion that must have been novel given his lament that "at present, our drums are chiefly employed for making noise and parade."[27] The Baron of Fourquevaux, in his *Instructions sur le faict de la guerre* (1548), likewise recommended that new conscripts be drilled to trumpets and drums to hold rank in battle, and Arbeau insisted that drums gave soldiers "heart, daring, and courage" and that "without them, the men would march in confusion and disorder, which would place them in peril of being overthrown and defeated by the enemy. This is why our Frenchmen are instructed to make the rankers and bondsmen of the

23. Wallhausen, *Art militaire pour l'infanterie,* 102. "Car moins aurez vous de soldats & plus d'officiers, & mieux seront ils dressez, en quoy il consiste beaucoup, sçavoir qu'un soldat sache habilement manier son baston, tirer en bon ordre & se rechanger, . . . car . . . il [le Prince] tient ses soldats si habiles que c'est un plaisir d'escrimer avec eux, ce qu'un chascun capitaine ayant une compagnie de 200 ou de 150 hommes pourroit tout aussi bien faire comme avec 300 s'il dressoit ses soldats."

24. Ibid., 65.

25. A revolution that would have to wait for the Swedish commander Gustavus Adolphus to demonstrate its full potential in battle in the 1620s. See Geoffrey Parker, *Military Revolution,* 16–25, and Delbrück, *Dawn of Modern Warfare,* 173–183.

26. See O[1] 915 (32, 184, 187, 188) and O[1] 917 (211), Archives nationales.

27. Machiavelli, *Art of War,* 76. Also consider his observation that the drumbeat is "a direction to the whole army, which acts and moves in a certain measure and pace according to the different notes and sounds so that the army may know how to keep due time and order" (75–76). Machiavelli's recommendations on drill are threaded throughout his treatise (cf.. 61, 63–64, for instance), though they are never more precise as far as the musical component is concerned.

squadrons march to certain rhythms."[28] As modest as Arbeau's instructions were, disciplined marching may have remained something of an ideal in his time. At the turn of the seventeenth century, French military experts such as Louis de Montgommery still took pains to explain how to march correctly: soldiers should march to the drumbeat, begin with the left foot, end with the right, feet turned slightly out, the heel touching the ground first, and so forth, and there should be two drums, one to mark the beat and the other to play more elaborate patterns.[29] Soldiers should be exercised in formation, always with drums. It would seem that even these basics could not be taken for granted.

Regular marching gave Nassau's companies a mobility far surpassing that of the Swiss or *lansquenets,* and this, in turn, expanded the range of maneuvers he practiced with his troops, which could break into small formations, close ranks, and attack in a series of shallow echelons in what eventually became the standard mode of infantry deployment. The greater fluidity of movement likewise fostered cooperation between companies of harquebusiers, musketeers, and pikemen. The integration of forces was a problem in the era of deep squares, for the pressure from the rear that gave squares their strength left no room for marksmen, who needed time at the front to fire and space into which to retreat after firing. The solution was to form separate companies of marksmen, but unless they fired from under cover or with support, they were often overrun by horsemen, and even foot troops with close-combat weapons could rout them with little trouble. Schemes of rotation devised in the sixteenth century had helped the situation somewhat, but orderly cycles in which those in front fired and then stepped back to reload worked better in theory and in military displays than they did in battle.

By 1600 the harquebus had been around for almost two centuries, and the musket, with its longer barrel and armor-piercing lead ball, had been in use since 1523, but neither had fully turned the tables on chivalric combat, despite the conventional wisdom that the invention of gunpowder rewrote the rules of knightly war-

28. Raymond de Beccarie de Pavie, Baron of Fourquevaux, *Instructions sur le faict de la guerre* (Paris: Galliot du Pré, 1548). With subsequent editions in 1549, 1553, and 1592, and translations into Italian, Spanish, English, and German, this was a very influential treatise. It owes a large debt to Machiavelli's *Arte della guerra,* on which see the introduction to G. Dickinson, ed., *The* Instructions sur le Faict de la Guerre *of Raymond de Beccarie de Pavie, Sieur de Fourquevaux* (London: University of London, Athlone Press, 1954). For Arbeau see *Orchésographie,* fol. 7v; 20. "Le bruict de tous lesdicts instruments, sert de signes & advertissements aux soldats, pour desloger, marcher, se retirer: & à la rencontre de l'ennemy leur donne coeur, hardiesse, & courage d'assaillir, & se deffendre virilement & vigoureusement. Or pourroient les gens de guerres marcher confusément & sans ordre cause qu'ils seroient en peril d'estre renversés & deffaicts, pourquoy nosdicts françois, ont advisé de faire marcher les rencs & jougs des escouades avec certaines mesures."

29. Louis de Montgommery, *La milice françoise reduite à l'ancien ordre et discipline militaire des légions,* 2nd ed. (Paris: François Rousselet, 1610), 113–114.

fare.[30] Firearms were simply too unhandy. All had to be loaded with powder and a bullet, which required a good bit of coordination, and then the firing device had to be prepared by placing igniting powder in the small pan near the matchlock and threading the match—a smoldering piece of cord—into the mouth of the hammer. Dampness could nix the entire process, and marksmen had to be careful to keep the burning match away from the powder while loading. It was Nassau's weapons drills in particular that modernized his army, bringing discipline to the use of the musket and harquebus and exploiting a tactical potential that had had to wait for these innovations in training before they could be realized. The human factor proved crucial to the advancement of gunpowder warfare, far more so than the introduction of guns in and of themselves.

This new technology inspired a flurry of treatises that hawked Nassau's secrets to a hungry public in the early years of seventeenth century. The first was Jacob de Gheyn's large book of engravings with command words for each movement, which was commissioned by Maurice's cousin, John of Nassau, and published in The Hague in 1607 in Dutch, German, Danish, English, and French, under the title *Maniement d'armes, d'arquebuses, mousquetz, et piques en conformité de l'ordre de Monseigneur le Prince Maurice;* in 1608 second editions were printed in The Hague, Amsterdam, and Frankfurt.[31] All carried privileges from the emperor, the king of France, and the States-General of the Netherlands.[32]

Although sumptuous, the folio engravings of Gheyn's treatise actually comprise a field manual for the captains who had to train inexperienced soldiers.[33] Each element of the drill is illustrated with an engraving keyed to commands such as "tenez le mousquet & enjouez" (hold the musket and aim, see fig. 5.4, although in this

30. See Delbrück, *Dawn of Modern Warfare,* 23–52.

31. I cite the second French edition (The Hague, 1608). On books emulating Gheyn's treatise and for a history of its publication see the commentary in Jacob de Gheyn, *Wapenhandelinghe,* facsimile ed. with a commentary by J. B. Kist (Lochem: De Tijdstroom, 1971).

32. It is worth mentioning the unusual set of privileges and the format of the book's title page, which left blank the spaces that were to be filled with dedications and coats of arms, for at the very least they presume a high level and pan-European spread of interest in Maurice's ventures. So, too, the pirating of Gheyn's treatise by Wilhelm Hoffmann (1609) and Adam van Breen (1618), the anonymous knock-off printed in a pocket format under the title *Mars, his Field, or the Exercise of Armes* [1611], the anonymous *The Military Discipline wherein is most Martially showne the order of Drilling for y[e] Musket and Pike* (London: Roger Daniell, 1623), and Wallhausen's many manuals on Nassau's infantry practices attest to the universal renown of Netherlandish weapons drills. Even Jacques Callot published a lovely little set of *Les exercices militaires* (1635). For a bibliography of Wallhausen's publications, see Anglo, *Martial Arts of Renaissance Europe,* 374.

33. It surely interested the captains of local militias as well. Masters of pike drills seem to have made tours through France instructing civilians in the new techniques. See Pussot, *Journalier,* 206.

FIGURE 5.4. Musket drill "Tenez le mousquet & enjouez." In Jacob de Gheyn, *Maniement d'armes* (1608). By permission of the British Library, shelfmark 61.h.19

period aim can be spoken of only in the loosest sense of pointing the gun in a general direction; indeed, in sixteenth-century portrayals one regularly sees marksmen with their heads turned away from the gun). The command words are preceded by instructions in which the motions of every limb, foot, and finger are described, making it possible to recreate the drill with total accuracy. The harquebus drill contains forty-two commands that specify the position of the gun, its point of contact with the body—shoulder or leg—the hands used to hold it and the match, and the position of the feet, which exhibit dancelike turn-outs, positions, and equilibrium. Gheyn's choreography is infinitely more sophisticated than anything in Arbeau, and with good reason, for not only was he attempting to teach new choreographies to a group of readers who had probably not already seen the drills, but the whole was performed with the risk that at any moment the match might ignite the powder by accident.

When the sequences had been learned, captains could drill for speed and accuracy using the brief command words. Here we return to the realm of the musical, for the command words created a verbal tempo that elicited coordinated responses of the sort that one sees in present-day military inspections. We can imagine the regular cadences of these drills and rhythms of the commands, which probably marked the beat upon which each movement was to be performed with preparatory rhythms like the modern command "company, present arms" (♫♩ 𝄾𝄾𝄾♪♩ ♩).

Gheyn's treatise contains an explicitly musical protocol in the commands for the pike drills, where the number of *temps,* or movements, required to perform each command is clearly indicated (see fig. 5.5). Each *temps* is shown in a separate engraving, yet some commands such as "carry the pike aloft" include three movements—three numbered engravings—that go together. For example, "portez la Pique hault" requires three *temps* to complete (see fig. 5.6, movement no. 4). Other commands, such as nos. 19, 20, and 21, require only one *temps.* Not only do the smallest components of the drill in this way refer to musical beat, but the pike drill in its final form uses an abbreviated set of command words "sans temps de reprises" that bear a regular rhythm of repeated imperatives: "portez la Pique hault," "plantez la Pique," "baissez la Pique," "remettez la Pique," "trainez la Pique," "posez la Pique contre le pied." They must have been spoken in time just as their execution was measured out in regular *temps.*

Gheyn's pike drills enhanced the tight ranks required by pike squadrons and enabled pikemen to handle longer spears—perhaps as long as fifteen feet—by which they acquired a significant advantage over infantry armed with shorter weapons.[34]

34. See Delbrück, *Dawn of Modern Warfare,* 53–54. Also see Wallhausen, *Art militaire pour l'infanterie,* 82–83.

Et comme il ſe trouue, que le Soldat ne peult promptement & bien replanter ou mettre bas la Pique, ſans l'experience de la ſcauoir bien manier & reprendre, pour l'eſclaraſſement de cecy, ſont ordonnez certains mots de commandement auec leurs temps de repriſes, qui correſpondent en ſuyte ſur les figures, a fin que les Capitaines puiſſent faire aprendre à leurs jeunes Soldatz & inexperimentez, ce maniement de la Pique autant quils en auront de beſoing, mais ayant l'experience, le Soldat ſera exerçé, par les motz de commandement cy apres mis à part, & ſans temps.

Motz de commandement.

Auec leurs temps de repriſes.

1. La Pique eſtant plantée, portez la hault en trois temps.

2.le premier
3.le ſecond } Temps.
4.le troiſieſme

Remettez, ou plantez la Pique.

5.le premier
6.le ſecond } Temps.
7.le troiſieſme

Mettez la Pique ſur l'eſpaule & la portez platte.

8.le premier
9.le ſecond } Temps.
10,le troiſieſme

11. Portez la Pique debiais.

Baiſſez ou preſentez la Pique par la porte.

12.le premier
13.le ſecond } Temps.
14.le troiſieſme

Remettez ou plantez la Pique.

15.le premier
16.le ſecond } Temps.
17.le troiſieſme

Portez la Pique haulte en trois temps.

18. Ces trois temps n'ont qu'une figure marquée nombre 18. pour ce que les deux aultres ſont monſtrez cydeuant.

Baiſſez ou preſentez la Pique en vn temps.

19. Vn ſeul temps.

20. Prenez la Pique à la pointe.
21. Trainez la Pique &

Preſentez la Pique en palmiant.

22. Il n'eſt beſoing d'obſeruer cecy.
23. En trois temps, comme ſe monſtre.
24. Par l'enſeignement nombre 24.

25. Poſez la Pique contre le pied droit, & tirez l'eſpée.

26. Ayant la Pique ſur l'eſpaule, preſentez la en ariere en trois temps.

27.le premier
28.le ſecond } Temps.
29.le troiſieſme

b Remettez

Figure 5.5. Command words with and without the *temps de repriſes.* In Jacob de Gheyn, *Maniement d'armes* (1608). By permission of the British Library, shelfmark 61.h.19

FIGURE 5.6. Pike drills. In Johann Jacobi von Wallhausen, *L'art militaire pour l'infanterie* (1615). By permission of the British Library, shelfmark 534.m.14

But all had to march with the same alternation of left and right—Montgommery and Arbeau are quite specific on this point—and move their weapons in unison. Gheyn's special list of commands with the *temps de reprises* for the pike drills was meant to ensure that the soldiers could perform the movements "promptement & bien" (swiftly and well) and points up the importance of rhythm to the use of spears in close combat. Indeed, the rhythmic component of the pike drills made marching a natural way to practice them. Drums may well have provided the *temps,* for each company of one hundred men included at least two drummers. Wallhausen describes an exercise in which the troops march in close formation "toute bellement, ce que le tabourin monstrera," moving faster and faster according to the drumbeat and finally lowering their pikes as if to enter combat.[35] And Montgommery, who also describes the Dutch drills, insists that the soldiers should exercise to the order (*ordonnance*) of the drums.[36] Gheyn's pike drill is of particular interest because it was not a static choreography in the manner of the gun drills but a vocabulary of motions keyed to command words that could be employed in any sequence. Gheyn recommends that once conscripts know the routine, the captains practice giving the commands in a variety of orders, "according to what the occasion requires." The drill, then, rehearsed a basic vocabulary for battle, to be executed on the move to the cadence of field drums.[37]

The same was true of the exercises in which musketeers practiced firing by rank. After learning to load and fire safely in close quarters, musketeers mobilized to the beat of drums, fired simultaneously by rank or file, and then retreated to the rear of the squadron to reload while the next line of soldiers advanced (see fig. 5.7, upper right-hand quadrant). "Take care," Wallhausen admonishes readers, "that they [the musketeers] aim together, fire together, and move off together so that no confusion follows."[38]

The musical point is that this choreography was performed in time: drums beat the *ordonnance,* forming the troops and ordering their movement according to rhythmic commands. The social point that follows from this is that drummers formed a crucial link in the chain of command, for which reason Wallhausen and Montgommery devoted whole chapters of their treatises to the fifes and drums.[39] Montgommery recommends two drummers for each company, one an ordinary player who works with a fife player and the second a *tambour colonnel* who serves as their captain. The drum colonel, who has a valet or assistant drummer to carry his drum, carries a wooden baton with which he beats the drums himself only when necessary. He must know several foreign languages, be cunning but not quarrelsome or talkative, and be faithful above all. He lodges with the sergeant major, to

35. Wallhausen, *Art militaire pour l'infanterie,* 83.

36. Montgommery, *Milice françoise,* 113–114. Wallhausen published a translation of Montgommery's treatise under the title *Militia Gallica,* which may account for some of the consistency between Wallhausen's work and Montgommery's. But clearly both men admired Nassau's army: Wallhausen makes this explicit in his title, Montgommery on p. 103 and ff. of his treatise ("Les evolutions & les exercises qui se font en la Milice de Hollande").

37. Ibid., 108. "Et de mesmes apres les avoir presentees soit en avant ou en arriere les faut aussi accoustumer de les pouvoir remettre en toutes les façons cy dessus mises: soit marchant ou pied ferme, ou en portant leurs picques qu'ils les baissent toutes à l'hauteur de trois pieds de terre ou peu moins."

38. Wallhausen, *Art militaire pour l'infanterie,* 53. "Prenez garde aussi qu'ils enjouënt ensemble, tirent ensemble, s'en aillent ensemble, afin qu'il ne s'ensuive aucune confusion." Also see Montgommery, *Milice françoise,* 111–112.

39. Wallhausen, *Art militaire pour l'infanterie,* 151–152; Montgommery, *Milice françoise,* 37–40.

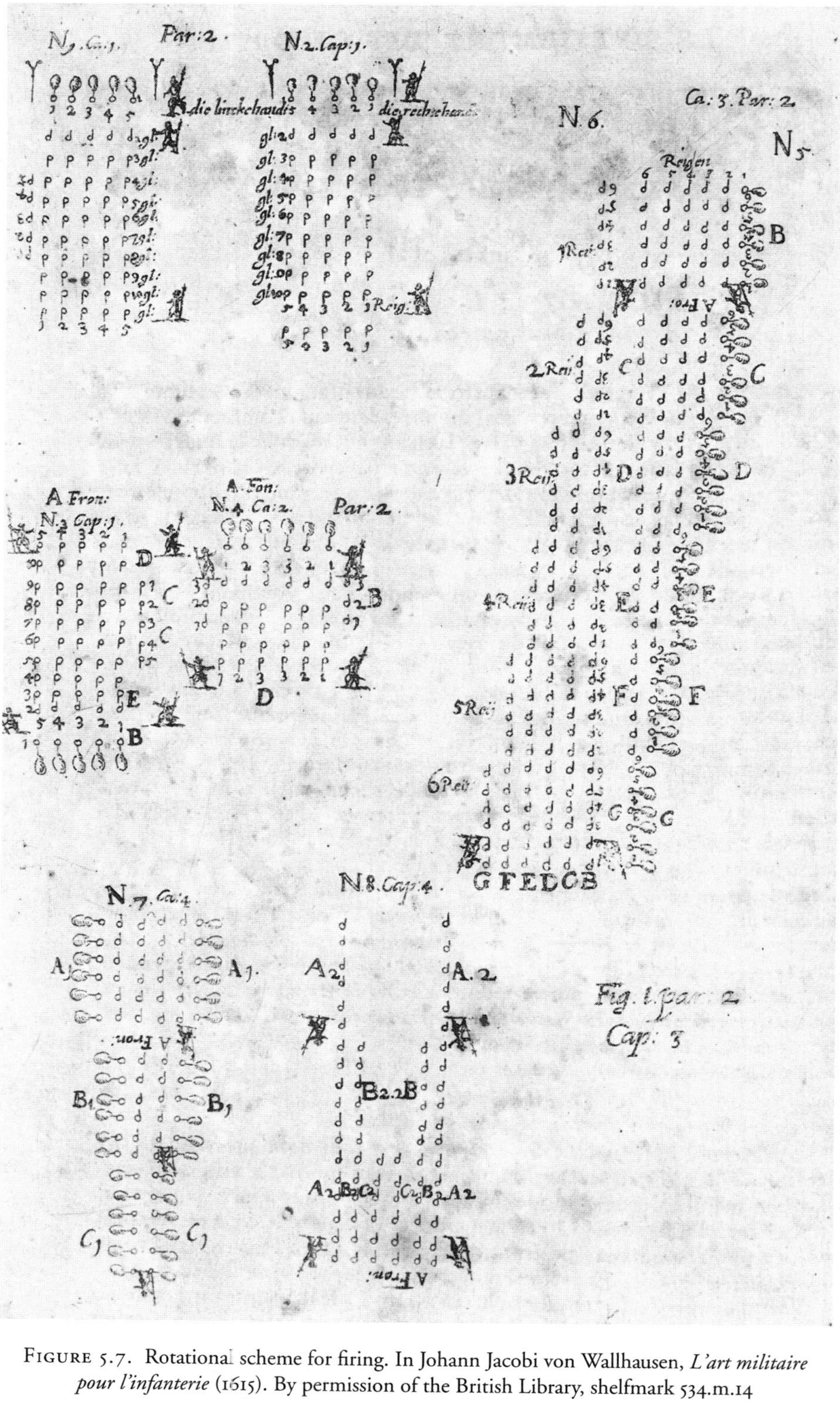

FIGURE 5.7. Rotational scheme for firing. In Johann Jacobi von Wallhausen, *L'art militaire pour l'infanterie* (1615). By permission of the British Library, shelfmark 534.m.14

whom he reports morning and evening, for not only does he relay orders from the sergeant major, he acts as the sergeant major's right-hand man. In fact, the lead drummer functioned as a sort of noncommissioned officer or aide-de-camp, operating independently from the rest of the company—he had to be adept at spying and reconnaissance and an able envoy (hence the need for foreign languages).[40] The drum colonel's visceral command of the company—like a conductor with his baton—was only the most audible manifestation of his elevated rank.

We should also note how seriously Wallhausen and Montgommery treat the patterns played by the drummers. In the first place, the drums beat out an acoustic badge of national identity that was understood by all. "As for the drumming, the French way is better than that of any other nation, even though the Spanish mock it and say that it seems that the French drums play a branle sooner than a march," Montgommery remarks, "but anyone who listens closely to its cadence will find that it distinctly marks a slow step."[41] No doubt the Spanish found the ornaments favored by French drummers a bit over the top. If Arbeau's eighty-some drum rhythms are any indication, French military drumming was an elaborate combination of beats, rolls, and rim-shots, something corroborated by Montgommery's instruction to use two drums, one to beat the "marche simple" and the other to "faict les fredons."[42] Despite its potential impracticality, the French style clearly elicited a good deal of Gallic pride. Montgommery's contemporary, Nicolas Bergier, praised French military drumming quite explicitly in his treatise *La musique speculative.*[43] Soldiers, Bergier tells us, march to a pyrrhichianapaestus rhythm (four shorts and a long), and when they assault, the drummers switch to a strict pyrrhic rhythm, which, with its redoubling of short beats, quickens the soldiers' ire and inspires courage. The use of the pyrrhic meter by French armies, he continues, is the true mark of their inflamed spirits, for drummers in other armies have chosen other rhythms—clearly the *wrong* rhythms—that reflect their particular (lesser) natures. In sum, the pyrrhic is an emblem of French military virtue.

Of course, all armies used distinct drum tattoos to issue orders.[44] Not only did drums coordinate troop movements and communicate over great distances, but the

40. On spying see Wallhausen, *Art militaire pour l'infanterie,* 151.

41. Montgommery, *Milice françoise,* 38. "Pour la batterie, la Françoise est meilleure que de nulle autre nation: bien que les Espagnols s'en mocquent, & disent qu'il semble que les Tambours François sonnent plustost un branle, que la marche. Mais qui considerera bien sa cadence, trouvera qu'elle marque distinctement le pas grave, tel que le soldat doit observer marchant en bataille ou entrant en garde."

42. Ibid. It is probably safe to presume that the *fredons* Montgommery speaks of are the rolls played by the drummers, to which Arbeau assigns the syllable "fre."

43. Bergier, *Musique speculative,* 192–194. Bergier may be citing *Orchésographie.*

44. See the long passage in Wallhausen, *Art militaire pour l'infanterie,* 152.

signals operated according to codes usually known only to those meant to receive them. This is one reason drummers were sent on reconnaissance missions, for the best among them knew the enemy's commands.[45] Only the Germans, Wallhausen laments, employed the same drum call for everything, which meant that orders to decamp and so forth had to be cried, making it ridiculously easy for spies to discover their plans.[46] French armies, by contrast, used a series of calls and relayed the signal to break camp with a cloth over the drum to dampen the sound.[47]

In a general sense, then, drums conditioned soldiers to respond at levels both muscular and mental. They regulated daily routines as well as engagements, serving as the army's timepiece, alarm, and engine. William Barriffe sums it up nicely when he says that "the Drum is the voice of the Commander, the spurre of the valiant, the heart of the coward, and by it they [the soldiers] must receive their directions when the roring Canon, the clashing of armes, the neighing of horses, and other confused noise causeth, that neither Captaine nor other Officer can be heard."[48] Above the great din of battle, enormous side drums measuring two and a half feet deep, the same in diameter, and built to be heard a league away created order where otherwise only confusion would have ruled.[49]

Field drums clearly impressed the young René Descartes during his tenure with the army of Maurice of Nassau, for in the *Compendium musicae* (a treatise written in the army's winter quarters at Breda) he lauds "that Military Instrument, the *Drum*" as being able to stand alone and provide "a certain *Delectation,* . . . so great is the force of *Time* in Musick."[50] Descartes's description of the way "a sound doth concusse, or shake all circumjacent bodies" seems inflected by his experience of military drumming; likewise, his idea that strong beats excite the body to motion aptly captures the effect of Nassau's weapons drills.[51] Indeed, in Descartes's paradigm the reflexes conditioned by music have the same effect on animals, whose base passions react in kind when struck by the strong beats of dance music:

> We may well affirm . . . by consequence [of music's concussing effect], that *Beasts* may dance to number, or keep time with their Feet, if they be taught and accustomed

---

45. Wallhausen expects that the regiment's top drummer will know "bien jouër, à la façon de toutes les nations" (ibid., 151).

46. Ibid., 152.

47. Montgommery, *Milice françoise,* 39.

48. William Barriffe, *Military Discipline: or the Yong Artillery Man* (London: Thomas Harper for Ralph Mab, 1635), 11.

49. On side drums see Arbeau, *Orchésographie,* fols. 6v–7v. For an earlier view, see the painting *The Three Soldiers* by Pieter Breughel the Elder, 1518 (Frick Collection, New York).

50. See Descartes, *Oeuvres,* 10:95. I quote the first English edition, *Compendium of Musick,* 7.

51. Descartes, *Compendium of Musick,* 6; *Oeuvres,* 10:94–95.

> thereto; because to this, nothing more is required, then only a mere naturall *Impetus*, or pleasant violence.[52]

We might see in this remark a hint of Descartes's coming fascination with mechanism, but it may also be that the young philosopher posits a higher truth about physical education. For what dance and drill did so effectively was to train the body to react consistently to regular stimuli, turning animals and raw conscripts into automata.

Descartes asserted that the repetitive impression of sensations on the brain caused predictable reactions to the same stimulus. Writing to Mersenne, he said he supposed that "if a dog were whipped five or six times to the sound of a violin, as soon as it heard this music again it would begin to cry and run away," attributing to physical "training" an element of cognitive memory.[53] In man the process worked the same: sensory stimulation produced a cognitive process as well as an immediate motor response (this was different from a strictly reflexive motor response that did not involve any recognition whatsoever, such as recoiling from pain). Hence, in the letter just cited, Descartes remarks that "for those who have taken pleasure in dancing to a certain tune, the desire to dance returns when they hear the same music again."[54] But the human mind is not consciously engaged. This conditioned reflex works "as if the mind is elsewhere" in humans. In such cases, the mental impressions made by stimuli "cause our limbs to make various movements, although we are quite unaware of them. In such a case we too move just like automata."[55] Of course, the point of military drill was to get soldiers to set aside their emotions and individual desires and simply to respond mechanically to commands.

The repetition of regular motions to the sound of drums did indeed produce a sort of mindlessness in conscripts even as it heightened esprit de corps and instilled courage. In *Keeping Together in Time: Dance and Drill in Human History,* William McNeill theorizes that drill produced feelings of exaltation, well-being, and social

---

52. Descartes, *Compendium of Musick,* 6; *Oeuvres,* 10:95.

53. Descartes, *Oeuvres,* 1:134. "La mesme chose qui fait envie de danser à quelque-uns, peut donner envie de pleurer aux autres. Car cela ne vient, que de ce que les idées qui sont en nostre memoire sont excitées: comme, ceux qui ont pris autrefois plaisir à danser lors qu'on joüoit un certain air, si-tost qu'ils en entendent de semblable, l'envie de danser leur revient; au contraire, si quelqu'un n'avoit jamais oüy joüer des gaillardes, qu'au mesme temps il ne luy fust arrivé quelque affliction, il s'attristeroit infailliblement, lors qu'il en oiroit une autre fois. Ce qui est si certain, que je juge que si on avoit bien foüetté un chien cinq ou six fois, au son du violon, si-tost qu'il oiroit une autre fois cette musique, il commenceroit à crier & à s'enfuïr." See Augst, "Descartes's Compendium on Music," esp. 129–132.

54. Descartes, *Oeuvres,* 1:134.

55. See the discussion of *L'homme* and of this letter to Plempius in Stephen Gaukroger, *Descartes: An Intellectual Biography* (New York: Oxford University Press, 1995), 276–290. I quote from Gaukroger's translation (282).

cohesion among the participants, creating an emotional resonance that made it safe for Nassau to "arm even the poorest classes, pay them a pittance, and still expect and secure obedience."[56] McNeill's human argument is that drill produces what he calls "muscular bonding" or "the primal upwelling of collective solidarity."[57] Essentially, dancing or marching together to music breaks down boundaries between individuals: the rhythmic exertion induces a trancelike state in which social bonds are established and strengthened.

In McNeill's analysis, Nassau drilled his soldiers with the goal of beating automatic, unthinking obedience into them and unwittingly succeeded in producing social discipline as a side effect. But contemporary treatises make it clear that military men understood the moral implications of rhythmic drill, which should not surprise us in an age so deeply concerned with the nature of the passions and the effects of music and dance upon them: "Obedience . . . conserves virtue and good morals, maintains friendships, prevents all sorts of evils and pillaging . . . [and] maintains all *ordonnances*. . . . It teaches the beautiful virtue of fidelity, and above all, it maintains the ranks and orders that are the only cause of victories."[58] These are the words of Jean de Billon, the most prolific French military theorist of the early reign of Louis XIII and an avid student of Nassau's methods, particularly as developed by Justus Lipsius (to whom I will turn momentarily). Billon articulates the moral and social benefits of obedience in terms of *ordonnances,* which we could read as order or orders, as formations or commands, or even as the drum signals used to convey them. Billon, for one, understood the technological and social advantages produced by obedience, which were, in the end, inseparable from one another. Well did the proponents of the new military discipline recognize that the same muscular bonding that enabled soldiers to stick together under attack helped them forge peaceful friendships back in camp. For the first time since the fall of the Roman empire, European powers amassed standing armies so vast and so well gov-

56. William H. McNeill, *Keeping Together in Time: Dance and Drill in Human History* (Cambridge, Mass.: Harvard University Press, 1995), 3; on Nassau's drills see 127–132. McNeill's book excited some controversy relating to the history of synchronized combat. See the review of John Keegan, "Keeping in Time: The Rise of Foot Drill and the Decline of the Minuet," *Times Literary Supplement,* July 12, 1996, 3–4, and the month of letters that followed in that journal. On musical synchronization see John Spitzer and Neal Zaslaw, *The Birth of the Orchestra,* 526–527.

57. McNeill, *Keeping Together in Time,* 132.

58. Jean de Billon, *Les principes de l'art militaire* (Rouen: Jean Berthelin, 1641), 12–13. "Elle conserve la vertu & les bonnes moeurs: Elle garde les amitiez: Elle empesche toutes sortes de malheurs, & pilleries: Elle faict estimer un homme, & luy cause des recompenses quand on a cognu sa patience à bien servir: Elle fait garder toutes Ordonnances"; "Elle apprend cette belle vertu de fidelité. Et sur tout elle fait garder les rangs & les ordres qui sont la seule cause des victoires."

erned that their social order rivaled anything found in civil society. Indeed, the research into Roman military practices that stood behind the pyrrhic revival and—as we shall see—behind rhythmic drill was inspired by an equally pressing desire to revive the strengths of imperial political organization.

LA MUSIQUE DES ANCIENS

In the marching and close-order drills just examined, we find the pairing of music and synchronized fighting that was so thoroughly mythologized as pyrrhic dance. Maurice and his sergeants probably did not think of marching in terms of the pyrrhic—though Arbeau and Bergier would have—but they certainly did associate it with the musical combat of the ancients. Machiavelli's *Arte della guerra* makes this connection clear.

The organization, training, and deployment of the infantry—a relatively new wing of the army—was largely conceived in imitation of the Macedonian phalanx and Roman legion. Beginning with Machiavelli, writings about the infantry referred to Xenophon's *Cyropaedia,* the *Histories* of Polybius (c. 200–c. 118 BCE), the *Tactica* of Aelianus (1st–2nd century CE), and a work from the late imperial Roman author Vegetius, whose retrospective treatise on the Roman army, *Epitoma rei militaris,* had enjoyed unbroken popularity from the time of its writing (between 383 and 450 CE) throughout the Middle Ages and into the Renaissance.[59]

Machiavelli's primary interest was in foot soldiers, who he believed would form the military arm of the future. During the Florentine Republic he campaigned long and hard to establish a native militia, drafted a militia ordinance that was passed in 1506, and finally became the Secretary of the Nine of Militia, which allowed him to enlist and train ten thousand men. His passion for a strong infantry was less prescient than anachronistic, based as it was on his desire to revive the ancient phalanx, but despite the rose-colored spectacles through which he read Aelianus and other classics, and despite the military failures of his militia, the experience from which he spoke in his *Arte della guerra* made it a valuable reference, one declaring that the infantry would restore to Italy the greatness she had known under the Roman emperor Augustus. In numerous editions and translations, Machiavelli's praise of ancient military practices reached a wide readership well into the seventeenth century.[60] In France his ideas were propagated by Fourquevaux's *Instructions sur le faict*

59. On Machiavelli's sources and on his military career see Neal Wood's introduction to Machiavelli, *The Art of War,* which I gloss here.

60. The *Arte della guerra* was the only one of Machiavelli's major works to be published in his lifetime, and it enjoyed numerous editions throughout the sixteenth century, as well as translations into French (1546), English (1560), and a pirate version of the text in Spanish by Diego de Salazar (1536).

de la guerre, the anonymous *Institution de la discipline militaire* (1559), and Montgommery's *Milice françoise* (1603), the full title of which makes the importance of ancient military discipline clear: *La milice françoise reduite à l'ancien ordre et discipline militaire des légions telle et comme la souloyent observer les anciens François à l'imitation des Romains et des Macédoniens.* Machiavelli may also have inspired Guillaume Du Choul's vividly illustrated treatise on Roman military discipline, *Discours sur la castrametation et discipline militaire des Romains* (1555).[61]

Machiavelli describes three sorts of exercises in which the Romans drilled their infantry: the first instilled endurance and agility and included sports such as racing, leaping, and wrestling; the second trained soldiers to handle arms and included the martial arts, especially a form of practice sparring in which the soldiers' arms and armor were weighted with lead to maximize strength training; and the third drilled soldiers to keep their ranks and obey orders, whether marching, in battle, or in camp. This third sort includes the close-order marching seen across the board in successful infantries from the Macedonian phalanx to the Swiss squadron, but the particulars of marching together receive scant mention from Machiavelli, who even suggests that the Romans left new soldiers to learn the physical order of marching thorough a process of osmosis that happened on the hoof (58–59). He seems to have believed ancient military music worked directly on the soul in an almost mechanical process that—unlike Descartes's characterization—did not require special training to produce its results. Rather, the drums caused the army to move "in a certain measure and pace according to the different notes and sounds" as a result of modal ethos and its power over the passions (75–76).

Machiavelli rehearsed several Renaissance commonplaces about modal ethos, relating how the Dorian mode fills men with resolution and the Phrygian excites fury in warriors, and even how Alexander the Great was so ravished at the sound of a Phrygian march that he drew his sword as if about to charge an enemy. What distinguishes Machiavelli's telling of these stories is the rhythmic dimension he adds to the diatonic qualities of the fabled modal melodies. Dorian and Phrygian songs become marches in his telling, complete with ancient "measures" or rhythmic modes. He calls for the revival or reinvention of these varied marches and their instigation in drills in order to expand the range of military "affects" available to infantry commanders.

A fascination with Phrygian marches might seem a far cry from the humanistic projects that would ultimately give birth to opera in Italy and *ballet de cour* in France, but at their deepest level of intent, all those who hoped to harness the pow-

61. Louis XIII enjoyed the *Castrametation des Romains* (Lyons: Guillaume Rouillé, 1555) from the age of four. Though it must have filled in as a picture-book for the young king, it put the images of Roman discipline on his horizon from a very early age. See Héroard, *Journal,* 1:147–154.

ers of the Greek modes were drawn to them in a search for ways to discipline the body and its passions. Machiavelli correctly perceived that discipline made the imperial Roman army great and gave the Swiss—"the only people with any trace of the ancient military institutions left among them"—their edge in battle. For discipline creates order, and from order springs courage: "good order makes men bold; confusion makes them cowardly" (61). Machiavelli concluded that "for this purpose [to keep due time and order], the ancients had their pipes, fifes, and other sorts of military music perfectly adapted to different occasions; for just as a man dancing keeps time with the music and cannot make a false step, so an army properly observing the beat of its drums cannot be easily disordered" (76). In this analysis, marches proved to be the secret behind the power of Greek and Roman infantry, marches cued so carefully to a variety of musical modes that one might well call them dances of war. Between Machiavelli, Arbeau, and Bergier on the one hand, and Descartes on the other, we see how music came into play both in old systems of thought (such as Neoplatonism and language-based reasoning) and new ones (such as mechanism and number-based reasoning), occupying a place at once classic and modern. In the first paradigm, music possessed extraordinary affects that struck diversely upon the soul and could be used to direct the body; in the second, the pulses of music could condition the body's reflexes and elicit predictable responses on command. To some extent, then, Descartes is not so new, particularly when one adds to the scientific viewpoint the likeness of this conditioning to the repetitive style of instruction employed by the educators who taught dance, handwriting, fencing, music, and manners. Civility itself made the same transition music did from being a deeply moralized system to a more mechanical one.

The physical order regulated by the drumbeat represented only the most palpable manifestation of an entire culture of discipline that Machiavelli proposed for his citizen militias, and it was this attention to discipline, to the human factor, that made the *Arte della guerra* still well worth reading at the end of the sixteenth century. Although sixteenth-century European weaponry vastly exceeded that of the Romans as far as firepower was concerned, ancient texts by Vegetius, Polybius, Xenophon and others continued to offer evidence of a human technology based on discipline that had barely been tapped by Renaissance generals. Indeed, Machiavelli's reforms remained just as imaginary as the battle he staged in the *Arte della guerra* until the establishment of Prince Maurice's army in the Netherlands, which, because it was a standing army, made possible not only physical discipline but also the regularized self-governance and military law that gave discipline its social dimension.

This is not to say, however, that Maurice and William Louis used the *Arte della guerra* as the blueprint for their new armies. Rather, their organization represented a wholesale rethinking of warfare—another Renaissance, if you will—based on ancient models. Maurice had studied with Justus Lipsius at the University of Leiden, whose *Politicorum libri sex* (1589) and *De militia Romana* (1596) turned the humanist's love of classical texts to the most pragmatic ends in a program geared to equip the soldier-prince with an armory of political, intellectual, and military abilities. According to Anthony Grafton and Lisa Jardine, Lipsius aimed to instill in young aristocrats "the self-control to endure religious war without despairing, the art of maintaining one's influence over subordinates, and technical mastery of the most vital craft of all, the soldier's."[62] Lipsius's *Politicorum libri sex* instructed the novice ruler with classical axioms conveniently updated to reflect contemporary political needs and included one book, titled *Military Wisdom,* that articulated how one might pursue them.[63] Lipsius received a handsome sum of money from the States-General after its publication. *De militia Romana,* which followed, offered a philological study of sources on the Roman army even as it called for greater military discipline in modern regiments. This was to be achieved through exercise, drill, order, laws, and *opus,* or hard labor such as digging trenches, which the ancients believed was part and parcel of military service.[64] The whole aimed to reproduce the strength, toughness, and courage of ancient soldiers, which Machiavelli associates with manly *virtù* (*vir* = man). These books became wildly popular—the *Politicorum* went through over fifty editions in Latin alone—but beyond their evident emphasis on the military craft, their importance rests in the place Lipsius secured for discipline more generally in contemporary political science.[65] Lipsius's recuperation of ancient military practices in this way elucidated the political benefits of discipline at court as well, reinforcing the reciprocality between civility and the control of armed force.

De militia Romana did not end Lipsius's project, for he was keen to see his theories into action by corresponding with the princes and their colonels. William Louis was himself an especially avid scholar of Greek and Roman sources and together with Maurice had the information contained in ancient texts translated into practice down to the very letter, for even the *motz de commandement* for the drills were based on translations of Greek and Latin commands.[66] We find the command

62. Anthony Grafton and Lisa Jardine, *From Humanism to the Humanities: Education and the Liberal Arts in Fifteenth- and Sixteenth-Century Europe* (Cambridge, Mass.: Harvard University Press, 1986), 198.

63. See Gerhard Oestreich, *Neostoicism and the Early Modern State,* ed. Brigitta Oestreich and H. G. Koenigsberger, trans. David McLintock (Cambridge: Cambridge University Press, 1982), 50–55.

64. See especially the fifth book of *De militia Romana,* in which Lipsius compares the Roman manner of waging war to contemporary methods.

65. See Grafton and Jardine, *From Humanism to the Humanities,* 198–199.

66. Delbrück, *Dawn of Modern Warfare,* 159.

words repeated verbatim in Montgommery's *Milice françoise,* a book that not only records the French interest in ancient military practices, but one that—along with the treatises by Gheyn, van Breen, Wallhausen, Billon, and so many others—promoted Nassau's new style of command.[67] Small wonder that European educators added Lipsius and a bit of drill and gunnery to their curricula,[68] or that young aristocrats such as Descartes flocked to the Netherlands to spend some time in the orbit of the great Prince of Orange, whose army had become a finishing school for those who wished to study the new discipline firsthand. As one of the most effective institutions of its age, it was a natural model for Europe's future leaders.

The intense scholarship behind all of these efforts, from Machiavelli's militia to Montgommery's reorganization of the French army after Roman and Macedonian models, shows how advanced the "science" of arms was in the sixteenth century—even as an object of philology—and how connected it was with broader intellectual and political currents. It also suggests that it is time to retreat from the field of battle in order to examine other sites in which humanists reconstructed, rehearsed, and renewed the practices of the ancients in performance. For while before 1600 one does not find pitched battles with scores of soldiers advancing to the pyrrhic feet of drums, one does indeed find experiments with music and arms being tried out in military displays and entries. It is to the festive birthplace of the new military discipline that we now turn.

CITY CULTURE AND THE PYRRHIC REVIVAL

In France the pyrrhic was first revived around 1550 as part of royal entries, where it called to mind the triumphs of Alexander the Great and Julius Caesar. It can be distinguished from the dance of the *bouffons, matachins, morescas,* and other native sword dances by the attire of the dancers, who wore the short tunics and light armor of ancient warriors.[69] But it represented far more than just a change of costume for traditional sword dances, for in it we first sight the Roman discipline cultivated in military drills at the end of the century. Based perhaps on readings of Machiavelli and certainly indebted to classical sources, these performances elucidate the role humanists played in advancing the science of arms. Indeed, the tremendous sums of money, armies of craftsmen, and months of rehearsal often expended on royal en-

67. Montgommery, *Milice françoise,* 103–107, 124–127.

68. Grafton and Jardine, *From Humanism to the Humanities,* 198–199.

69. In 1556, Guillaume Paradin (*Le blason des danses* [Pairs: Didot, 1830], 16) likened the pyrrhic to the dance of the buffoons with its acrobatics and fancy swordplay in time to the music; Arbeau would of course make the same equation in *Orchésographie,* titling his sword dance the "Danse des bouffons" even while introducing it as a pyrrhic.

tries made mounting such a performance not unlike mobilizing for war, and the command of these vast human and economic resources gave the entries' poet-generals an opportunity to demonstrate the strengths of a humanistic regime. Mustering the city's leaders, guilds, youth groups, collegians, religious, and militia in their finest array, entries were a form of civic self-fashioning, in which humanists reinvented cities in utopian forms that did not neglect the virtues of military might.

In 1548 the poet Maurice Scève organized a pyrrhic for the entry of Henry II into Lyons.[70] The twelve performers were members of the Enfants de la Ville, a high-status youth group made up of the sons of municipal magistrates. In addition to supporting a confraternity, the Enfants regularly turned out for weddings and civic ceremonies, and for this entry they had purchased sumptuous gladiatorial costumes and rehearsed daily for four months.[71] The large group of which they were a part was divided into an *avant-garde,* a *bataille,* and an *arrière-garde* like contemporary armies; the Enfants who performed the pyrrhic made up the *bataille,* while their colleagues in the *arrière-garde* performed one of the horse ballets studied in the next chapter.[72] The pyrrhic lasted for half an hour, during which the gladiators waged furious attacks that, according to all accounts, frightened spectators with their force and amazed them with their coordination. What appeared on the surface to be an aggressive melee was, in fact, a choreographed dance.

Margaret McGowan believes that Scève relied on Dionysius of Halicarnassus (c. 54–57 BCE) as a source for his reconstruction.[73] In the *Roman Antiquities* VII.72, Dionysius relates specific information about the instruments used for the dance, its costumes, and the arrangement and movements of the dancers.[74] According to Dionysius, the Romans performed the dance in the processions that opened major games. Numerous bands of men, youths, and boys paraded in three divisions—no

70. See *La magnificence de la superbe et triumphante entrée de la noble & antique cité de Lyon* (Lyons: Guillaume Rouillé, 1549), fols. B4v–C4r; the analysis by McGowan, "Renaissance War Dance"; and Richard Cooper's introduction to Maurice Scève, *The Entry of Henry II into Lyon, September 1548* (Tempe, Ariz.: Medieval and Renaissance Texts and Studies, 1997), 19–20, 37–40 (this work is a facsimile of the 1549 print). The entire visit and the music provided for it are discussed in Frank Dobbins, *Music in Renaissance Lyons* (Oxford: Clarendon Press, 1992), 109–117.

71. See Jean Tricou, "Les Enfants de la Ville," *Bulletin de la Société littéraire, historique et archéologique de Lyon* 14 (1937): 107–138, and Jean Guéraud, *La chronique lyonnaise de Jean Guéraud, 1536–1562,* ed. Jean Tricou (Lyons: Imprimerie Audinienne, 1929), para. 46. Also see Davis, *Society and Culture,* 111, 113–114.

72. On such battle formations for entries see Graham and Johnson, *Paris Entries,* 20–21.

73. McGowan, "Renaissance War Dance," 33.

74. Dionysius of Halicarnassus, *The Roman Antiquities,* trans. Earnest Cary, 7 vols., Loeb Classical Library (Cambridge, Mass.: Harvard University Press, 1937–1950), 4:365–371.

doubt a ceremonial version of the actual division of hoplites into three divisions according to relative age and their deployment in three echelons in the Roman legion. The bands carried short spears and wore scarlet tunics girded with bronze cinctures, swords at their sides, and helmets adorned with "conspicuous crests and plumes" of the sort gladiators donned in the blood sports so popular in Rome in that day and age. Small flutes, ivory lyres of seven strings, and other stringed instruments called *barbita* accompanied the dancers, who followed the movements of their leaders in the same way Roman soldiers obeyed their centurions. They danced with "warlike and rapid movements, usually in the proceleusmatic rhythms," which is to say, in feet of four short syllables known to incite action.

There is no mention of music for the pyrrhic in Lyons, nor much information about the costumes, but the chronicle provides specifics about the choreography. The combat *à l'antique,* the account explains, "was not antique because of the arms, but because of the order, of knowing how to support each other and how to move between their ranks without breaking them."[75] Certainly the variety of arms—which gave the spectacle a distinctly gladiatorial flair—complicated their attacks and must have demanded extra rehearsal, for spears fought against two-handed swords, two swords against one sword and a shield, one sword and a Bolognese dagger against a sword and a Barcelonan shield, and so forth. Facing off in four ranks, the first ranks fell back after fighting, allowing those behind them to renew the attack. Alternating "en ceste ruse d'ordre" (in this crafty order), spears broke to bits and swords flew to pieces. Even though the battle never became disordered, the whole scene frightened the spectators, who were unaware of the training involved.

The king was well pleased with the display, finding in it "une nouvelle mode de combatre & si dangereuse" (a new and very dangerous mode of combat) and asking for it to be repeated six days later (which it was, indoors).[76] Although "antique," the dance struck the king as new, for it posed a solution to one of the great problems facing infantry commanders: the coordination of forces. Not only did the pikemen need to hold together in battle, but if their squadrons could be made small and dependable enough, they could be deployed more flexibly, used to defend marksmen against cavalry onslaught, and dodge enemy artillery. The gladiators smashed very real arms to bits with great *fureur* and yet swung back into their original formation at the end, ready to attack again. Certainly the tactical implications of the variety of arms and the great destruction, combined with the composure of

75. *Magnificence de la superbe et triumphante entrée,* fol. C1r. "Là commencerent un combat à l'antique, non quant aux armes, mais quant à l'ordre de se scavoir secourir, & entrer les rangz les uns dans les autres sans se rompre."

76. Ibid., fols. C1v, L3v.

the dancers, impressed Henry. The dancers' ability to deploy in multiple formations, fight full out, and continually regroup made them worth three times as many soldiers who, after engaging the enemy, fell back in disarray. Through the revival of the ancient Roman inheritance of which Lyons was so proud, Scève rediscovered a truth about Roman military technology—and did so some forty years before Lipsius.[77] The living descendants of those conquering legions whose ancient edifices were still visible in the city reenacted the military virtue of their ancestors and the power by which Romans secured peace in civil society.

Here we should note that Roman power was twofold, legal and military, and that many of the Enfants' fathers practiced jurisprudence, which was still based on Roman law. The Enfants de la Ville represented not just imperial might but the Roman code at a time when the turn to litigation, rather than violence, as a means of settling local disputes brought with it greater respect for lawyers, judges, and officers of the royal courts. Given their parentage, we might also see in the Enfants' performance an arming of justice. Their participation in the entry emphasized the prominence of legal professionals and the Roman legal inheritance. "Gladiators" henceforth reappeared occasionally in royal entries in Lyons: in 1595 "les Gladiateurs & maistres d'escrime" fought before Henry IV, in 1600 they marched just before the queen in the entry of Marie de' Medici, and in 1622 Louis XIII beheld a pyrrhic at the Jesuit college as part of his reception in the city.[78]

Eight years after Henry II witnessed the pyrrhic in Lyons, Guillaume Paradin published his *Blason des danses* (1556) in nearby Beaujeu.[79] Its lengthy chapter on the pyrrhic confirms the dance's currency among the Renaissance scholars who read Plato and Xenophon, but Paradin's *Blason* is more than an encyclopedia entry of the sort pulled together by Hieronymous Mercurialis and Johannes Meursius. Like Henry II's reaction to the pyrrhic in Lyons, Paradin seeks the relevance of historical accounts to contemporary military practice. Most pointed is his identification of a common infantry maneuver in the classical choreography: "the Romans often saw this pyrrhic dance, in which two bands of young men ran, dancing against each other in arms and fighting, making snails [*faisoient limaçons*] and turnings in the

77. McGowan, "Renaissance War Dance," 33–36.

78. See [Matthieu], *L'entrée de tres-grand . . . Prince Henry IIII . . . en sa bonne ville de Lyon,* 18, 21; [Pierre Matthieu], *L'entrée de . . . Marie de Médicis . . . en la ville de Lyon* (Rouen: Jean Osmont, 1601), 47; and *Réception de très-chrestien, très-juste et très-victorieux monarque Louis XIII, roi de France et de Navarre, premier comte et chanoine de l'église de Lyon . . . par Messieurs les Doyens, Chanoines & comtes de Lyon en leur Cloistre & Eglise, le xi Decembre, 1622* (Lyons: Jacques Roussin, 1623), 49.

79. Guillaume Paradin, *Le blason des danses* (Beaujeu, 1556). I use the Parisian edition of Didot, 1830.

manner of a skirmish."[80] The *limaçon* or "snail" was a Swiss maneuver used to compact the loose ranks of cross-country marching into the close formation required for battle. The maneuver got its name from the impression it gave of a snail out of its shell, expanding and contracting as it inched along, and it seems to have been used to impress dignitaries in parades, though infantries also must have rehearsed "making the snail" for battle.[81] The capacity to move in and out of close order allowed the mercenaries to make, break, and reorganize ranks in developing situations. Arbeau tells drummers to shift to the intense shorts of the pyrrhic foot (˘ ˘) when approaching the enemy in order to get the soldiers to close ranks for battle, in this way equating pyrrhic meter with the music for the *limaçon.*[82] Machiavelli says outright that ancient militias practiced the *limaçon.*[83]

It is unsurprising, then, to find the *limaçon* in Renaissance pyrrhics, where this simple maneuver was inscribed with the same classical import attributed to it by military theorists.[84] During her entry into Tournon in 1583, Magdeleine de la Rochefoucaud beheld three martial dances performed by the students at the Jesuit college there, who made their second entrance "marching to the cadence of the lutes, sometimes joining one another, sometimes separating," just as in a *limaçon.*[85]

---

80. Ibid., 15. "Les Romains frequentoient fort ceste danse Pyrrhique, en laquelle deux bandes de jeunes hommes couroient, dansans les uns contre les autres armez, et combatans, faisoient limaçons et tournoiemens à la mode d'une escarmouche."

81. François de Rabutin, *Commentaires des guerres en la Gaule belgique* (Paris, 1555), ed. Charles Gailly de Taurines, 2 vols. (Paris: Henri Didier, 1932–1942), 1:8–9. Also see Delbrück, *Dawn of Modern Warfare,* 10–11, 17.

82. Arbeau, *Orchésographie,* fol. 17r; 37–38.

83. Machiavelli, *Art of War,* 65.

84. Guillaume Du Choul, *Discours de la religion des anciens Romains* (Lyons: Guillaume Rouillé, 1556), 74, and Claude Guichard, *Funerailles, & diverses manieres d'ensevelir des Rommains, Grecs, & autres nations, tant anciennes que modernes* (Lyons: Jean de Tournes, 1581), 180, both describe Roman pyrrhics as including the *limaçon.* Although the terms are the same, their descriptions—both equestrian—suggest not the infantry *limaçon* defined by Machiavelli, but a maneuver used by marksmen on foot or mounted that was most often called the *caracole* by cavalrymen. See figure 5.7, the upper right-hand quadrant of which depicts this maneuver on foot. Some descriptions of *limaçons* used in ballets do seem to resemble the rotations of the caracole. See, for example, that in Beaujoyeulx, *Balet comique de la Royne,* fol. 22v. "Elles se tournoyent aussi, faisans le limaçon au rebours les unes des autres, tantost d'une façon, tantost d'une autre, & puis revenoyent à leur premiere marque."

85. *La triomphante entrée de tresillustre dame Madame Magdeleine de la Rochefoucaud, Espouse de hault & puissant Seigneur Messire Just-Loys de Tournon, Seigneur & Baron dudict lieu, Comte de Roussillon, &c., Faicte en la Ville & Université de Tournon le dimenche vingtquatriesme du moys d'Avril 1583* (Lyons, 1583), facsimile ed. ed. Maxime Gaume, Images et témoins de l'âge classique 4 (Saint-Étienne: Presses de l'Université de Saint-Étienne, 1976). The description of the pyrrhics can be found on pp. 25–34 and 78. For the second entrée see p. 33 and n. 103 below.

In the Lyons entry of 1548 and the Tournon entry of 1583, the pyrrhic refracted old into new (Roman dance into "a new and dangerous mode of combat") and new into old (the *limaçon* into classical dance), and rarefied the most spectacular aspects of war by setting them to more elaborate music and enhancing their dancelike elements. The *limaçon* became ballet in a musicalization of combat that, far from abstracting it, enhanced its destructive capabilities. Though missing the captives, spoils of war, sacrificial beasts, and liberated soldiers who joined their Caesars on the way to the capitol, the ritual triumph of entries preserved the ceremonial reenactment of the primal battle upon which the monarchy was founded. In them pyrrhics staged the original triumph in a form that took music to be both a civilizing element in society and a source of enhanced military might.

One difference from those Roman triumphs, however, is that the players had changed. The Enfants de la Ville in Lyons and the Jesuit students in Tournon complicate our picture of the pyrrhic revival, for they were not soldiers. It is hardly surprising to find pyrrhics in royal entries, for their amalgam of music, classicism, and martial arts fit well with triumphal themes. But infantry did not perform them. Although Machiavelli praised the infantry and for a time Francis I called his battalions "legions," a social gulf separated foot soldiers from the noblemen who served on horseback. The term "infantry" itself is not unrelated to the names of youth groups such as the Enfants de la Ville, which sometimes formed the basis of urban militias in which townsfolk served.[86] In some intrinsic sense, then, the pyrrhic was revived by a culture of arms located in cities. It was an upper-class culture, certainly, for the Enfants paid handsomely for their costumes, and it did not exclude nobles, who boarded at Jesuit colleges before going on to study weapons drills at military academies and captain infantry companies. But it was civic in allegiance. Like Machiavelli, the aldermen and college rectors who supported these experiments with military discipline must have thought in terms of policing the city—not, as Henry II did, of the royal army.

Just as the French nobility became more militaristic during the religious wars, a variety of civic bodies took to arms in those years. Cities bucked up their militias, Catholic confraternities multiplied in order to enforce bans against practicing Protestantism within city walls, and—in the 1580s—cells of the radical League formed in Paris and other cities.[87] The League spearheaded Catholic militancy at the time of *Orchésographie*'s publication, and it is instructive to consider Arbeau's links to the League, which was particularly strong in his hometown of Dijon. Indeed, *Orchéso-*

---

86. Davis, *Society and Culture,* 111.

87. For an excellent investigation of piety and League violence in another provincial capital—Rouen—see Benedict, *Rouen during the Wars of Religion,* esp. chap. 8. In Rouen as in Dijon, Jesuit and League sentiments often conjoined.

graphie was dedicated to the son of his nephew, Estienne Tabourot des Accords, a Dijonnais author of some note with strong League sentiments that offer context for the militarism of his uncle's dance treatise.[88]

Early in his career, Tabourot was made captain of the militia for the parish of Saint Jean and, eventually, parish alderman. As militia captain, he helped guard the city and conduct the watch and could imprison anyone who disturbed public repose. It was a position he would hold until his death in 1590, and one in which Arbeau's instructions on marching would have come in quite handy when training guards. But this was only the beginning. At the height of his career, Tabourot became king's councilor and *procureur du roi,* positions that drew him to the inner circles of civic power, where he allied with others who wished to assure the Catholic purity of Dijon, the provincial seat of ultra-Catholic Burgundy.[89] His publications, for example, show a strong loyalty to the Duke of Mayenne, who led the League beginning in 1588. Tabourot wrote a distich for Mayenne's entry into Dijon in 1574 and served as secretary for the assembly of the Burgundian Estates convoked by Mayenne in 1589 to recognize the Cardinal of Bourbon as king of France.[90] The city militia swore an oath of fidelity to Mayenne and promised to guard Dijon in the name of God, Roman Catholicism, and the crown.[91] Thus the instructions on marching in *Orchésographie* seem intended to aid the militias that mobilized in the name of the League and the *ecclesia militans.*

Tabourot's affiliations also extended to Burgundian confraternities. As early as 1567, the lieutenant governor of Burgundy, Gaspard de Saulx-Tavannes, founded the Confraternity of the Holy Ghost, the first of a series of quasi-chivalric orders that would multiply and eventually feed the League. These "shock troops of the church militant," as Mack Holt has described the Confraternity of the Holy Ghost, forced hundreds of Huguenots to abjure.[92] Though we do not know if Tabourot was a member, he published a long poem celebrating the baptism of the son of Jean de Saulx-Tavannes, who founded the Ligue de Saint Esprit, and his militia served their cause when Mayenne came to power in 1589.

Finally, Tabourot's Catholic zeal played itself out in associations with the Jesuits:

---

88. For details of Tabourot's biography see Pierre Perrenet, *Estienne Tabourot, sa famille et son temps* (Dijon: Éditions du Raisin, 1926) and, more recently, Estienne Tabourot, *Les bigarrures du Seigneur des Accords (premier livre)* (2nd ed., Paris, 1588), facsimile ed., ed. Francis Goyet (Geneva: Droz, 1986), viii–xvi.

89. Burgundy was particularly resistant to Protestantism, on which see Mack P. Holt, "Wine, Community, and Reformation in Sixteenth-Century Burgundy," *Past and Present* 138 (1993): 58–93.

90. Tabourot, *Bigarrures, premier livre,* xi, xv, xviii, 140B, 139F.

91. Holt, *French Wars of Religion,* 146.

92. On the confraternity see ibid., 67–68; on Dijon under the League see ibid., 145–147.

in 1581, Tabourot and Pontus de Tyard were made executors of an estate that endowed the establishment of the Jesuit college in Dijon. Thus Tabourot proved instrumental in bringing the militant order to the city.[93] At the time Arbeau dedicated *Orchésographie* to his grandnephew, Guillaume, both Guillaume and his younger brother Pierre (aged fifteen and twelve or thirteen in 1588) were enrolled at the Jesuit college in Dijon, the very college their father, Estienne, had helped establish.[94] The *abécédaire* quality of *Orchésographie* made it a perfect textbook for Jesuit colleges, where dancing was sanctioned and where the book's all-male pyrrhic would have been particularly appreciated.

In some sense Arbeau dedicated his treatise to Jesuit collegians everywhere and to those who would teach young Catholics to raise the sword against heresy. The military exercises it contained perfectly supported the Catholic mission of the brothers as well as their ethos of Christian soldiering. It would be nice if we knew more about the college in Dijon where Arbeau's great-nephews studied and where *Orchésographie*'s pyrrhic might have been taught, but in the absence of such information, we can turn to the Jesuit college in Tournon, about which more is known and where pyrrhics were certainly performed. In fact, the pyrrhic in *Orchésographie* bears a striking resemblance to one described in the *livret* from the 1583 Tournon entry, which may have been one of Arbeau's choreographic sources. And Arbeau's interest in the activities at Tournon should not surprise us, for next to the Collège de Clermont in Paris, Tournon's was one of the most distinguished Jesuit colleges in France.[95] In 1584 it had chairs in logic, physics, and metaphysics, and two in theology, not to mention instruction in Greek and Hebrew; fifteen hundred students, of families both noble and bourgeois and from as far away as Ireland and Milan, studied there.[96]

Alongside rigorous instruction in the classroom, the brothers at Tournon carefully directed recreation time, some share of which was devoted to the performing arts. As at other Jesuit colleges, students celebrated Twelfth Night, Carnival, and Saint John's Night with masquerades, tragedies, and comedies that drew avid crowds (which in 1588 led to the banning of such entertainments in Paris). They

93. Tabourot, *Bigarrures, premier livre,* xii.

94. Ibid., xi–xii. On Arbeau's relationship with Guillaume, whose first instruction he oversaw in Langres, see Perrenet, *Estienne Tabourot,* 22.

95. On the Jesuit colleges in France see Huppert, *Public Schools in Renaissance France,* 104–115; on Tournon see 106–108. The college was evacuated in 1562, 1568, and 1570.

96. See Compère and Julia, *Collèges français,* 1:696–700. It is difficult to know about adjunct faculty at the colleges because the students often paid for lessons in dance, music, drawing, arms, and even writing themselves. See Camille de Rochemonteix, *Un collège de jésuites aux XVIIe et XVIIIe siècles: le collège Henri IV de La Flèche,* 2 vols. (Le Mans: Leguicheux, 1889), 2:23–24.

wrote verse, prepared costumes, set choreography, and rehearsed.[97] The program for the entire 1583 entry was designed by two students at Tournon, the budding poet Honoré d'Urfé and his brother, who took part in the performance as well.

In Tournon, moreover, unusual allowances were made for music.[98] The rector was Michel Coyssard, whose four-part settings of catechistic texts were considered in the previous chapter. The students at Tournon learned how to sing in order to participate in liturgical services, and they sang "figured" music as well, which we know because they performed a four-part ode during the 1583 entry.[99] Students were probably also recruited to play lute for the entry, for the pyrrhics took place in a theater erected in the courtyard of the college, where a massed lute band would have been needed to project over the swordplay *en plein air.* With a large number of lutenists using plectra and small lutes doubling larger ones at the octave, the ensemble would have made a nice, martial racket.[100]

In addition to these more artistic activities, the Jesuits encouraged physical recreation, which included fencing, dance, ninepin, bowls, lawn tennis, and prisoner's base, an outdoor team game upon which one of the dances from 1583 was modeled.[101] Such sports clearly played into the Jesuit's love of pyrrhics, as we can

---

97. See ibid., 2:10ff., esp. 47–48, on the visit of Father Lorenzo Maggio to the Collège de Clermont in 1588 and the manuscript instructions prepared at that time.

98. In many Jesuit colleges, liturgical singing was not allowed, but in Lyons and Tournon, it was preserved during a review in 1571, because it had become a tradition in those colleges. See Culley, "Musical Activity in Some Sixteenth-Century Jesuit Colleges," 6, and on Jesuit attitudes toward training students to sing liturgical music see Kennedy, "Jesuits and Music," 73–81.

99. The ode is "Puis que de Just de Tournon le renom," in *Triomphante entrée de tresillustre dame Madame Magdeleine de la Rochefoucaud,* 28. One should not dismiss the possibility that students in this period learned the lute or viol at college. Descartes clearly knew something of how to play the lute and flute, which he may have learned at La Flèche, and the painting of Étienne Bergerat leading a student viol consort in the performance of a motet for Louis XIII (Musée des Beaux-Arts, Troyes) likewise suggests that some instrumental performance was encouraged by the brothers. At the German College in Rome, it was decided in 1564 that "those who will know something about singing or about playing will be able to practice" during recreation time (Culley, "Musical Activity in Some Sixteenth-Century Jesuit Colleges," 4). But some Jesuits discouraged the study of instruments, on which see the report from Prague cited in Kennedy, "Jesuits and Music," 80. Coyssard seems to have been personally involved in the spectacles mounted for La Rochefoucaud, for the published account of the entry closes with a sacred ode that subsequently appeared in his *Hymnes sacrez.*

100. Small lutes are very bright and penetrating and would have carried well in a courtyard. The music may also have been scored for lutes of various sizes, for doubling at the octave increases the volume dramatically.

101. There was a field for prisoner's base at La Flèche, where students played this game, ninepin, bowls, and lawn tennis. See Geneviève Rodis-Lewis, *Descartes: His Life and Thought,* trans. Jane Marie Todd (Ithaca: Cornell University Press, 1998), 232–233, n. 19.

see from their choreographic similarities, yet they were only one aspect of an intellectual climate that encouraged combat at every turn. During their weekly and monthly debates students sparred verbally in teams named for the Romans, Spartans, and Trojans and led by decurions. The *Ratio studiorum* regulating all the Jesuit colleges laid out guidelines for these "jousts," "contests," or *concertationes,* and Estienne Tabourot describes them with glowing admiration in an essay on the education of children.[102]

In sum, the combative attitudes fostered by the Jesuit debates found a festive counterpart in the ancient sword dance. Jesuit colleges probably specialized in pyrrhics because of the prohibition against women and female characters in their theater pieces, it is true, but the militaristic ethos of the Jesuits also encouraged the religious lessons that might be learned from dramatizing strife. The 1583 production, for example, thematized the very credo of the Society of Jesus, which took its name from a military order founded in 1450 to fight the Turks and spread the Christian faith. After a band armed with Turkish bows and arrows performed their musical version of prisoner's base, bands of Moors, savages, and satyrs armed with swords marched forward to the cadence of the lutes and began a *moresca,* "hitting each other in time to the sound of the instruments, now singly, to the whole measure, high and low in a square against two at a time, now interwoven at the half-measure in a circle, against six at a time."[103] Like Arbeau's pyrrhic, this one included an entrance march, squares in which the dancers wheeled from right to left in order to engage two opponents in turn, and the stances and blows of fencing. But the pyrrhic in Tournon emphasized the clash of opposing forces by employing exotic

---

102. See Estienne Tabourot, *Les bigarrures du Seigneur des Accords, quatriesme livre* (Rouen: Jean Bauchu, 1591), fols. 6r–7r, 23r–25r; Allan P. Farrell, *The Jesuit Code of Liberal Education: Development and Scope of the Ratio Studiorum* (Milwaukee: Bruce, 1938), 274–275; Aldo Scaglione, *The Liberal Arts and the Jesuit College System* (Amsterdam and Philadelphia: John Benjamins, 1986); Rochemonteix, *Collège de jésuites,* vol. 2; Henri Fouqueray, *Histoire de la Compagnie de Jésus en France des origines à la suppression, 1528–1762,* 5 vols. (Paris: Bureaux des Études, 1910–1925), 2:691–720; and Gaukroger, *Descartes,* 38–45.

103. *Triomphante entrée de tresillustre dame Madame Magdeleine de la Rochefoucaud,* 33–34. "Les Mores, Sauvages, & Satyres s'en viennent l'espée au poing, faisans une fort plaisante entree, démarchans à la cadance des luts, tantost s'accouplans, tantost se separans: puis tous ensemble commencent à jouër la Moresque, se frappans d'accord au son des instrumens, maintenant simple, à mesure entiere, haute & basse en carré contre deux à la fois: maintenant entre-lassee à demy mesure en rond, contre six à la fois, tantost de taille, tantost de revers: & à la parfin d'estocade, se meslans avec une merveilleuse dexterité les uns avec les autres." On Christian triumph as a traditional theme of the *moresca* see Daniel Heartz, "Un divertissement de palais pour Charles Quint à Binche," in Jacquot, *Les fêtes de la Renaissance,* 2:329–342. The satyrs probably came from Dionysius, who relates that in Roman pyrrhics the soldiers were often followed by mocking bands of satyrs. On the similarity between Arbeau's pyrrhic and the *moresca* at Tournon, see Heartz, "Divertissement à Binche," 337.

costumes that visibly divided the dancers into teams or armies. This was not the fencing of individuals, but a war.

The Jesuits cast their battle in a land beyond Christian sectarianism, which is hardly surprising given that they had been driven from Tournon more than once by angry Huguenots and were forced to accept Huguenot students in 1576.[104] Nonetheless, the cultural alterity implied by the *morescas* could not hide their celebration of the very immediate and very Catholic victories of the Baron of Tournon, who, as announced in d'Urfé's panegyric ode, fought with the royal army at the sieges of Issoire and La Mure in the 1570s and, closer to home, suppressed the peasant uprising in the Dauphiné in 1580. It took little imagination to relate fighting Turks and Moors to Europe's liberation of Jerusalem and the local crusades to rid France of heretics. The pyrrhics acted out the theme of an *ecclesia triumphans* that fit well with the Counter-Reformation militancy inciting local confraternities and city militias in league with the Holy Union.

TIMELY as such pyrrhics were, the usefulness the brothers found in the ancient sword dance outlived the Edict of Nantes. In fact, pyrrhics gained even greater prominence in Jesuit theater of the seventeenth century.[105] During state visits, they demonstrated the good formation of collegians, and by 1622 they were such a standard of Jesuit ceremonial that when Louis XIII made his entry into Lyons that year, he asked the rector of the Jesuit college if there would not be a pyrrhic—a spectacle they produced a few days later.[106] The subtext of a Christian army remained, but the integration of pyrrhics into the disciplinary programs of the order, particularly

104. See Fouqueray, *Histoire de la Compagnie de Jésus en France,* 2:31.

105. On ballet in seventeenth-century Jesuit theater, see McGowan, *Art du ballet de cour,* 205–227.

106. In 1614 Louis stopped at La Flèche, where the students greeted him with a Parnassian spectacle. Apollo, Pallas, nymphs, and sylvans recited felicitations before his throne on a grassy knoll, after which the sylvans—who could hardly have been much younger than their king—"exécutent une danse champêtre et simulent un combat à coup de massue" (Fouqueray, *Histoire de la Compagnie de Jésus en France,* 3:337). In 1622 at Avignon, Jesuits designed Louis's entire entry, for which their college staged a musical "Duel of Just Rigor and Clemency" that we can presume featured pyrrhics. The airs were written by a Monsieur Intermet, a local canon, and they so ravished the king that the parts were torn from the hands of the musicians in order that copies could be made for him. See *La voye de laict ou le chemin des heros au palais de la gloire ouvert à l'entrée triomphante de Louis XIII . . . en la cité d'Avignon le 16. Nov. 1622* (Avignon: J. Bramereau, 1623), 8–9 on the organization of the entry, 221–222 on Intermet (about whom nothing else is known), and 265 on the pyrrhic.

For the pyrrhic in Lyons see *Réception de très-chrétien, très-juste et très-victorieux monarque Louis XIII,* 49. An overview of theater and ballet at the college is given in Pierre Guillot, *Les Jésuites et la musique: le Collège de la Trinité à Lyon, 1565–1762* (Liège: Mardaga, 1991). On the cramped conditions in the hall where the pyrrhic took place see 192–193. For a discussion of the ballet within the context of the other spectacles staged for the royal visit see McGowan, *Art du ballet de cour,* 217–218.

in the "postwar" era, had much more to do with the civilizing process and the goals of a new French state than it did with Catholic apologetics. It is to the Jesuit brand of civility that we now turn.

JESUIT CIVILITY

At the outset, the Jesuits' arch-Roman bias limited their influence in France. Clermont, Tournon, and Dijon counted among the few colleges they were able to open before 1600. One obstacle to their mission was the French system of municipally administered colleges, which was healthy enough that French cities had less need for Jesuit educators than cities in other countries. Another was anxiety about the brothers' connections with Spain and Rome and their version of Catholic politics. The Jesuits were known to incite violence—one notorious preacher was so incendiary that he "tore his garments into shreds, threw down his hat, and drooled from his mouth as he got more and more excited" during his political sermons—and after the coronation of Henry IV they gained a bad reputation for preaching against the king.[107] Following the assassination attempt against him in 1595, Henry accused the Jesuits of complicity in the plot and expelled the order from the country.

But in 1603 a remarkable turnaround occurred. Henry realized that he would never be able to gain control over the provincial colleges as long as they were run by municipalities, which tended to resist royal intervention.[108] And so he invited the Jesuits to return to France and administer a new series of colleges that would answer directly to the crown. The Jesuits agreed. Their flagship institution was the College of Nobles at La Flèche, which Henry established in one of his family châteaux. So tight were the ties between Henry and La Flèche that he asked that his heart be buried there when he died; following his assassination in 1610, his heart was interred at the college in an elaborate ceremony that the brothers continued to commemorate annually. The Jesuits were thus quite literally close to the heart of the Bourbons in their fulfillment of royalist educational aims.

The Jesuits were a perfect choice to head educational reform, for their pedagogical style brought military discipline directly into the service of a monarchy seeking greater control over the nobility. The papal bull that established the society in 1540, the *Regimini militantis ecclesiae,* made explicit the bellicose model at its foundation. Loyola the soldier took a military structure for his organization, one whose discipline, not incidentally, impressed Justus Lipsius as a model for that of the new army devised by Nassau. Loyola's spiritual exercises thus found their way into the mod-

107. Huppert, *Public Schools in Renaissance France,* 108n. 14, from which this quotation is drawn.
108. Ibid., 104–115.

ern army, and the Jesuits formed a link between collegiate, military, and ecclesiastical life.[109]

The discipline of the Jesuits depended on the regulation of every aspect of life in college according to the *Ratio Studiorum.* The hours of teaching, recreation, meals, and rest were strictly maintained, courses of study never varied, and even the games that could be played during free time were controlled. By setting up a system of student monitors (a highly desirable position), no action, no utterance, no moment of the day or night escaped surveillance, and it was in this way that Jesuit colleges achieved a kind of civility that proved effective in both civilian and military institutions. Although in one respect this created a highly theatrical environment in which every act constituted a performance, theatrical productions and ceremonies were still the ultimate showcases for the physical and rhetorical polish practiced at every turn, and in them we can observe the contribution of military discipline to a civilizing process that employed musical means.

In 1622 La Flèche pulled out all the stops to celebrate the canonization of Saint Ignatius of Loyola and Saint Francis Xavier.[110] Crowds gathered in the courtyard to see the students perform a tragedy enlivened by "excellent music" and "ingenious and nimble pyrrhics" that brought great applause.[111] Even more illustrative of the Jesuit style was the great bonfire set the following evening:

> Fifty students, richly attired in their theatrical costumes, came out of one of the halls, each carrying a white wax candle, and in the middle of a great number of people made the *limaçon* around the fire in the order [*ordonnance*] of an infantry company, having their captain, lieutenant, ensign, and sergeants and halting at various points to say some verses in praise of Saint Xavier.[112]

The pensioners marched as soldiers of Christ in a display so explicitly designed to project military might that they not only adopted the outward form of an infantry

109. See Oestreich, *Neostoicism and the Early Modern State,* 54.

110. *Le triomphe des Saints Ignace de Loyola fondateur de la Compagnie de Jésus, et François Xavier apôtre des Indes, au collège royal de la même compagnie, à La Flèche* (La Flèche: Louis Hébert, 1622), reprinted in Rochemonteix, *Collège de jésuites,* 2:219–250. Also see ibid., 2:47–48.

111. Ibid., 2:241, 244. "Jamais on ne vit mieux faire; l'éclat des habits tout chargés d'or, la musique excellente, les pyrrhiques ingénieuses et lestes, ne furent rien au prix de la naïveté de ceux qui représentaient, desquels les larmes furent accompagnées de celles de leurs auditeurs et de leurs applaudissements" (244).

112. Ibid., 2: 247. "Cinquante pensionnaires, richement vêtus de leurs habits de théâtre, sortirent d'une des salles, ayant chacun un flambeau de cire blanche en la main et au milieu d'un monde de peuple, firent le limaçon autour du feu, en ordonnance d'une compagnie de gens de pied, ayant leur capitaine, lieutenant, enseigne, sergents, et faisant halte en divers endroits pour dire quelques vers à la louange de saint Xavier."

company but also executed the *limaçon.* Likewise, at the pyrrhic ballet performed for the entry of Louis XIII into Lyons later that year, Jesuit students snapped from one formation into another at bullet commands like those used in infantry drills. "En coing . . . En croissant . . . Divisez-vous . . . Frappez . . . En fleur de Lys," their captain called out, producing a sequence of geometrical formations only slightly more fanciful (and demanding) than the experimental forms being tested by Nassau's army (see fig. 5.8).[113] The choreography closely approximated the tactics of the new infantry, which relied on swiftness and cohesion. It also pointed toward the massive infantry formations that would be used in seventeenth-century field warfare. Gone were the unruly frays that Blaise de Monluc claimed typified war in his day, the "fights, encounters, skirmishes, ambushes, an occasional battle, minor sieges, assaults, escalades, captures, and surprises of towns" of the early civil wars.[114] Louis no doubt saw in the pyrrhic the technology that would support his own predilection for field battles, a style of warfare that was later advanced by Turenne, who began studying with the Princes of Orange when he was only thirteen. "Make few sieges, and fight plenty of battles," declared Turenne, who mobilized French forces in open combat.[115]

The *limaçons* and geometrical formations bring home the extent to which the Jesuits actually practiced cutting-edge military discipline. The hierarchy of officer-instructors and pensioners, the system of rewards and punishments, the exercises both physical and mental, the constant activity, and above all the self-discipline promoted by the Jesuits comprise far more than a superficial resemblance to the order of the modern army: both aimed to achieve excellence and obedience using the same methods. Indeed, Gerhard Oestreich has observed that Lipsius's years under the Jesuits decisively influenced his interpretation of Roman discipline and the ways it might be instituted in a standing army.[116] Taking the two basic elements of the Jesuit structure, *exercitia* and *constitutiones,* Lipisius incorporated them into the classical concept of military discipline as *exercitia* and *ordo.* The *ordonnance* of the

113. *Réception de très-chrétien, très-juste et très-victorieux monarque Louis XIII,* 51–52.

114. Quoted in Geoffrey Parker, *Military Revolution,* 41. See his chapter 1 for a concise account of the revolutions in siege and field warfare between 1500 and 1650, the shift from horsemen to infantry, and the effectiveness of weapons drills. By contrast, David Parrott, *Richelieu's Army: War, Government, and Society in France, 1624–1642* (Cambridge: Cambridge University Press, 2001), argues that whereas battles were more common in the Wars of Religion, siege warfare dominated the French army's practices during Richelieu's ministry (56–59). Nonetheless, both Turenne and Condé preferred battles to sieges, and overall, combat in the open field predominated before 1650. See Lynn, *Giant of the* Grand Siècle, 530–532.

115. Quoted in Lynn, *Giant of the* Grand Siècle, 531.

116. Oestreich, *Neostoicism and the Early Modern State,* 54.

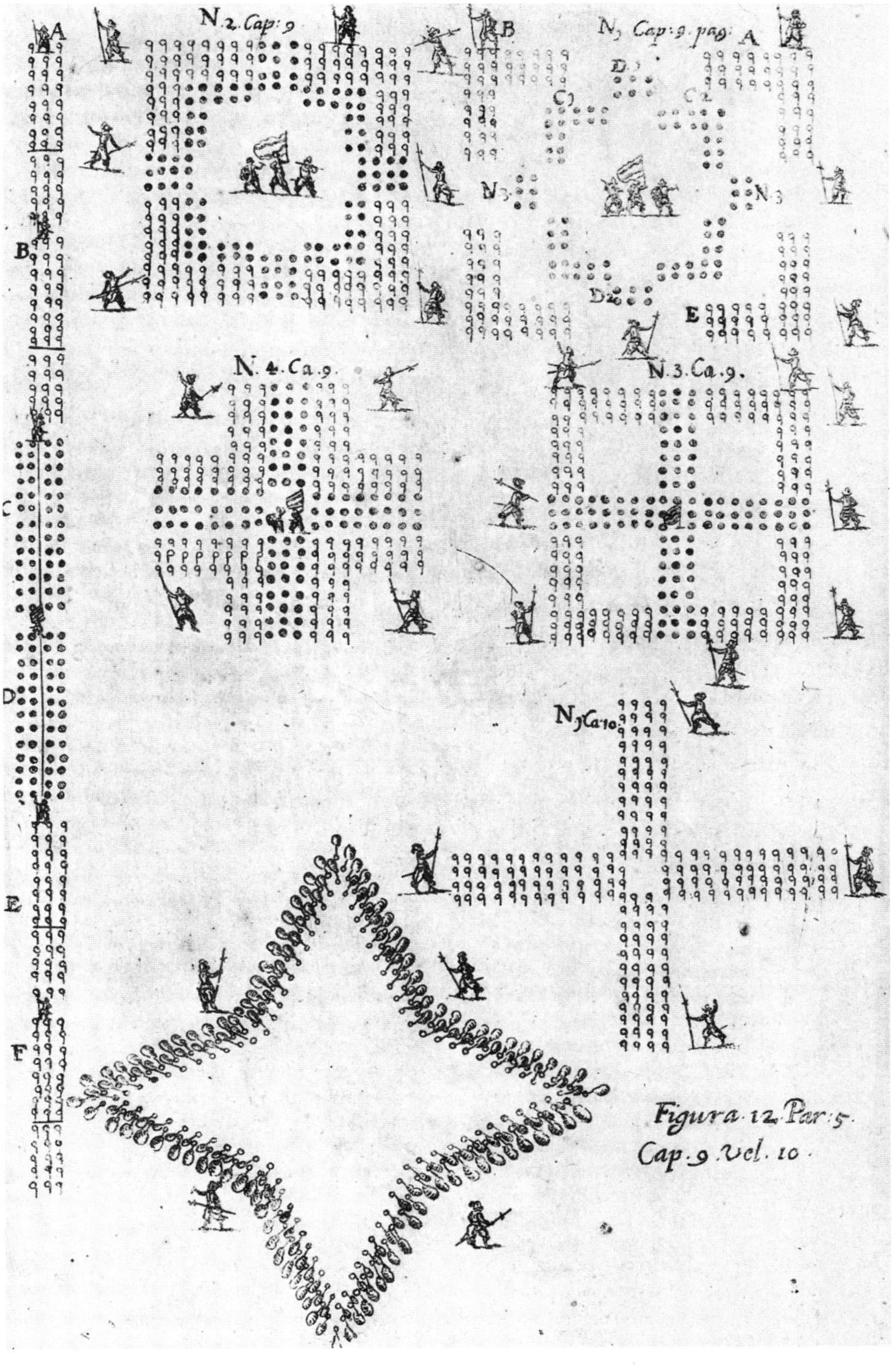

FIGURE 5.8. Geometrical formations for battle. In Johann Jacobi von Wallhausen, *L'art militaire pour l'infanterie* (1615). By permission of the British Library, shelfmark 534.m.14

Jesuit procession—its chain of command, its order of march, and the *limaçon* undertaken in it—taught students to serve well as civilians or officers, to know their place and to fulfill their duties with grace and self-restraint. Whether they made their ways from the College of Nobles to Pluvinel's academy, to the army of Maurice of Nassau (as Descartes did), or to the court of Louis XIII, the Jesuits equipped them to prevail in a system demanding intellectual, spiritual, and physical poise and musical discipline.

After the geometrical exercise described above, three battalions armed *à la romaine* waged a frightful battle. To music that was "royalement martiale" provided by six trumpets, the students fought aggressively, steel flashing and crashing, and soldiers falling from wounds so convincing that some spectators rushed to finger the blunted edges of the cutlasses afterward. "All were so skillfully hardened that one saw neither slack courage nor foolish step."[117] Like the other accounts of pyrrhics, both ancient and Renaissance, this one draws attention to the physical mettle of the dancers, their battle-worthiness, the containment of great violence within a musical order, and their moral fiber.

Yet this was not a real battle. Before the fight began, the captain acknowledged that they were not the king's battalions from the powder-filled field, but students living in the happy pavilions of Minerva. The goddess of wisdom, war, and the liberal arts certainly exemplified the bellicose aspects of the Jesuit enterprise, as did the combination of heroic verse, musical performance, and dancing in the pyrrhic. But they were not warriors. Little younger than their twenty-one-year-old king, the students were restless to graduate to real battles. Louis had himself just finished off three seasons of military forays to the south that culminated in impressive victories over the Huguenots. His success in the field proved that he could rule with the same military authority as his legendary father. The military campaigns, the succession of entries, and the pyrrhics all fit together as rites of passage to manhood for the king. To paraphrase the *livret* from Lyons, the pyrrhic took His Majesty's prowess as its subject, magnifying the first consummation of his military skill.[118]

---

117. *Réception de très-chrétien, très-juste et très-victorieux monarque Louis XIII,* 53. "La Musique, à vray dire, fut Royalement Martiale; le combat ne fut point un jeu d'enfant, car en ces trois bataillons fort espais d'un costé & d'autre, on ne vid aucun desordre. Tous estoient si adextrement aguerris, qu'on ne vid oncques ny courage lasche, ny pas niais. Plusieurs des spectateurs eurent peur voyans des vrayes espées nues en un si rude chamaillis, de sorte que cela donna occasion à quelques personnes des plus signalées, d'arrester un Acteur pour manier son espée, lors on cognut qu'elle estoit faicte exprez. Quelques uns tomboient si à propos, que vous les eussiez tenus pour blessez à mort. On ne voyoit que briller l'acier, & n'entendoit-on que des coups parmi l'effroy de six trompettes."

118. Ibid., 49. "Un bataillon en Pyrrique dressé à la Romaine, avec le quatrain apres chasque figure sur les prouësses de sa Majesté, servit d'avant-jeu à la guerre & la victoire suivante."

The students exemplified the fighting force of the future, and perhaps a utopian one at that, for the establishment of a substantial peacetime army in France was still decades away.[119] Henry IV's reduction of the royal army after 1598 and destruction of fortresses may have helped stabilize France's economy, but it did not empower his son. Louis was forced to cobble together his army from mercenary forces, royal troops, and private armies that did not necessarily respect royal authority as their first allegiance. Dissatisfied with this system, Louis instigated a great change in the organization of armed force in France, and in 1635 he began to issue commissions to officers to raise and train regiments in his name. The state-commission army transformed the nature of the officer corps itself, for henceforth officers served the crown.

It was only in this army, a large state-run institution that was maintained in peacetime and in war, that discipline and drill would be implemented. Louis personally admired the "Roman" drills from the Netherlands, and some exemplary regiments began to practice them as early as 1624.[120] In 1629 he ordered weekly drills in garrisons and made a point of personally reviewing them in scenes probably quite like the visit to the Jesuits described above. This intense concern with military exercises was shared by Louis XIV, who realized that the disciplinary order they projected was more powerful than armed force itself: "Many more battles are won by marching in good order and making a good show than by blows of the sword and musket. This good order shows confidence, and it seems that it is enough to look brave, because most often our enemies do not wait for us to approach near enough to have to show whether we are in fact brave."[121] Whereas the more outrageous formations proposed by Wallhausen (fig. 5.8) would never have succeeded in battle, they seem to have been quite effective at securing victories nonetheless. The remarks of Louis XIV suggest a political truth about the *ballet de cour* as well, for its laby-

---

119. Lynn, *Giant of the* Grand Siècle, 61–64. Henry IV showed little interest in maintaining large peacetime garrisons, and he destroyed many fortresses out of fear that city militias or the personal retainers of the great lords he relied on to man them might use them against him. This reduced the need for royal troops to garrison fortified places, which Henry allowed to dwindle to alarmingly low levels during times of peace, as had the Valois (Wood, *King's Army*).

120. On Louis's involvement with the drills see Lynn, *Giant of the* Grand Siècle, 517–524. Parrott believes that the drills implemented by Louis were far more rudimentary than Dutch drill, on which see *Richelieu's Army*, 38–44. Whether or not the army fully adopted the Netherlandish art of drill, we should observe that musical discipline was instilled in the future officers who studied with the Jesuits.

121. Louis XIV, *Mémoires de Louis XIV*, ed. Charles Dreyss, 2 vols. (Paris: Didier, 1860), 2:112–113. "Beaucoup plus de batailles se gagnent par le bon ordre de marche et par la bonne contenance que par les coups d'épée et de mousquet. Ce bon ordre fait paraître l'assurance, et il semble que ce soit assez de paraître brave, puisque le plus souvent nos ennemis ne nous attendent pas d'assez près pour nous donner lieu de montrer si nous le sommes en effet."

rinthine choreographies and military details evinced the external control of a well-ordered court through a performance of internal control. In this way it might effectively persuade those who came from outside—such as the Polish ambassadors who witnessed the *Ballet des provinces*—that the king's sovereignty was complete. Ballet insisted upon a political order that, whether or not it was ever tested, was in some sense beyond question and in which the dancers participated as though mobilized by its eternal forms.

Modern military discipline was an attribute of the state army, for the drills, surveillance, and extensive command structure by which it was achieved played no role in old-fashioned forces. The time and energy weapons drills required exceeded the resources of small armies, which provided recruits with basic training but nothing more. In any case, this older sort of army disbanded as soon as possible, which negated the long-term benefits of drill. With the rise of the state commission army, discipline belonged to the king alone, and music became an arm of the state, for the entire disciplinary technology that advanced the physical obedience of Jesuit students and soldiers was beyond the capabilities of the private armies mustered by France's *grands.*

Thus the royalist objective of Jesuit education bore fruit in a new style of aristocrat whose refined sense of discipline began with games of prisoners' base, pyrrhics, and the desire to become a decurion at school and progressed as if by natural extension to "classical" drills, military academy, a career at court and in the army, and, eventually, the glory of a commission as a captain or colonel. Rhythm asserted its disciplinary empire at every turn. If the young warriors put on stage by the Jesuits pleased Louis as much as they seem to have done, perhaps it was due to the promise they held of a vast royal army under his direct command. The extent to which the Jesuits were able to practice military discipline and instill allegiance to the king—both so spectacularly dramatized in the pyrrhic ballet—showed that the students of the royal colleges would soon help him achieve and consolidate military gains.

6

"DRESSER L'HOMME"

The *Ballet à Cheval*

No nation clung to the military institution of mounted knights more tenaciously than the French, for the *gendarmerie* epitomized a noble culture defined through heroism in battle. Equestrian statues were raised of Henry IV, and Rubens famously depicted him leading a charge on horseback at Ivry, his scepter a blaze of light (see plate 3).[1] These images of chivalric valor were not wholly fabricated, as Henry had indeed won the battles of Coutras (1587) and Ivry (1590) with cavalry charges. But Coutras and Ivry were aberrations. By the end of the sixteenth century most western European armies had reduced the number of their horsemen to a mere 10 percent, and these were primarily scouts and light horse—heavy horse were abandoned almost entirely.[2] The Great Horse of the French and the honor they preserved were unusual.[3]

As early as 1515 Machiavelli, in his *Arte della guerra,* dismissed *gens d'armes* as of secondary importance, correctly predicting that infantry would be the wave of the future.[4] The Battle of Pavia (1525) bore him out when Spanish and German infantrymen armed with muskets routed the French knights. The superior strength of the French in horsemen and cannons notwithstanding, imperial troops captured Francis I and many of his knights, marking Pavia as a turning point in the value accorded both to foot soldiers and hand firearms. In 1548 Raymond de Fourquevaux advised the king of France that "it is not as necessary to have good horsemen as it

1. Also see Giuliano Giraldi, *Esequie d'Arrigo IV, Cristianissimo re di Francia e di Navarra, celebrate in Firenze dal Serenissimo don Cosimo II granduca di Toscana* (Florence: Sermartelli, 1610), engravings by Aloisio Rosaccio, and the series of paintings at the Chateau de Berny (now lost) described in Godefroy and Godefroy le Jeune, *Cérémonial françois.* On these and other equestrian depictions of Henry IV, see Sara Mamone, *Firenze e Parigi: due capitali dello spettacolo per una regina, Maria de' Medici* (Milan: Amilcare Pizzi, 1987), chap. 10.

2. Geoffrey Parker, *Military Revolution,* 69.

3. For the English perspective, see *Military Discipline,* 45–46.

4. Machiavelli recommends only 150 heavy horse and 150 light horse per regiment, compared to 5000 infantry (Machiavelli, *Art of War,* 82).

is to have good foot soldiers, for the foot soldiers are the ones who win or lose a war, and not the horsemen, or at least, not unless by great chance."[5] Forward-thinking military men such as François de La Noue counseled those horsemen who remained to discard the "anvils" they strapped on in the name of armor, the "iron pots" they wore as helmets, and their cumbersome lances, arming themselves like the light horsemen who had traditionally supported their star performances in battle.[6]

Despite the obsolescence of heavily armored knights, old habits died hard. The French invested the Great Horse with tremendous honor. It was honor and bravery, in fact, that gave Coutras and Ivry their particular complexion, for these battles were fought between rival factions of nobles who continued to take mounted combat as one raison d'être of their estate.[7] As we shall see, notions of individual honor proved very obstructive to deploying knights effectively in battle, and it was this cultural obstacle that Henry of Navarre overcame with a new formation for charges.

At Coutras the royal forces were led by Henry III's favorite, the Duke of Joyeuse, who made up in bravura what he lacked in tactical savvy. His troops included numerous members of the royal court—warriors who were relatively unseasoned but nonetheless so festively armed that Agrippa d'Aubigné could not refrain from quipping that the counts, marquises, barons, and lords in the front line were "the cavalry the most covered with tinsel and gold that had ever been seen in France."[8] This noble vanguard of twelve hundred chevaliers formed up in the old fashion, in a broad line two deep, and charged with their long lances to meet Navarre's forces. Their formation, known as a "hedge" or "wing," was the most practical way to put lances into battle, but in the second half of the sixteenth century military strategists began to question its value. La Noue, who criticized the use of heavy armor and lances, likewise found problems with the hedge:

> When a troop is arranged in a wing, the good ones, who are ordinarily the fewest, march gaily into combat, nonetheless the others who hardly wish to bite (who feign

5. Beccarie de Pavie, sieur de Fourquevaux, *Instructions sur le faict de la guerre,* fol. 109r. "Il n'est pas tant requis que lon ayt bonne Chevallerie, comme il est necessaire avoir bons Pietons: car les Pietons sont ceulx qui peuvent donner gaignee ou perdue une guerre, & non pas les Gens de cheval, si ce n'est grand' aventure."

6. François de La Noue, *Discours politiques et militaires,* 331.

7. A good account of these battles can be found in Charles Oman, *A History of the Art of War in the Sixteenth Century,* 2nd ed. (London: Greenhill Books; Novato, Calif.: Presidio Press, 1991), 470–480, 495–505. It is largely based on d'Aubigné, *Histoire universelle,* 7:123–144, 8:164–173.

8. D'Aubigné, *Histoire universelle,* 7:131. "La cavallerie la plus couverte de clinquant et d'or battu qu'autre qui ait esté veuë en France."

> a bloody nose, a broken stirrup, or a missing horseshoe) lag behind in such a way that two hundred feet underway one sees this long line thin out and great breaches appear in it, which give a marvelous courage to the enemy.[9]

In short, discrepancies in relative bravery had a dire outcome during the charge.

The alternative was to form up in compact squadrons of horsemen, and it was this formation that Henry adopted at Coutras.[10] He deployed his horsemen in blocks interspersed with detachments of harquebusiers five deep. Instructing the gunmen not to fire until the enemy came within twenty yards (lest they certainly be trampled), he managed to weaken Joyeuse's line with gunfire before his blocks of horsemen charged into the duke's rash affront. The royal knights were massacred. Joyeuse himself was killed quite mercilessly while fleeing; his brother Claude also died in the fight, along with three counts and a number of other courtiers.[11] The charge of the hedge gave greater latitude for individual heroics, but the cooperative cavalry square was victorious. By deploying his horse in squadrons, Henry deserves credit as the first general to mobilize what might properly be called a cavalry in the modern sense of the word.[12]

Light, fleet, and collective, cavalry stayed the tide of the infantry revolution and preserved the role of horsemen in battle. But battles themselves had fallen somewhat by the wayside. War offered fewer and fewer opportunities for gentlemen on horseback to display their military virtue.[13] And the cavalry, as a new institution, lacked the mystique of the old *gendarmerie.* Far from transforming cantankerous and independently minded knights into a cooperative mounted branch of the army, the cavalry ultimately replaced them with squadrons descended from the light horse: nobles still served in the cavalry, as they did in all branches of the military, but in new ways.

Although the Wars of Religion did revive knighthood in France, warfare had lost much of its chivalric character. Fifteenth-century combat had unfolded as something of a duel on horseback—battles often resulted only in the capture of noble-

9. François de La Noue, *Discours politiques et militaires,* 337. "Mais quand une troupe est ordonnee en aisle, les bons, qui sont ordinairement le moindre nombre, encor qu'ils marchent gaillardement au combat, neantmoins les autres qui n'ont gueres d'envie de mordre (qui faignent seigner du nez, avoir une estriviere rompuë, ou leur cheval desferré) demeurent derriere; en sorte qu'en deux cens pas de chemin, on void esclarcir ceste longue file, & apparoissent de grandes bresches dedans. Ce qui donne un merveilleux courage aux ennemis."

10. See Ronald S. Love, "All the King's Horsemen: The Equestrian Army of Henri IV, 1585–1598," *Sixteenth Century Journal* 22 (1991): 510–533.

11. D'Aubigné, *Histoire universelle,* 7:143.

12. See Delbrück, *Dawn of Modern Warfare,* 136.

13. See Geoffrey Parker, *Military Revolution,* 41.

men (as with Francis I at Pavia), and codes of honor refereed the blood sport of war. Gunpowder warfare was another matter, as we have seen, for infantry armed with armor-piercing muskets could decide the outcome of battles. All the more important, then, was the forum provided by jousting tournaments, where knights could prove themselves with honor. Jousts persisted well into the seventeenth century despite the impending obsolescence of lances and notwithstanding the ban Catherine de' Medici placed on them after Henry II died from wounds suffered during a tourney in 1559. Here we should note that even at their instigation circa 1100, tournaments constituted both training and a substitute for war, and by the time of the fabled tourneys staged by the Burgundians, jousting exuded the decadence of an expiring crusade culture, commemorating forays against the infidels that would never come to pass with ceremonial challenges among Christian chivalric orders.[14] In the Renaissance, too, jousting for rings and unhorsing challengers taught young courtiers skills required in social gatherings just as surely as they trained men to survive in battle. This is to say that court life required its complex of military skills even as actual battles began in waves of color and spectacle.

The main point to be made about the dual function of court festivities and the way tournaments mediated between warfare and upper-class sociability is that bellicose fetes were continually evolving under martial as well as social influences. Tournaments, entries, mock combats, and naval battles created and responded to the mutable rules of social commerce *and* the revolutions of military technology. It should therefore come as no surprise that coincident with the success of tight cavalry formations at the end of the sixteenth century, a new equestrian art form arose that showed off the latest mode of battle. This was the *ballet à cheval.* Horse ballet culminates this study of music and arms with special clarity because it bound ancient myths of music and horsemanship so effectively to that other spectacular humanistic creation from those years, the *ballet de cour.* Music projected refinement across equitation, harmonizing the relationships among the riders, "civilizing" a martial art, and bringing a new political dimension to horsemanship. Moreover, because horsemanship quite literally defined knighthood for the French (*cheval, chevalier*), horse ballet registered the whole matrix of classicism, heroism, honor, chivalry, and military capability it purveyed at the highest social level, one popu-

14. See the classic work of Johan Huizinga, *The Autumn of the Middle Ages,* trans. Rodney J. Payton and Ulrich Mammitzsch (Chicago: University of Chicago Press, 1996) and Malcolm Vale, *War and Chivalry: Warfare and Aristocratic Culture in England, France, and Burgundy at the End of the Middle Ages* (Athens: University of Georgia Press, 1981), which revisits many of the issues raised by Huizinga. On the techniques employed in jousting and other forms of mounted combat, see Anglo, *Martial Arts of Renaissance Europe.*

lated by the *grands* whose traditional role had been to ride into battle in defense of the kingdom.

MUSICAL DRESSAGE

Before the establishment of the French military academies, young French nobles went to Italy for a year or two to study at any one of a number of several famous riding schools. Antoine de Pluvinel had himself attended the school of Giovanni Battista Pignatelli in Naples. To Pignatelli goes credit for first training horses with pillars, an innovation that marks the new aesthetic impulse, which refined the blunt utility of medieval riding through meticulous training that entailed fixing the horse almost fast between pillars that restricted its movement so that horsemen could concentrate on details. Restraint, equilibrium, and artistry—the triumph of artifice over nature—came together in a delicate symbiosis of horse and rider, and students cut striking figures when they returned home. At mid-century, Naples was the European center of equitation: another Neapolitan, Federico Grisone, had opened a school there in 1532, and he published a treatise on the new method entitled *Gli ordini di cavalcare* (1550). It was quickly translated into French and German and remained a standard reference work well into the seventeenth century.[15] Indeed, the first substantial French treatise on riding, *Le cavalerice françois* by Salomon de La Broue (1602), recommends Grisone to its readers at the outset, and both La Broue and Pluvinel, France's foremost horseman until his death in 1620, laud Pignatelli as the *fons et origo* of the new art.[16] Naples maintained its precedence in the late sixteenth century with the school of Cesare Mirabbello.

The other Italian capital of dressage was Ferrara, a city celebrated for the *feste a cavallo* of the Estense and the school of Cesare Fiaschi.[17] In Fiaschi's hands, the musical inclinations of dressage, with its attention to *tempo* and *misura giusta,* were made explicit. Grisone had told his riders to move in conformity with the horse's

15. Federico Grisone, *Gli ordini di cavalcare* (Naples: G. P. Suganappo, 1550), first published in German in 1570 (Augsburg: M. Manger), Spanish in 1568 (Baeza: J. B. de Montoya), English in 1560 (London: W. Seres), and French in 1559 (Paris: C. Perier). For a modern edition of sections of Grisone's treatise see Carlo Bascetta, ed., *Sport e giuochi: Trattati e scritti dal XV al XVIII secolo* (Milan: Il Polifilo, 1978), 205–225.

16. La Broue, *Cavalerice françois,* 1:5, 2:5–7; Pluvinel, *Instruction du roy,* 30.

17. To mention just two such Ferrarese tourneys from the time, see Ercole Estense Tassone, *Il monte di Feronia* (Ferrara: Valente Panizza, 1562) and idem, *L'isola beata* [Ferrara: Francesco Rossi, 1570]. On Italian tourneys in general see *La società in costume: Giostre e tornei nell'Italia di Antico Regime* (Foligno: Edizioni dell' Arquata, 1986), esp. chap. 7 on music for tourneys (167–181). Also see Paolo Fabbri, ed., *Musica in torneo nell'Italia del Seicento* (Lucca: Libreria Musicale Italiana, 1999). On sixteenth-century equestrian festivities, see the articles in Jacquot, *Fêtes de la Renaissance,* 2:283–436.

FIGURE 6.1. "Un passo e un salto." In Cesare Fiaschi, *Trattato del modo dell' imbrigliare, maneggiare e ferrare cavalli* (1628). Courtesy of the Music Library of the University of California, Berkeley

gait "with no less agreement than if there were music," and Fiaschi took this one step further, codifying the musical method with a series of short tunes corresponding to each of several different steps.[18] Fiaschi's *Trattato del modo dell'imbrigliare, maneggiare e ferrare cavalli* (1556) thus offered the first comprehensive musical vocabulary for dressage (see fig. 6.1).[19]

---

18. Grisone, *Gli ordini di cavalcare,* in Bascetta, *Sport e giuochi,* 216. "Quando salta o veramente para, e a qualunque cosa, lo accompagnarete a tempo, conforme al motivo ch'egli farà, così come egli a tempo risponde al vostro pensiero e in ogni richiesta, per che bisogna che il vostro corpo con la schiena vadi giusto e gli sia corrispondente e ordinato, con non meno concordanza che se fusse musica."

19. Cesare Fiaschi, *Trattato del modo dell'imbrigliare, maneggiare et ferrare cavalli* (Bologna: A. Giaccarelli, 1556). Selections reprinted in Bascetta, *Sport e giuochi,* 227–240.

Far from relegating music to the most advanced lessons of his treatise, Fiaschi built his entire method upon a foundation of musical training. "And because it might seem strange to some horsemen that I have wanted to include music in my second book, thinking that it is not necessary," he announced, "I respond that without measure and time one cannot make something good, and I have shown [the measure] in this way."[20] In Fiaschi's airs, pitch, rhythm, and vowel sounds interlock in musical commands keyed to the pace and exertion of the horse's movements. For example, *un passo e un salto,* illustrated in figure 6.1 utilizes three short "ah"s at a lower pitch for the preparation, a higher, longer "ahi!" for the leap, and a rest for the recovery. By contrast, the less demanding *galloppo raccolto* requires "la voce somessa," rendered musically with a lower pitch (B-flat), longer note length (semiminim), and the soothing syllable "ah."

Cavaliers were directed to learn enough music to be able to sing while riding because music communicated more directly than speech. Singing, tongue-clicks, soothing tones, and encouraging cries became the primary tools of contact between horse and rider as proponents of the new style of dressage employed the voice to instruct and command. Grisone specifically lists the voice first in a succession of disciplinary methods, followed by those of increasing invasiveness: the voice, the crop, the bridle, the thighs, the stirrups, the spurs, and pulling the horse up short (the bit).[21] Likewise, he says that speaking is the best encouragement for a horse and, when one adopts "pleasing and low" tones, the best reward.[22]

The voice thus introduced music into dressage, laying the groundwork for the pairing of loud bands and equitation in the *ballet à cheval.* Vocal cues heightened the magic of a beautifully managed horse, as they allowed the rider to command almost invisibly and to avoid any extroverted physical gestures. The same corporeal restraint encouraged in social settings *à pied* could thus be maintained *à cheval,* where it seemed, moreover, to prove the universal value of good grace by smoothing communication between chevalier and his steed. If civility attempted to dictate a code of behavior facilitating social commerce, then good horsemanship trumped even etiquette with its invisible and metalinguistic discourse based on musical sounds. Whether the terms *arie* and *airs* applied to the various gaits referred to the general visual sense of these words or their musical one, it seems clear that the best riders achieved the airs they wished with the musical sorts of cues found in Fiaschi's treatise.

20. Fiaschi, *Trattato del modo dell'imbrigliare, maneggiare et ferrare cavalli,* in Bascetta, *Sport e giuochi,* 231. "E perché potrebbe forsi parer strano a qualche cavaliero che io abbia voluto inserir in questo mio secondo trattato musica, giudicando forsi essi non esser necessaria, rispondendo dico che senza misura e tempo non si può far cosa buona e io così lo mostro."

21. Grisone, *Gli ordini di cavalcare,* in Bascetta, *Sport e giuochi,* 222.

22. Ibid., 223–224.

The attention to sound and hearing in equitation fit well with the Neoplatonic mythology of musical ethos, encouraging writers on horsemanship to mine classical sources for affirmation that horses were responsive to music. In 1563 François de Provane, an early French translator of Fiaschi's treatise, rehearsed all the standard myths of musical suasion in his dedication, from the premises of Pythagoras through the stories of Amphion, Orpheus, and Arion.[23] As for horses, not only was it possible to train them "par cadences & mesures," but musical training was easier on the horses than other methods. Horses possess, Provane suggested, an almost human sensibility for music. In a world where music divined inner humanity and served as a touchstone of refinement, horses registered so high on the scale of musical affinity that they stood in a different class from other domesticated animals. Like the generous, magnanimous, and natural spirits Pierre de Ronsard praised in his preface to the *Livre de meslanges,* horses are sweetly ravished by music in a way that implicated them in the *harmonie universelle* that human souls yearned to regain.[24] Although Provane never says so, the conclusion that musical dressage might improve the human condition is very much on the horizon. Fiaschi himself exhorts the good chevalier to imitate as much as possible the good musician: the horse must be trained to the same point of perfection as a musical instrument, and the rider, like the performer of "rare and excellent music," must show this sublime and sovereign control not through knowledge, but in the perfection of the musical performance.[25] The pairing of sensitive and well-schooled horse with a virtuous rider evinced an essentially musical harmony of movement.

Despite the bold claims made for the new style of dressage by Frenchmen such as Provane, the continual disruptions caused by the civil wars—and, no doubt, the construction of increasingly massive armor for knights—prevented a properly French school from coming into being. Throughout the latter half of the sixteenth century, the French published edition after edition of Grisone and Fiaschi in translation without producing a native work on equitation, Frenchmen traveled south to drink from the source, and even the royal court imported Italian coaches. Fancy dressage was an Italian art. Only the publication of La Broue's *Cavalerice françois* in 1602 initiated a French school of dressage in print. It was followed by the treatises of René de Menou (*La pratique du cavalier,* 1612), the posthumous treatises of An-

23. "Epistre à Jacques de Silly, Chevalier de l'ordre du Roy," in Cesare Fiaschi, *Traicté de la manière de bien embrider, manier, et ferrer les chevaux,* trans. François de Provane (Paris: Charles Perier, 1564), fols. A2r–A4v.

24. Ronsard, *Oeuvres complètes,* 2:1171.

25. Fiaschi, *Trattato del modo dell'imbrigliare, maneggiare et ferrare cavalli,* book 2, chap. 3. The French edition of Auvray (Paris, 1578) embellishes Fiaschi's original text with the notions of the sublime and sovereign (fol. 59v).

toine de Pluvinel (*Le maneige royal,* 1623, and *L'instruction du roy,* 1625), and the treatise of Pierre de La Noue (*La cavalerie françoise et italienne,* 1620).[26] These publications reflected the establishment of the French academies of Pluvinel and others. Nonetheless, before that time the French seem to have been excellent students: Claudio Corte, who spent the years 1567–1573 at the French court, praised the French for being infinitely accomplished in all manner of equestrian pomp (except the distinctly Italian *palio*), meanwhile dedicating the second edition of his treatise *Il cavallerizzo* to Charles IX in 1573.[27] Likewise, Fiaschi dedicated the first French translation of his treatise to Henry II.[28]

For tourneys and parades, the French did indeed put on wonderful shows of horsemanship, particularly in entries, where the unique amalgam of musical humanism, royalist allegory, and military display helped to create favorable circumstances for the development of horse ballet in France. One finds all these elements present as early as 1548 in the entry of Henry II in Lyons. Riders costumed "à la mode de l'antique Cavallerie Romaine" amazed spectators with their dexterity on horseback as they made caprioles and circles, leaped, and redoubled their jumps into the air, the whole to the sound of small bells attached to the caparisons of the horses, which were "so pleasantly resonant that the harmony of their sweet sound did not tickle the spirits of the astonished people any less than the flash of the gleaming gems dazzled their eyes so that those watching did not know if they dreamt or lived" (see fig. 6.2).[29] Although simple, the tinkling silver *timballes* made the performance opulent to the ear as well as to the eye, enough so to evoke the Neoplatonic terms of musical ravishment from Maurice Scève, the commentator. His description of "resonance," "harmonie," "doux son," and "chatouillement des esprits" tells us that the music charmed the listeners, just as the cavorting horses

26. Menou, *Practique du cavalier;* Pierre de La Noue, *La cavalerie françoise et italienne* (Lyons: Cl. Morillon, 1620); Pluvinel, *Le maneige royal où lon peut remarquer le défaut et la perfection du chevalier en tous les exercices de cet art* (Paris: G. Le Noir, 1623); idem, *Instruction du roy.*

27. Claudio Corte, *Il cavallarizzo* (Venice: G. Ziletti, 1562); 2nd, expanded ed. published as *Il cavallerizzo: nel qual si tratta della natura de' cavalli, della razze, del modo di governarli, domarli e frenarli.* (Lyons: Pierre Roussin, 1573). On Corte and for selections from the 1573 edition see Bascetta, *Sport e giuochi,* 241–256; on France see ibid., 245.

28. See Cesare Fiaschi, *Traicté de la manière de bien emboucher, manier, et ferrer les chevaux* (Paris: Guillaume Auvray, 1578), probably a reprint that preserved the original dedication of an earlier edition.

29. *Magnificence de la superbe et triumphante entrée,* fol. D1r. The caparisons of the horses were embellished with "petites timballes d'argent si plaisamment resonantes que l'harmonie de leur doulx son ne chatouilloit moins les esperitz du peuple estonné, que l'esclair des pierreries reluisantes esblouissoit les yeulx de tel, qui en les voyant ne scavoit s'il songeoit ou vivoit. Car à la verité c'estoit plus tost une droicte faerie, que chose vraysemblable." On the riders, see Cooper's introduction to Scève, *Entry of Henry II into Lyon,* 40–41.

FIGURE 6.2. "Captain on Horseback." In *La magnificence de la superbe et triumphante entrée de la noble & antique cité de Lyon* (1549). By permission of the British Library, shelfmark 811.g.33

seemed to dance to the music their leaps produced and in this way to have been as moved by it as the spectators were.

The Roman costumes added yet another dimension to the show, freighting the balletic dressage with historical import. Classical texts contained a treasure trove of equine lore, including much concerning music, in which humanists such as Provane sought authorization for the use of music in dressage. "I know well," he said, "that the Sybarites and Libyans . . . trained . . . their horses to the sound of instruments."[30] And everyone knew Xenophon's description of mock combats in the Hippodrome replete with trumpet calls (*Hipparchicus* 3.10–13) and his comparison of prancing horses with dancers (*De re equestri* 9). Descriptions like Xenophon's fueled the design of entries and fulfilled many of the aesthetic preconditions necessary for the emergence of equestrian ballet, which bore the seal of a classical art recovered. Similarly, historical studies such as Guillaume Du Choul's *Discours sur la castrametation et discipline militaire des Romains* (Lyons, 1555) provided visual and textual material upon which to model recuperations of Roman practices, whether festive, military, or both (see fig. 6.3).[31] Du Choul, in fact, helped organize the 1548 entry.[32]

Like all entries, this was a civic pageant meant to receive the king in splendor and to impress him with the largesse mustered by the city. Motivated by competition with royal majesty as much as by awe of it, city coffers bore tremendous expenses, plans were elaborate, and preparations often took months. So it was that the equestrian display laid claim to a certain nobility that seemed—at least to the king and his entourage of noblemen—beyond the grasp of the "notables, bourgeois & citoiens" of Lyons. For the culture of horses and chevaliers had little to do with urbanity. Rather, horsemanship predominated in life on the seigneuries, both for defense and for hunting or traversing one's land. At first take it seems surprising that the Cavallerie des Enfants de la Ville had achieved such artistry in dressage. The king, princes, and other gentlemen were "dumbfounded to see them—people from the

30. François de Provane, "Epistre," in Fiaschi, *Traicté de la manière de bien embrider, manier, et ferrer les chevaux,* fol. A3r. "Je sçay bien, que les Sybarites & Lybiens lors qu'ilz guerroient contre les Crotoniens domtoient & galloppoient leurs chevaulx au son des instrumentz." Claude François Ménestrier attributes this remark to Pliny (see Ménestrier, *Traité des tournois, joustes, carrousels, et autres spectacles publics* [Lyons, 1669; facsimile reprint, Lyons: Editions Horvath, 1978], 169–170).

31. Du Choul, *Discours sur la castrametation et discipline militaire des Romains.* See Orest Ranum, "Strengthening the Noble Male Body: Guillaume Du Choul on Ancient Bathing and Physical Exercise," in *Politics, Ideology, and the Law in Early Modern Europe: Essays in Honor of J. H. M. Salmon,* ed. Adrianna E. Bakos (Rochester: University of Rochester Press, 1994), 35–51.

32. *Magnificence de la superbe et triumphante entrée,* fol. A2r. Cooper gives a detailed account of the preparations for the entry in Scève, *Entry of Henry II into Lyon,* 14–31.

FIGURE 6.3. "Man-at-Arms from the Time of the Ancient Romans." In Guillaume Du Choul, *Discours sur la castrametation et discipline militaire des Romains* (1555). By permission of the British Library, shelfmark 143.e.2

city, and not of this calling—so adroit that it would be almost impossible to do better."[33] Whereas the pyrrhic combat on foot staged by the Enfants de la Ville impressed Henry II as a demonstration of a new form of combat, his appreciation of the fine dressage came from visceral understanding of the effort required to master it. The beautiful mounts, ringing caparisons, and caprioles made a point about the lineage of the Enfants, who came from the top ranks of Lyonnaise society, reminding us that many officers of the royal courts owned land and even seigneuries. If it was land, rather than noble title, that separated social orders in sixteenth-century France, as James Collins has suggested, then horsemanship emblemized their connection to the land upon which social distinctions turned.[34] Such displays were usually made by noblemen and musical praise reserved for them. One account reports that during his entry into Metz in 1569, for example, Charles IX was the only rider to have kept his horse under control despite the harquebus and artillery salvos, "handling it with such good grace that riding along he seemed to dance to the sound of the clarions, trumpets, drums, and fifes that (in a great consonance of harmony) sounded in all the main streets and squares of the city."[35]

Spectacular horsemanship was, of course, common at court, where competitive equestrian games had originated centuries before. The jousting tournament at which Henry II suffered his mortal wounds was but the most famous of the perpetual tourneys mounted for special occasions that climaxed annually during Carnival season. The well-documented accounts of the 1564 fetes give us an idea of the occasions for *combats à cheval:* on Jeudi Gras the Cardinal de Bourbon invited the court for dinner and a *combat à cheval* at his abode, and on Mardi Gras the king hosted "tourneys and the breaking of lances."[36] Thus the *bergeries* and mascarades that have justly made the Fontainebleau fetes of 1564 so famous took place during a month filled with martial games as well.

---

33. *Magnificence de la superbe et triumphante entrée,* fol. D1r. "Et qui accroissoit merveille sur merveille, cestoit de veoir le Capitaine, Lieutenant, Portenseigne, & bonne part des autres si dextres à cheval, & si bien le scachant manier, faire pennades, bondir, voltiger, & redoubler le sault en lair: qui ne pouvoit donner que grand plaisir au Roy, aux Princes & autres Gentilz hommes, non sans s'esbahir de les voir (pour gentz de ville, & non appellez à celà) si à droictz, qu'il seroit presque impossible de mieulx faire."

34. Collins, *Classes, Estates, and Order,* 2–13.

35. *La celebre et magnifique entrée de Charles neufiesme, Tres-chrestien Roy de France, faicte en sa Ville & Cité de Metz* (Paris: Jean Dallier, 1569), fols. B4v–C1r. "Il [le Roy] entra tout de cheval, le manyant de si bonne grace, que chemynant il sembloit danser au son des Clerons, Trompettes, Tabours, & Phifres, qui (en grande consonance d'armonye) sonnoient par toutes les principales ruës & places de ladicte Ville & Cité, ou passa sa Majesté."

36. For descriptions of these and the following combats see Graham and Johnson, *Royal Tour of France,* and Brantôme, *Recueil des dames,* 53ff.

In May of that year, Gaspard de Saulx-Tavannes held *combats à cheval* in honor of the royal party as it passed through Dijon. Tavannes was the lieutenant governor of Burgundy, a strongly Catholic province, and in 1563 he established the Confraternity of the Holy Ghost in Dijon, whose members swore to take up arms with him to defend the faith. Thus we should not see Tavannes's knightly combats as mere entertainment fit for a king, nor the king's trip to Dijon as a casual excursion to the provinces. Indeed, the entire purpose of the royal tour was to reassert monarchic control in the wake of the first religious war, and Tavannes's combat showed that Burgundy stood ready to fight. By the end of 1568 a dozen Catholic confraternities of the Holy Ghost had sprung up across the land as the Protestant threat revived the crusade mentality at the heart of medieval tournaments and knightly orders.[37]

Moreover, Tavannes's *combats à cheval* were the divertissement of one of France's greatest commanders. He was made political and military advisor to the young Henry III (then the Duke of Anjou), and he led the royal army to victory at Jarnac (1569) and Moncontour (1569) at the side of his teenage protégé. What separated *combats à cheval* from the headlong cavalry charges those same knights made in battle was thus not a matter of kind but of degree. Indeed, combats at the barrier may have encouraged the recklessness that played well before the ladies but had variable results in battle. Tavannes knew enough to save flamboyant sorties for tourneys. He managed to lure the Prince of Condé into a rash charge at Jarnac (one not unlike the fatal charge Joyeuse would make at Coutras): unhorsed and with his leg broken, Condé was captured and shot in the back of the head by a captain currying favor with Anjou. The Huguenots lost their most inspiring leader that day, and chroniclers parlayed the victory into a glorious triumph for the future Henry III.

The Royal Tour progressed south for the winter of 1564–1565, ending at Bayonne, where the Valois received a delegation from the Spanish court. The Spanish presence heightened the French lust for ostentation and the display of arms, precipitating so many races, jousts, *combats à cheval,* and indoor combats on foot that the Spanish were said (by the French) to have marveled at the great number of *gens d'armes,* their adroit horsemanship, and confident lance blows.[38] Alongside the informal contests that included the Spanish (which they reportedly lost quite invariably), the festivities built to a crescendo across the fortnight with a series of costumed combats on fantastical themes "performed" by the French alone.

The apotheosis of the fabulous suite of *magnificences* was a combat between Virtue and Love, in which the king (Virtue) and his brother (Love) each commanded a band of knights. Music took a leading role in the spectacle. Chariots bearing sing-

37. On the confraternities, see Holt, *French Wars of Religion,* 67–69.

38. Graham and Johnson, *Royal Tour of France,* 309.

ing goddesses and Mercuries, Cupids, and children introduced the tourney, and after all the knights had jousted one on one, two on two, three on three, and so forth, the horsemen regrouped in companies at each corner of the field and made a succession of carefully choreographed and perfectly paced charges, "passing by one another without mixing with, troubling, or merging with each other and passing to their place at full speed, just to the spot from which their enemies came."[39] This was not an equestrian ballet, per se, as the trumpets seem to have "sounded with great melodiousness" only at its end.[40] But it did feature a high degree of "dexterity and deftness" that relied on the "perfect timing" of the riders, for it was timing that allowed them to strike one another safely and to give the impression of fighting pell-mell without true confusion.[41] As Xenophon says of his very similar choreography for the Hippodrome (*Hipparchicus* 3.10–13), charging through the opposing forces in this way was more warlike and novel than the usual cavalry line-ups.

Probably in order to demonstrate the battle-worthiness of the knights and their mounts, the spectacle came to an apocalyptic end when "balls of fire" were tossed in among the riders, who vanished into a chaos of noise and smoke, a denouement significant for its visceral evocation of war. If riding together, as one source explains, was meant to show how Virtue and Love are brother and sister, and how the king and his brother were in accord with one another, this *lieto fine* was undone by the sulfer of battle. In this story, military Virtue—masculine, bellicose, royal—wins the day. The famous depiction of the tournament by Antoine Caron tells all, compressing events into a single spectacular image that included the deus ex machina fireworks, trumpet calls, and the frightful clash of arms, man, and beast (see plate 4). Whatever "musicality" might have been demonstrated by the coordinated charges proved but a military exercise preparatory to war; the French meant for the Spanish to leave with a taste of their host's firepower and knightly ferocity, one flexed and strengthened by control.

COURTLY AND MILITARY CHOREOGRAPHIES

Horse ballet required conditions different from those of the bellicose occasions for which tourneys were mounted. It was more suited to festivities in which the ten-

39. Ibid., 287–288, 371–372. "Lesquels ayans ainsi combattu, se mirent en croix en quatre endroits, distans egallement les uns des autres: lesquels venoyent quasi à un mesme temps à se rencontrer et frapper au milieu du Camp, passans à travers les uns des autres, sans se mesler, troubler ny confondre, passans de leur place à toute bridde, jusques au lieu dont partoyent leurs ennemis. Ce qu'ils firent par plusieurs fois, avec si grande dexterité et addresse, qu'il n'y en eut un seul qui faillist à rencontrer et frapper son ennemy aussi à propos, comme s'ils eussent combattu en foulle, et s'entrechargerent, se meslans pesle mesle, jusques à ce qu'ils furent separez par une grande quantité de balles de feu" (372).

40. Ibid., 288.

41. Ibid., 372.

sion between the court and an external force did not create such obvious drama between the monarch and his protagonists. In this respect it resembled the *ballet de cour,* which aimed to assimilate antagonists and prolong the fiction of its harmonic narrative by drawing king, queen, and the rest of the nobility into a lengthy grand *bal* at its end. *Ballet à cheval* framed its militaristic exhibition of balance with narratives of social concord, harnessing them to the tourney's masculine virtue even while steering the whole away from the combustible imagery of knightly jousting.

The first *combat à cheval* called a ballet by contemporaries celebrated a happy, "private" affair at court: the marriage in 1581 of Anne de Joyeuse to the half sister of the queen, Marguerite de Lorraine. Joyeuse entered the court after distinguishing himself in the sixth civil war (1577), and Henry III quickly pulled him into its innermost circles, where his polished manners and the humanistic education he had received at the Collège de Navarre facilitated his social ascent. The marriage was an arranged one, part of a series of alliances organized by the king, who surrounded himself with a close network of minions whom he rewarded with favorable engagements, dukedoms, and other gifts—Henry's other favorite and general, Jean-Louis de La Valette, became the Duke of Épernon the same year.

The best-documented event of the Joyeuse wedding festivities was Beaujoyeulx's famous *Balet comique de la Royne,* but this was just one of the *magnificences* staged for the marriage. Beaujoyeulx tells us in his introduction to the *Balet comique* that there was also a "horse ballet after the manner of the ancient Greeks . . . accompagnied by excellent consorts of music,"[42] and it is surely for this ballet that Pierre de Ronsard wrote his *Cartel pour le combat à cheval en forme de balet.*[43] Cartels were poetic challenges that explained the conceits of tourneys, whether between happy and unhappy lovers, between Love and his detractors, or between factions from Arthurian romance.[44] But Ronsard's cartel for the ballet takes a didactic tack, pos-

42. Beaujoyeulx, *Balet comique de la Royne,* fol. 1r. "Des balets aussi à pied & à cheval, prattiquez à la mode des anciens Grecs, & des nations qui sont aujourdhuy les plus esloignees de nous: le tout accompagné de concerts de musiques excellentes & non encores jamais ouyes." The ballet was also remarked by L'Estoile, *Mémoires-journaux,* 2:34. "Le jeudi 19e, pour fin des carousels et ballets, fut fait le ballet des chevaux, auquel les chevaux d'Espagne, coursiers et autres du combat, en combattant, s'avançoient, se retiroient et se contournoient, au son et à la cadence des trompettes et clairons sonnans, y aians esté aduits et instruits cinq ou six mois auparavant."

43. Ronsard, *Oeuvres complètes,* 2:271–272. On the rich symbolism of horses in French verse and on this cartel in particular, see Hélène Naïs, *Les animaux dans la poésie française de la Renaissance* (Paris: Marcel Didier, 1961), 481–488, 545–556, and Marie Madeleine Fontaine, "Nager, voler," in *Ronsard et les éléments,* ed. André Gendre (Geneva: Droz, 1992), 67–124, esp. 77–78.

44. Cartels were a standard genre of occasional verse, and all of the poets who served the court in the second half of the sixteenth century—Mellin de Saint-Gelais, Ronsard, and Philippe Desportes—turned out numerous cartels, sometimes on short notice. See Mellin de Saint-Gelais, *Oeuvres poétiques*

sibly because ballet was so novel. It turns around three subjects: the ancient history of horse ballet, its choreography, and its royalist message. It is noteworthy that Ronsard hit upon these subjects right away, as they point up precisely the matters that distinguished the horse ballet from combats: proponents of the new style of dressage had already given it a classical pedigree, which Ronsard elaborates at the opening of his cartel, going on to describe the choreography in terms of the geometrical figures one would have witnessed in *ballets de cour* and finishing with an equestrian moral of good governance that moved the horse ballet away from tourneys (which displayed individual heroism) and into the humanistic arena of the *ballet de cour* (which manifested social harmony).

The poem beautifully introduces the ethos of equestrian ballet with Ronsard's comparison of the chevaliers to benevolent centaurs:

> Ces Centaures armez à nostre âge incognus,
> Au bruit d'un si haut Prince en France sont venus
> Pour les peuples instruire, et les rendre faciles
> Autant que sous le frein leurs chevaux sont dociles.
> Et faire de son nom tout le monde ravir,
> Afin que toute-chose apprenne à le servir.[45] (Lines 31–36)
>
> [These armed centaurs, unknown to our age,
> Are come at the renown of this high Prince of France
> To instruct his people and to render them as easy
> As their horses are docile under the bit.
> And to ravish the world with his name,
> To the end that all will learn to serve him.]

Half horse and half man, centaurs represented the uneasy coexistence of nature and culture in ancient Greece. The most famous myth involving them tells how Pirithous invited the centaurs to his wedding, and as soon as they tasted wine—the product of human artifice—they became unruly and assaulted the bride. Known for drinking, unbridled lust, and violence, centaurs represented the reversal of civilization, a challenge to humanity made all the more succinct given the role of horses in human culture. But Ronsard appeals to another side of the centaur's mythology—to the story of Chiron, who educated Asclepius, Jason, and Achilles

---

françaises, ed. Donald Stone Jr., 2 vols. (Paris: Société des Textes Français Modernes, 1993–1995), vol. 1; Ronsard, *Oeuvres complètes,* 2:228–294; and Philippe Desportes, *Cartels et masquarades; épitaphes,* ed. Victor E. Graham (Geneva: Droz, 1958). Their work was well remunerated: L'Estoile reports that "le roi donna à Ronsard et à Baïf, poètes, pour les vers qu'ils firent pour les mascarades, combats, tournois et autres magnificences des nopces, et pour la belle musique par eux ordonnée et chantée avec les instruments, à chacun deux mil escus" (*Mémoires-journaux,* 2:23).

45. Ronsard, *Oeuvres complètes,* 2:272.

(on the latter, see plate 1).[46] Chiron had himself been educated by Apollo and Diana and was renowned for his skill in hunting, medicine, music, and prophecy. In Chiron we find a species of natural virtue born of animal passions and tamed by humanistic studies—the head rules the lower bodily stratum. And in the chevaliers Ronsard likens to him we find the triumph of culture over nature, of obedience over willfulness, of restraint over chaos, and of the king over his subjects. The subtext includes potential violence, for this invests the chevaliers and the centaurs with their power, but it is a violence taken in hand.

The choreography Ronsard describes is strikingly like the geometrical designs of early *ballet de cour;* indeed, the cartel revisits many of the images Ronsard employed in an earlier poem to evoke the turns and mutable geometry of court ballet. In that poem he said that "the ballet was divine . . . first it was round, then long, then straight, then pointed in a triangle in the way one sees a flock of cranes avoiding the frost."[47] Here the divinity of the ballet comes not only from the disposition of the dancers in these planetary geometries, but from its artifice and its "contre-imitation" of nature. So, too, Ronsard sees in the *ballet à cheval* the figures of *ballet de cour* and their power to "contrafact" and contain something beyond themselves. Not only does Ronsard compare the horse ballet to dolphins and a flock of cranes, he sees the ballet as making war into peace:

Tantost vous les voirrez à courbettes danser,
Tantost se reculer, s'approcher, s'avancer,
S'escarter, s'esloinger, se serrer, se rejoindre
D'une pointe allongée, et tantost d'une moindre,
Contrefaisant la guerre au semblant d'une paix,
Croisez, entrelassez de droit et de biais,
Tantost en forme ronde, et tantost en carrée,
Ainsi qu'un Labyrinth, dont la trace esgarée
Nous abuse les pas en ses divers chemins. (Lines 15–23)

[At one moment you will see them dance in *courbettes,*
At others pull back, come close, advance,
Move apart, retreat, close ranks, rejoin,
From an elongated point, and then from a shorter one,
Contrafacting war in the semblance of peace,

46. Brantôme, *Recueil des dames,* 831–832.

47. Ronsard, "Le soir qu'Amour vous fist en la salle descendre," in *Oeuvres complètes,* 1:403–404. "Le ballet fut divin . . . ores il estoit rond ores long or'estroit, or' en poincte en triangle en la façon qu'on voit l'escadron de la Grue evitant la froidure." The ladies-in-waiting of Catherine de' Medici, who were known as the *escadron volant,* performed many early *ballets de cour.* On this poem's Neoplatonicism, see McGowan, *Ideal Forms,* 223–225.

Crossing, intertwining straight and at angles,
First in circles and then in squares,
Just as a Labyrinth, in which the straying path
Misleads our steps in its diverse turns.]

The labyrinth charges this passage with a very specific set of meanings, recalling Theseus's battle with the Minotaur, a half-human monster of great violence restrained by human cunning within the labyrinth Minos built on Crete. The labyrinthine dance performed by the horsemen in this way reenacts Theseus's heroic venture into the maze and the ritual dance of Delos in which the dancers' twists and turns imitated his escape from it.[48] Not only does the horse ballet evoke the blood-battle of the legendary Athenian king—who, incidentally, also helped to fight the centaurs when they disrupted the wedding of Pirithous—but its figures "imitate war in the semblance of peace," bringing the battle closer to home. Like a joust or ordinary combat, the ballet is mimetic, though its imitation cuts both ways, "contrefaisant" war with its arms and vocabulary of battle and "semblant" peace with the collective effort of the horsemen. In this way it engaged the horsemen in a performance of political deference of precisely the sort Henry III must have hoped for when he elevated faithful courtiers such as Joyeuse to prominent positions in those years.

Ronsard lards his cartel with classical allusions in order to establish a historical lineage for the new art form, joining it to the projects of classical recovery associated first and foremost with lyric and *ballet de cour.* But while song and dance did have recoupable Greek and Roman antecedents, those for equestrian ballet were thinner on the ground, leading Ronsard to fabricate a history of horse ballet almost of whole cloth. He turned to the well-substantiated accounts of pyrrhic dance, which he mounts on horseback, giving Pyrrhus and Pallas, who is said to have invented the pyrrhic, a cavalry to train:

Pyrrhe en celle façon, sur le tombeau d'Achille
Feit une danse armée, et aux bords de Sicile
Enée en decorant son pere de tournois,
Feit sauter les Troyens au branle du harnois,
Où les jeunes enfans en cent mille manieres
Meslerent les replis de leurs courses guerrieres.
Pallas qui les conduit, a de sa propre main
Façonné leurs chevaux, et leur donna le frein,
Mais plustost un esprit, qui sagement les guide
Par art, obeissant à la loy de la bride. (Lines 5–14)

48. See Wright, *Maze and Warrior,* esp. chap. 5.

[Pyrrhus, in this fashion, on the tomb of Achilles,
Made an armed dance; and on the shores of Sicily,
Aeneas, honoring his father with tourneys,
Made the Trojans jump to the chafing of the harness,
Where the young children in a hundred thousand ways
Intermingled the parries of their martial combats.
Pallas, who led them,
Trained their horses with her own hand and gave them the bit
And especially breathed into them a spirit which guides them well
With art, obeying the bridle's command.]

Ronsard may have picked up the pyrrhic reference from Claude Guichard's study of Greek and Roman funeral ceremonies, published the same year, in which Guichard describes a type of cavalry display "that the Greeks call pyrrhic."[49] But whether or not Ronsard could substantiate his reference is less important than the way it exemplifies the genuine proximity of pyrrhic dance and equestrian ballet as projects of classical recovery tied to military innovation. If, as we saw in the previous chapter, legends of pyrrhic dance produced a positive impulse toward more dancelike infantry maneuvers, so too this legend of an equestrian war dance encouraged spectators to rethink the traditional deployment of knights and light horse in battle. That same year François de La Noue was penning his *Discours politiques et militaires* and calling for cavalry squadrons to replace the old hedges of lances; less than six years later Henry of Navarre would defeat the royal forces at Coutras with his blocks of horsemen. This was the battle in which Joyeuse met his death, a rather ironic twist of fate given that the horse ballet performed for his wedding hinted at the tactical advantages of cavalry reform. But it was, in fact, precisely the personal bravura of Joyeuse and other noblemen and its association with aristocratic virtue that kept cavalry squadrons so long in coming, for each knight wished to be the first to charge and fight. La Noue saw this class-based predilection as a national disability that could be overcome only with firm authority.[50] The significance of cavalry squadrons had already been realized, but their instigation required a new order among *gens d'armes* accustomed to acting on their own. Thus the horse ballet of 1581 rehearsed not only the outward form of the new technology (tight blocks or

49. Guichard, *Funerailles, & diverses manieres d'ensevelir,* 180. "Apres on se mettoit à faire une course à cheval autour du bastiment de charpenterie là dressé pour brusler, où les chevaliers Rommains, avec les autres gents de guerre couroyent en rond, & se contournoyent comme en limaçon, gardans bonne ordonnance, & se mouvans d'une prompte & hastive desmarche, que les Grecs appellent Pyrrhichee. Les pietons en faisoyent tout autant, & environnoyent le bucher, mais à leur mode, faisans des courses feintes à l'entour fort plaisantes à voir." Guichard's source may have been Du Choul, *Discours de la religion des anciens Romains,* 74.

50. François de La Noue, *Discours politiques et militaires,* 338.

squares), but the cooperative means by which it might be achieved, promoting the sort of concerted effort that did not come easily to French nobles.

THE DUKE of Joyeuse cannot be faulted too severely for failing to see in the ballet the impending supremacy of a more collectivistic cavalry. The physical vocabulary of the ballet moved dressage away from the battle maneuvers that would have been familiar to him, and to many the steps of the *haute école*—the airs above the ground—appeared but useless show. For instance, the English translator of Claudio Corte's treatise, Thomas Bedingfield, complained that "now adaies nothing is almost used [in the schools] but sundrie sorts of superfluous dansing and pransing, which *M. Claudio* calleth *Corvette* & *pesate vaue.*"[51] Many of these elevated postures derived from natural movements such as rearing, kicking, and pivoting on the haunches, any of which could be useful in battle, but dressage transformed these useful moves into a vocabulary that was, if not wholly superfluous to the needs of an armored knight, at least far less applicable, especially given the massive weight of armor at the turn of the seventeenth century. Horses disposed to rear were not practical for battle, no matter how prized for dressage, making the separation between military training and the *haute école* complete. In his treatise of 1602 La Brou responded to critics of the *airs relevés* by pointing out that not all horses were born to serve in war: "It is a great displeasure for a knight armed to the hilt to be on a nervous horse naturally inclined to jumping . . . for a [small] number of jumps are sufficient to put the horse out of breath and [out of] combat."[52] Trying to turn a jumper into a fighter was just as great a mistake as trying to make a warhorse dance and prance.

Warhorses were meant to stay on the ground and learn the movements of the *manège bas* or low school. There the principal exercise was the *passade,* in which the horse passed repeatedly over a straight track, made a half-turn, and switched "hands" at each end.[53] Since horses tend to be right- or left-"handed," this exercise produced greater ambidexterity.[54] The turns themselves required the horse to keep the hind legs more or less in place while shifting its weight back, supporting itself on its hind legs, and turning on a small circle. In this way, the *passade* developed

51. Claudio Corte, *The Art of Riding.* trans. Thomas Bedingfield (London: H. Denham, 1584), 18.

52. La Broue, *Cavalerice françois,* 1:138. "C'est un grand desplaisir au chevalier armé de toutes pieces, qui est sur un cheval nerveux & naturellement sauteur, duquel il ne peut tirer quatre passades ou voltes de guerre, qu'il n'aye plustost souffert l'incommodité d'un nombre de sauts sur l'esquine, qui seront quelquefois suffisans de le mettre hors d'aleine & de combat."

53. Antoine de Pluvinel describes the *passade* and how to teach it in *Instruction du roy,* 86–90. Also see La Broue, *Cavalerice françois,* 2:36–50, and Pierre de La Noue, *Cavalerie françoise,* 56–76.

54. On horses favoring their right or left sides, see La Broue, *Cavalerice françois,* 1:64ff.

strength as well as suppleness. Horses destined for the *haute école* perfected the *passade* in a succession of low leaps called the *passade terre à terre* before progressing to the *airs relevés;* Pluvinel called the *passade* "the true proof of the horse's goodness, because by beginning one learns his speed, by stopping his good or bad mouth, by turning his skill and grace, and by beginning again several times, his force, vigor, and his loyalty."[55] But for dressage horses, no more than five turns would be attempted at a time, lest the horse tire and lose its good posture, whereas more were needed "to develop and maintain the breath of warhorses, so that in battle they do not come up short."[56] Innumerable *passades* required aggressive interventions (such as the crop, heavy hand, or spurs) that were rarely used in the *haute école* and that tested the strength and good nature of horses destined for battle, even if the fineness of their gait and turns deteriorated toward the end of the exercise. Thus, for warhorses, strength superseded precision.

The steps of which Bedingfield complains—the *pesade* and *courbette*—were both airs of the *haute école.* The *pesade* introduced a steed to the high airs by having him lift his forelegs off the ground while lowering his hocks, turning a rear into a more sustainable move. From the *pesade,* performed in place, students advanced to the *courbette,* a fundament of musical dressage, in which the horse raised the forelegs and hopped on the hind legs. Bedingfield's association of the *courbette* with dancing is entirely appropriate, as *courbettes* emphasized evenness of cadence and tempo. According to Pierre de La Noue's *Cavalerie françoise, courbettes* must be "diligently beaten," with "the hind legs lifting equally to the sound and the true measure of them [the *courbettes*], without having one leg or the other slow or rush by some unequal movement the even pace of the forelegs."[57] Performed in this way, *courbettes* were one of the most overtly musical of all the *airs relevés,* and given that they could be repeated in succession, they did indeed constitute not just prancing, but dancing, with all of its musical order. Moreover, in *courbettes* the horses' fore-shoulders moved up and down in an exaggerated cadence involving the whole body. The broad strokes of the *courbette* fit perfectly with the needs of the large spaces in

55. Pluvinel, *Instruction du roy,* 87. "Vostre Majesté a tres-bien jugé les passades estre la vraye espreuve de la bonté du cheval, pource qu'en partant on cognoist sa vitesse, en arrestant sa bonne ou mauvaise bouche, en tournant son adresse & sa grace; & en repartant plusieurs fois, sa force, sa vigueur, & sa loyauté."

56. Ibid., 90. "De gaigner, & de maintenir l'haleine aux chevaux de guerre, afin que dans l'occasion ils ne demeurent court."

57. Pierre de La Noue, *Cavalerie françoise,* 116. "Les courbettes sont plus basses du devant à la verité [than *pesades*], mais diligemment battuës, prestement avancées & poursuyvies de la crouppe ferme, & bien appuyée sur les jarets qu'il tient fort tendus, portant également les jambes de derriere au ton, & à la vraye mesure d'icelles, sans que l'une ou l'autre retarde, ou avance par quelque inegal mouvement la juste cadance de celles de devant."

which equestrian ballets were staged, and it is no wonder that they became a staple of the *ballet à cheval.* Ronsard names this air in his description of the 1581 ballet for the Joyeuse wedding, telling how the horses will "dance in *courbettes*" (line 15).

In light of the distinct vocabularies of movement developed in the low and high schools, it may seem unreasonable that Ronsard would claim battle-worthiness for the horse ballet of 1581, which he said imitated war. But the choreographic patterns of the ballet did indeed have a warlike quality. Renaissance horsemen, as we have seen, favored one of two battle formations—the squadron and the hedge. From these starting positions, the hedge held out few possibilities for special maneuvers. Indeed, after the charge, at best a captain might hope to rally one end of the line in order to carry out a flanking motion before the whole devolved into a general melee. But the compact and impenetrable squadron held its shape in battle, offering more possibilities for maneuvering during the engagement.

As we know, squadrons fit poorly with the individualistic modes of French knights, but not only for the cultural reasons stated above; technological differences between the lines and squares exacerbated the divide between the knightly hedge formation and the more egalitarian squadron. Unlike Navarre's squadrons of nobles at Coutras and Ivry, cavalry squares were more directly descended from the light horse than the heavy horse, and their horsemen were armed with pistols or harquebuses rather than lances, weapons which required a new set of tactics, which created their own military culture, and which registered lower on the social scale. These hurdles had to be negotiated before French knights could fight more successfully together, a negotiation instigated in part through the *ballet à cheval.*

Pistols evolved to cope with the constraints of handling firearms on horseback, where the rider had only one hand free to operate a weapon. Like matchlocks and harquebuses, pistols had a host of problems, namely a very limited range of three paces (some said that they were only effective when held against an opponent's belly) and a tricky wheel lock mechanism that often failed to ignite the powder.[58] Nonetheless, pistols could usually get off three or four shots, they required only one hand to fire, and horsemen made up for their unreliability by carrying more than one. In his *Regole militari . . . della cavalleria,* Lodovico Melzo recommended stashing two at the saddle pommel and two behind the saddle on the right side.[59] Some carried an extra one in their boot.

Among the first to switch to pistols were the German *reîtres,* who by the 1570s seem to have been arming themselves exclusively with pistols and sword.[60] Thus the

58. See Montgommery, *Milice françoise,* 186.

59. Melzo, *Reigles militaires,* 35.

60. On European mounted forces and their arms, see Delbrück, *Dawn of Modern Warfare,* 117–145.

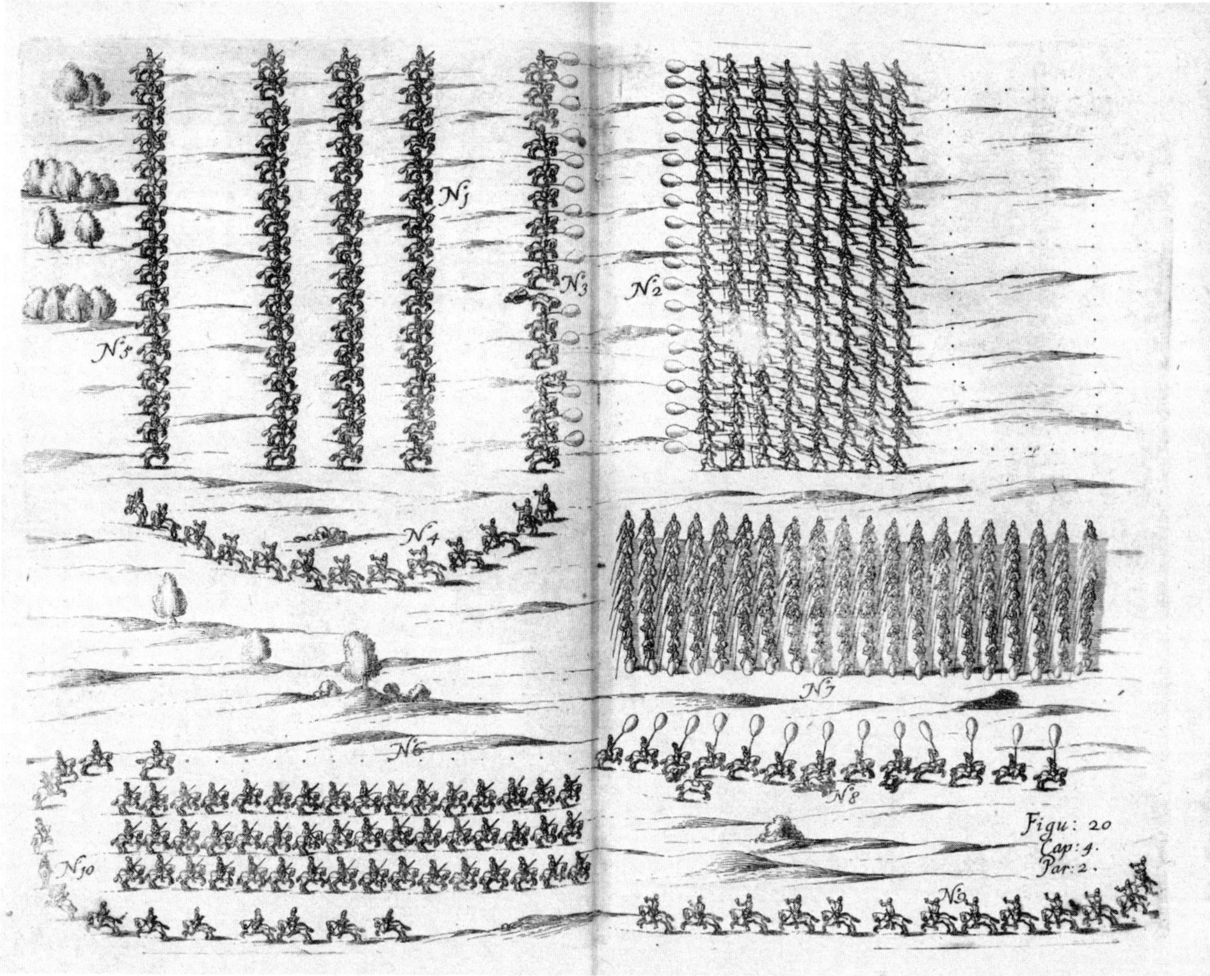

FIGURE 6.4. The Caracole. In Johann Jacobi von Wallhausen, *L'art militaire à cheval* (1616). By permission of the British Library, shelfmark ML.p.10 (2)

confrontation between the squadrons and the hedge formation was one of national style (German vs. French), of class difference (mercenary *reîtres* vs. knightly *gens d'armes*), and of contrasting arms (the pistol vs. the lance).[61] The French did arm groups of horsemen with firepower, turning their light horsemen into *cuirassiers* with pistols and mounting infantrymen with harquebuses on nags (they were called dragoons), but the lance remained, and as more than a symbol of knighthood.[62] Although the lance often shattered at the first blow, leaving its heavily armored bearer to fight as best as he could with a sword, its range of eighteen to twenty feet exceeded that of a pistol, preserving its advantage as an offensive weapon.

Pistols required new strategies for horsemen, the most innovative of which involved translating a rotational form of *limaçon* from an infantry maneuver into one for horsemen, who called it the caracole. What worked for the harquebus on foot worked in this case for the pistol on horseback, and the procedure of having a squadron advance, fire rank by rank, and peel off to retreat and reload gave the pistol greater effectiveness, at least theoretically (see fig. 6.4). Volley fire allowed the squadron to maintain its position in battle, despite the pistol's short range, and in this way the cohesive squadron could attack with persistence. In his memoirs Tavannes relates how the *reîtres* executed the caracole at the battle of Dreux (1562), one of the first reports of its use in battle:

> The [German] gentlemen were the first to arrange themselves in squadrons of fifteen and sixteen ranks, adopting a better means of fighting than the hedges of the French, and nonetheless the formations were imperfect, because, encountering resistance, the riders would not pass through their enemies and would engage them without plunging into the line. The first rank turns to the left, uncovering the second, which does the same, and the third similarly, one after the other, making a "limaçon" and retreating to the left to recharge.[63]

---

61. François de La Noue, *Discours politiques et militaires,* 355–362.

62. On the composition of the French cavalry see Montgommery, *Milice françoise,* 177–200. Montgommery relates that for a large battle, the cavalry should consist of 400 *gens d'armes* armed with a cuirasse and a blunderbuss or lance, three quadrilles of 100 light horsemen armed with pistols, and the *carabins* armed with blunderbuss and pistols.

63. Gaspard de Saulx-Tavannes, *Mémoires* (Paris, 1574), reprinted in *Choix de chroniques et mémoires sur l'histoire de France, Mémoires de Gaspard de Saulx-Tavannes, Mémoires de Boyvin du Villars,* ed. Jean-Alexandre C. Buchon (Paris: A. Desrez, 1836), 291. "Les gentils-hommes furent les premiers qui se rangerent en escadrons composés de quinze et seize rangs; prindrent une meilleure façon de combattre que celle des hayes des François, et néantmoins les ordonnances imparfaictes, parce que, trouvant resistance, ils ne passoient au travers de leurs ennemis, et tiroient à l'abordée sans enfoncer. Le premier rang tourne à gauche, descouvre le second qui tire de mesme, et le tiers semblablement, l'un après l'autre, faisant un limaçon et s'esloignant à main gauche pour recharger." Tavannes mentions the "limaçon accoustumé" of the *reîtres* again in the account of the battle of Moncontour (ibid., 377).

Tavannes admired the deftness of the *reîtres* and their squadron formation, one with which the royal army began to experiment in 1568.[64] Indeed, François de La Noue claimed that the royal victory at Moncontour owed to Tavannes's use of squadrons, calling it a victory of the squadron over the hedge (see fig. 5.3).[65] But despite Tavannes's interest in the squadron, he disliked the caracole because it evaded hand-to-hand combat and in this way appeared to be a tactic of cowards. The caracole also brought its own problems: it required a large amount of space to make the turns, which made it difficult to fit with infantry squares, and it could only be performed easily to the left (as most riders shot with the right hand). Furthermore, its regular pattern unraveled quickly if a charge could penetrate. La Noue found the caracole so unsuitable for war that he likened it to a game of prisoner's base—mere play.[66]

But La Noue's sarcastic dismissal of the maneuver ignores the significant benefits that resulted from the discipline it promoted long before the battle began. As the first identifiable drill for large numbers of horsemen, it was the caracole that taught the *reîtres* to stick together "like glue," to maintain their order in battle, and not to scatter helter-skelter in retreat.[67] Rehearsing the caracole taught riders to find and keep their places in the squadron, and it accustomed their horses to gunfire and to following complex directions. In this respect it is hardly surprising that contemporaries believed the caracole to have been practiced by the ancient Romans, for at a time when commanders raised on Machiavelli's *Arte della guerra* were busy disciplining their infantry with marching and drills in emulation of Roman legions, it made sense that the latest in cavalry discipline would likewise have a classical antecedent.[68]

No doubt most military horses and riders lacked schooling. When one considers that contemporary riding treatises regularly spent page after page telling how to accustom the horse to flapping standards, noisy trumpets, flailing drumsticks, the smell of gunpowder, the sight of fire, scurrying infantry, flashing armor, wooden bridges, and, of course, the terrors of cannon, musket, harquebus, pistol, lance, hal-

---

64. Ibid., 291–292, 329–330.

65. François de La Noue, *Discours politiques et militaires,* 336. Although La Noue participated in the battle, he overstates his case, since Moncontour was more than a simple confrontation of square and hedge. Indeed, it was probably the royal reserves of Cossé and Biron that saved the day, troops that, judging from the engraving of Perrisot and Torterelle, were not formed up in squadrons. For an analysis of the battle see Oman, *History of the Art of War,* 448–458. For Tavannes's account see Saulx-Tavannes, *Mémoires,* 375–378.

66. François de La Noue, *Discours politiques et militaires,* 360.

67. Ibid., 358–359.

68. Guichard, *Funerailles, & diverses manieres d'ensevelir,* 180.

berd, and pike, it becomes clear that just getting onto the field and lining up in any sort of formation would have been, for many horses, quite an accomplishment.[69] The most elementary schooling for a warhorse required at least a year of concentrated effort, and even then one could not expect to maneuver much or jump hedges and the like.[70] As for the riders, Pierre de La Noue observed that as soon as the lowest page felt his tail in the saddle, he set off in "confusions, turns, and curves to the right and the left" in an attempt to make *passades à la soldate* and show off his valor.[71] Such stunts were "contrary to the military art, which does everything with beautiful order and good measure"; to correct them required an intelligent and patient horseman. The caracole proved a wonderful cure-all: as a daily calisthenics, it exercised military horses for concentration, strength, and precision; it brought greener horses up to snuff by working them alongside seasoned mounts; it internalized regular formations; and it challenged the horseman to adopt the focused mindset required by the new dressage. Discipline—and this is the importance of the caracole—began with a new mentality for horse and rider.[72] As René de Menou said of French dressage, "The principal point of our science consists of judgment, to make war in the eye, to change an action from moment to moment as needed, and to work the minds of our horses sooner than their legs."[73] The choreographies engendered by the pistol raised the level of control needed in battle and forced horsemen to train together. Furthermore, the coordination required as each rank fired simultaneously and then turned and rode off demanded consistent tempo and

69. See, for example, La Broue, *Cavalerice françois,* 1:54–59, and Pierre de La Noue, *Cavalerie françoise,* 76–110.

70. La Broue, *Cavalerice françois,* 1:119–120.

71. Pierre de La Noue, *Cavalerie françoise,* 71. "Il n'y a si petit compagnon qui ne vueille paroistre bon gend'arme dés qu'il se voit le cul sur la selle, qui faict que dés aussi tost que telles gens sont dessus leurs chevaux, que c'est à eux à trouver leurs jambes, pour passader à la soldate, disent-ils, ne se souciant-pass'ils vont d'école ou non, ce leur est assez de les faire aller selon qu'ils l'entendent, & qu'ils se persuadent qu'il faut faire pour se monstrer vaillant & courageux: Mais attendu qu'ils ne considerent-pas que leurs confusions, tours, & détours à droitte & à gauche, sont contraires à l'art militere [*sic*], qui fait tout par bel ordre & bonne mesure; il faut qu'ils sçachent comme ceste passade, qu'ils appellent à la soldate, se doit faire pour estre parfaicte."

72. Delbrück stresses this point: "The significance of the often mentioned caracole is to be sought less in its direct, practical use than in the drill itself, that is, in the development of discipline that results willy-nilly from any regular drill. But it is precisely this development of discipline with which we are concerned at this moment of transition from knighthood to cavalry." *Dawn of Modern Warfare,* 124.

73. Menou, *Practique du cavalier,* 105. "Le principal poinct de nostre science consiste au jugement, faire la guerre à l'oeil, changer de moment en moment d'action, selon le besoin, & travailler plustost la cervelle de nos chevaux, que les jambes."

gait. While not nearly as musicalized as infantry drills, caracoles gathered disparate horsemen into a military corps with a good degree of mobility and a physical identity that had been internalized by each constituent.

By the first decade of the seventeenth century, caracoles were a choreographic norm in the repertory of European cavalries. Melzo, whose *Regole militari* was the most popular textbook for cavalry commanders, presumed that all mounted troops would execute caracoles in battle.[74] In addition to the cuirassiers armed with pistols, Melzo expected harquebusiers to be versed in the maneuver and even the lances, who carried pistols in reserve, to have practiced the caracole as a unit.[75] Yet true to form, the adoption of the caracole in France proved more complex. The last word in French regulations of the time, Montgommery's *Milice françoise,* assigns it only to a special detachment of light cavalry called the *carabins,* who specialized in quick and multiple attacks, hasty retreats, and skirmishes. Montgommery's treatise reveals that even though the French *gens d'armes* had begun to abandon the lance and to fight in quadrilles, the culture of the *gendarmerie* perpetuated a number of very practical distinctions between knights and light horse. Knights still charged into battle ready to fight to the death, "to vanquish or to die sooner than to turn back," which meant that the protected retreats enabled by the caracole continued to smack of the ignoble.[76] Indeed, the qualitative difference between firing a gun from a distance and braving hand-to-hand combat continued to suggest the difference between low-class soldiers looking only to save their skins and the heroism of noblemen. French generals found themselves caught between the juggernaut of improving firepower and immovable chivalric traditions that only became more entrenched in the face of the military revolution, and in this light it is small wonder that battles themselves fell out of fashion, putting the *gendarmerie* out of commission.

Those military strategists who championed the squadron validated the new formation by searching for its ancient precursors. François de La Noue claimed that the Romans never shied from trying out new formations, and even Montgommery advertised his delightfully practical handbook as being based on "the ancient order and military discipline of the legions such as the ancient French observed in imitation of the Romans and Macedonians."[77] At once newer and older than knightly warfare, Montgommery's revisions of French practice claimed to be modeled upon

74. The same was true for Wallhausen. See his *L'art militaire à cheval* (Frankfurt: Paul Jacques, 1616), 61–63.

75. Melzo, *Reigles militaires,* 26–36, esp. 30, 32, 35.

76. Montgommery, *Milice françoise,* 186, 192–193.

77. The full title of the treatise is *La milice françoise reduite à l'ancien ordre et discipline militaire des légions telle et comme la souloyent observer les anciens François à l'imitation des Romains et des Macédoniens.*

Greek, Roman, and Gallic precedents. Thus he likened the *carabins* to the famous cavalry of Parthia, which was renowned in the ancient world for its chain-mailed spearmen and horse archers.[78] It is true that horsemen of the eastern kingdom of Parthia were unusual, for cavalry played a limited role in Greek and Roman armies, and for this reason classical sources such as Xenophon's *Cyropaedia* were studded with stories of unfortunates who had come up against the Parthians in battle.[79] In addition to these Latin and Greek accounts, Montaigne treated readers of his *Essais* to an entire chapter titled "Des armes des Parthes" (2: ix). Parthians appeared in tourneys and carrousels as well, such as the 1606 carrousel, where twelve Chevaliers of Fire battled the other elements dressed as Parthians.[80] By evoking a Parthian mythology in his description of the *carabins,* Montgommery gave contemporaries a way to understand their style of skirmish and retreat as something more than cowardice. Fleet and effective, light horsemen had always born a stronger resemblance to ancient horsemen than the *gens d'armes* did; military humanism just provided strategists the historical clout they needed to bolster their claims for the new style.

At a time when the past had become a tool for reworking the present, descriptions of ancient horsemanship—however cursory—contributed to the active transformation of riding, dressage, and the cavalry, and most important, their reception. Drawn through late-sixteenth-century accounts of equestrian entries, *combats à cheval,* musical dressage, *ballets à cheval,* and the precision riding of the *carabins,* references to classical horsemanship connected all these activities with a common thread: ancient military discipline. The persistent association of military discipline with the ancients at this time shows up the broad current of renascent activity that gave birth not only to a military revolution "in imitation of the Romans and Macedonians," but to a new set of art forms calculated to promote military discipline in civilian society. When the *gens d'armes* and *chevaux légers* of the royal army staged a pageant in 1612, its "lesson of military art" aimed not only to reassure the young King Louis XIII of his army's strength, but also to give Parisians a glimpse of the beauty achieved by order and control. Gone, ostensibly, was the vicious rabble of civil war days—now Parisians had nothing to fear from a military that exuded beauty, good behavior, and order. The light horses made a caracole while the lances rode around them, passing again and again, "at which the rejoicing was very great among all the Parisians—French Greeks—who would shortly see all sorts of other

78. Ibid., 193.

79. On the Parthians see Hans Delbrück, *Warfare in Antiquity,* trans. Walter J. Renfroe, Jr., vol. 1 of *History of the Art of War* (Lincoln: University of Nebraska Press, 1990), 441–449.

80. See the description in Claude Malingre, *Les annales generales de la ville de Paris* (Paris: Pierre Rocolet, Cardin Besongne, Henry Le Gras, vefve Nicolas Trabouilliet, 1640), 539–540.

nations assembled" for the engagement of Louis XIII to Anne of Austria.[81] Overall, the reconstitution of ancient military practices provided occasions to declare that French horsemen were latter-day incarnations of the Parthians, that Henry the Great constructed the Place Royale in express imitation of the Roman amphitheater used for the Olympics, and that the Parisians themselves were "Grecs François," making Paris into a new Athens and France into a new Republic.[82]

Ronsard's assertion that *ballet à cheval* imitated war in a peaceful guise was not mere rhetoric, for it struck viewers owing not only to the sheer majesty of the dancing steeds, but especially to the discipline required to bring it off, a discipline shared both by an increasingly "civilized" military and the pacified circles in which softened manners grew so pronounced after 1600. Indeed, its refinement witnessed a symbiosis of chivalry and civility so complete that stylized training to warfare could double as training to stylization, enough so to turn the equestrian ballet into a practice for being a gentleman at large, even when one was not engaged in a military career. Military academies often staged carrousels, and some educators insisted that noble curricula feature *ballets à cheval.*[83] Pierre de La Noue titled a chapter of *La cavalerie françoise* "How to teach horses to dance in order to use them in carrousels," and Claude-François Ménestrier presented equestrian ballet as a standard genre of divertissement in his *Traité des tournois, joustes, carrousels et autres spectacles publics.*[84] Ballets and carrousels persisted because they also operated politically, advancing the needs of nobles, lords, and kings to reinforce the visibility of the armed force at their disposal. In them, the social, martial, and "harmonic" orders that have concerned us throughout this book merged quite perfectly, and in none more so than the *ballet à cheval* choreographed by Antoine de Pluvinel for the 1612 carrousel celebrating the engagement of Louis XIII, a spectacle that epitomizes the

81. Pierre Beaunis de Chanterain, sieur des Viettes, *La resjouyssance des Compagnies qui auroyent faict monstre devant le Roy & la Royne Regente,estans armez à Paris* [Paris, 1612], 6–7. "Les Trompetes Royales recreatives commencerent à claironner, la Cornette, & guidon desployez aupres des capitaines portez, qui plenierement en ordre faisoyent le limasson, Les lances y paroissoyent en entourant iceux cavaliers courtisans, passant, & repassant, dont la resjouissance estoit fort grande entre tous les Parisiens, Grecs François, qui en si peu de temps auroyent veu tant de sortes de nations assemblez, faisant le recit des autres compagnies, & Regimens qui sont aux champs."

82. *Histoire generale de tout ce qui s'est passé au Parc Royal sur la resjouïssance du Mariage du Roy avec l'Infante d'Espagne* (Paris: Anthoine du Brueil, 1612), 5. The Place Royale (now the Place des Vosges) was described as the "huictiesme merveille du monde . . . qu'Henry le Grand, comme il semble estre plus divin qu'humain, sçachant ces triomphes devoir estre, fist construire expres à l'imitation de l'Amphiteatre Romain, où se celebroient les jeux royaux, & olympiques: Mais toutesfois incomparable, pour estre plus admirable, & plus beau."

83. See Motley, *Becoming a French Aristocrat,* 145–148, and Saint-Germain, *Advis de l'établissement de quatre académies,* 26.

84. Pierre de La Noue, *Cavalerie françoise,* 141–145; Ménestrier, *Traité des tournois.*

relationships between music, discipline, and arms in early modern France, and to which we now turn.

DRESSAGE, ABSOLUTISM, AND THE 1612 BALLET

By 1600, civility had left its mark on French equitation and made the equation of beautiful riding with virtuous behavior complete. "It is certain that only the most beautiful minds are apt for the most beautiful exercises," remarked La Broue, adding that "the perfection of such knowledge can never be communicated to . . . weak and crude minds."[85] Here we find precisely the reflexive relationship between body and soul that made good manners so compelling. La Broue invested the body with signs of inner morality, locating virtue in "just stature" and graceful movement. Indeed, the definition of moral rectitude through upright posture was beautifully born out by his notion that the good rider required a still and balanced soul.

As an art perfected through careful attention to balance, number, and measure, riding was itself harmonic or musical. La Broue's cavalier must "have a lot of natural judgment, even of proportions: because all the airs and most beautiful exercises are composed of numbers and measures, and of many equalities that must be carefully observed."[86] He must know music: "It is most unlikely that he who cannot appreciate or comprehend the harmony, air, and measure of music, and consequently of instruments and of dance, could ever understand the airs and proportions of our schools well."[87] Little separated the physical geometry of the airs from the sounding proportions of music. This, of course, had an eminently practical origin, for, like Fiaschi, La Broue is simply saying that sensitivity to beat and tempo in a horse's gait will make it easier to ride well. Music is the key to coordination. But there is more to it than this, for La Broue insists upon a moralized aesthetic that freighted music with the ability to enlighten and harmonize. Measured or "musical" behavior shows a man's *honnêteté,* his sovereignty over his own inner life, and his suitability to govern those below him, beginning with his mount.

85. La Broue, *Cavalerice françois,* 1:4, 3. "Il est certain, qu'il n'y a que les plus beaux esprits, qui soient propres pour les plus beaux exercices." "Et pour ne point dissimuler, je pense, qu'il n'advient jamais que la perfection d'un tel sçavoir se communique à certains esprits foibles & grossiers."

86. Ibid., 1:3. "Il est aussi necessaire que le Cavalerice aye beaucoup de jugement naturel, mesmes aux proportions: d'autant que tous les airs & plus beaux maneges sont composez de nombres, de mesures & de plusieurs egalitez qu'il faut soigneusement observer."

87. Ibid. "Je puis asseurer qu'il est fort malaisé que celuy qui ne peut gouster ny comprendre l'harmonie, l'air & la mesure de la musique, & consequemment des instrumens & de la dance, puisse jamais bien entendre les airs & proportions de nos escoles. Tellement que voyant un Cavalerice sçavant en son art, sans doute on peut croire, qu'il est nay pour bien faire beaucoup d'autres choses honnestes."

The cooperative relationship between horse and rider demonstrated the smooth function of hierarchical orders, exemplifying the proper relationship between subject and sovereign. When amplified in the form of an equestrian ballet, dressage made a strong political statement. Ronsard said as much at the end of his cartel for the Joyeux ballet: "these armed Centaurs are come at the renown of this high Prince of France, to instruct his people and to render them as easy as their horses are docile under the bit . . . to the end that all will learn to serve him." This propagandistic feature of ballet, whether *à pied* or *à cheval,* was part and parcel of ballet from the very beginning, and it continued to play a vital role in the organization of seventeenth-century ballets, directing choices such as the timing, themes, and the participants in the productions. So it was that a highly political event inspired Pluvinel's famous horse ballet, which celebrated the engagement of Louis XIII to the Spanish Princess Anne of Austria and of Louis's younger sister Elisabeth to Prince Philip of Spain.

With the assassination of Henry IV in 1610, Marie de' Medici was forced into a shaky regency that would never have survived the war with Spain that was perennially on the French horizon. As peace was her next best option, she proceeded to negotiate a double marriage with the Habsburgs, even though Louis XIII was only ten and her daughter younger still. The prospect of an alliance with the arch-Catholic rulers in Spain threw French Protestants into a panic—Henry IV had sympathized with the Protestants, but it looked as though Marie would be less mindful of their needs. And so she came into conflict with many nobles. With infallible political intuition, she and her advisors decided to present the marriages as a fait accompli, with a huge public spectacle to celebrate the engagement. Thus, although the marriages would not take place for another three years, the carrousel of 1612 symbolically married the two realms in the eyes of public. It is of great significance indeed that the designers of the carrousel sought popular approval for the alliance, for contrary to the fetes held at court, this festivity hoped to awe the whole of Paris—if not of France—and it grew to magnificent proportions. One source reports that 200,000 witnessed the event.[88] It was held in the Place Royale, one of a series of royal squares begun by Henry IV and Sully in 1605 as part of a project to improve the city, generate rent for the crown, and glorify the monarchy. The stunning symmetry of each *place* would ultimately frame a centrally placed equestrian statue, of Louis XIII in the Place Royale and of Henry IV in the Place Dauphine, launching the tradition of urban development and statue campaigns furthered by Louis XIV with the Place des Victoires, the Place des Conquêtes, and provincial

88. Pluvinel, *Maneige royal,* 63.

royal squares.[89] Like the square itself, the carrousel was a royal event that took place in and for the city, not unlike a Roman circus.

The carrousel was organized as a knightly combat over a large "Palace of Felicity" built in the square, which symbolized the marital happiness of the royal couples. During the entries of the knights, a flaming Mount Aetna spouted lava, "giants" processed on stilts, and lions pulled carts, to the utter delight of the crowds (see fig. 6.5). Despite the carnival atmosphere the carrousel must have created, its chivalric narrative underscored the princely virtue it was meant to celebrate with its parade of chevaliers from France's most prominent families. Five Knights of Glory defended the Palace of Felicity against a series of knightly assailants: the Knights of the Sun, the Knights of the Lily, the Knights of Fidelity, the Knights of the Universe, and the Roman Conquerors. The narrative for the tourney came right out of Arthurian legend, with a hefty dose of classicism to include the glory of the Roman Republic.

By this account it would seem that not much had changed since the days of Philip the Good, when the Knights of the Golden Fleece met in Burgundy for banquets, tourneys, and vows to crusade to the Holy Lands. To some extent this was true, for the knights had come to face off against one another at the jousting barrier. But the entries of the knights, accompanied as they were by scores of footmen, mounted squires, dozens of trumpets, carriages, floats, ships, and elephants, were so spectacular and so long that the jousting barely began before nightfall. And not only was the jousting itself overshadowed by the elaborate parade, but the Knights of the Lily stole the show by stopping to perform a brief equestrian ballet before the royal pavilion (see fig. 6.5, upper left quadrant of the square). The printed accounts of the carrousel all lavish special attention on the ballet, describing it figure by figure and even including diagrams of its geometrical choreography.[90] The spectators packed into the Place Royale seem to have found it a stunningly new and wonderful thing to behold, and the queen herself requested a repeat performance in the courtyard of the Louvre three days later.[91]

---

89. See Michel Martin, *Les monuments équestres de Louis XIV: Une grande entreprise de propagande monarchique* (Paris: Picard, 1986).

90. Descriptions of the ballet can be found in François de Rosset, *Le romant des chevaliers de la gloire* (Paris: Veuve Bertault, 1612), fols. 69r–74v; Honoré Laugier de Porchères, *Le camp de la place Royale* (Paris: Jean Micard and Toussaint Du Bray, 1612), 116–147, which includes diagrams of the ballet figures; Pluvinel, *Maneige royal*, 59–63, which includes engravings of the ballet and the riders' costumes; and *Histoire generale de tout ce qui s'est passé au Parc Royal*, 14–16. See also Jacques Vanuxem, "Le Carrousel de 1612 sur la Place royale et ses devises," in Jacquot, *Fêtes de la Renaissance*, 1:191–203.

91. Laugier de Porchères, *Camp de la place Royale*, 142.

FIGURE 6.5. Carrousel in the Place Royale (c. 1612). Engraving by J. Ziarnko. Bibliothèque nationale de France, Cabinet des estampes, Qb1-1612

Pluvinel choreographed the ballet and rode in it with the director of another military academy, Benjamin, the Duke of Vendôme, the son of the Duke of Épernon, La Valette, the Baron of Pontchasteau, and Jean Zamet, a captain in the king's guard and son of the very rich Italian financier and friend of Henry IV, Sébastien Zamet. La Valette and Pontchasteau were probably students or recent graduates of Pluvinel's academy, and Vendôme certainly was; they were, in any case, all young *grands.* Zamet and Benjamin were not, but Benjamin was clearly an expert rider, and Zamet probably was (if not noble, Zamet's participation exemplified the ennobling quality of money—when once asked by what title he should be announced, his father is said to have responded, "The seigneur of seventeen hundred thousand *écus*").[92]

The organization of the entry as a triumphal procession reinforced the social rank of the knights with emblems of military valor. Fourteen trumpets led the way, their brassy tones the proper sign of royalty and the instrument of the cavalry. Twenty-four horses followed, led by footmen, followed by twenty-four mounted pages, all sumptuously attired in crimson, black, and white (the queen's favorite colors) with yards of braided trim, silvered spurs, and a multitude of feathers in their hats. Then came the machine—a cart supporting a great pedestal surmounted by two golden crowns representing the marriage of France and Spain. All about the cart hung swords, lances, javelins, targets, drums, and branches of laurel and myrtle, much the same as those one finds tumbling down the margins of frontispieces in books. Below were helmets covered with flowers and eight boys dressed as Cupids who tossed flowers, fruits, and perfumes from the cart and trampled the fallen arms under their feet, a sign of love's triumph over war (or love's triumph *after* war). In their midst sat a Trojan Venus on a throne. Then came a mounted orchestra, six squires carrying the knights' battle lances painted in crimson, black, and white, and six more squires with their shields emblazoned with devices, some fictitious, some real.[93] The knights wore helmets covered with plumes and feathers, jeweled belts, ruffs, silver cloaks, and white boots embroidered with silver to match the silver fleurs-de-lis embroidered on their skirted doublets. Their Spanish mounts were caparisoned with long saddlecloths to match, and they sported white plumes and feathers on their heads and at their tails. This synopsis, by the way, reduces page after page of description in sources that hoped to capture the magnificence of the event with detailed attention to the velvets and silks used, the braid, embroidery, and jewels that encrusted them, and the egret feathers, ostrich plumes, silvering, and other metalwork that embellished the costumes.

---

92. Quoted in Pillorget and Pillorget, *France baroque, France classique,* 2:1239.

93. Vendôme, La Valette, and Zamet used their own devices; the others were fabricated. See Rosset, *Romant des chevaliers,* fol. 74r. Clearly, the accounts vary. In general, I rely on Rosset with some details drawn from Laugier de Porchères.

For all this splendor, however, it was the musical equitation that set this entry apart from the others, giving it a sumptuousness and register of meaning surpassing that of other elements in the carrousel.[94] One might even say that while the rest of the carrousel took its cues from traditional tourneys, the entry of the Knights of the Lily attempted to negotiate the terrain between the Neoplatonic spectacles held at court and the more commonplace parades, floats, and processions of civic ceremonies. Like a royal entry, with its elaborate succession of triumphal arches, open-air theaters, and carts that served as ephemeral stages for lyric poetry and song, the entry of the Knights of the Lily brought an extremely humanistic commodity—*ballet de cour*—out of the court.

When the cart arrived before the royal pavilion, the Trojan Venus sang three songs of praise written in so-called ancient meters, the first "anapestic," the second "paeonic," and the third "choriambic."[95] Their unremarkable texts express the knights' hunger for battle and laud the queen, who shines like a sun over the French empire; they were penned by Guillaume de Baïf, a minor poet and son of Jean-Antoine de Baïf.[96] Yet no matter how outwardly ordinary, the classical meters cited in Honoré Laugier de Porchères' description refer unmistakably to the work of Baïf and Courville's Academy of Poetry and Music and the *musique mesurée* produced there.

The music to which the horses danced was also described as *mesurée à l'antique,* perhaps owing to the fact that the music for the dance and that for the songs was one and the same. Laugier de Porchères says that after the songs, the horses formed a triangle facing the king and queen and began to dance "to the cadence . . . of metric airs composed according to ancient numbers or measures."[97] This remark iden-

94. By Rosset's account fifteen shawms and cornetti and four drums played for the ballet; according to Laugier de Porchères (who says they marched on foot) the band included twenty-seven sackbuts, shawms, cornetti, violins, fifes, and drums; and the anonymous *Histoire generale de tout ce qui s'est passé au Parc Royal* says twenty drums, shawms, and fifes.

95. Laugier de Porchères, *Camp de la place Royale,* 130–131. The three *airs* are conserved in two musical sources, one polyphonic (*Airs à quatre de différents auteurs [troisième livre]* [Paris: Pierre Ballard, 1613], fols. 49v–51r) and the other for voice and lute (*Airs de différents auteurs, mis en tablature de luth, quatriesme livre* [Paris: Pierre Ballard, 1613], fols. 53r–54r). The texts are attributed there to Guillaume de Baïf. The chansons are not attributed to any composers, but they appear in the polyphonic print in the middle of a series of *airs* by Charbonnières.

96. Also see Nicolas Métru, *Recueil des vers du sieur Guil. de Baïf, mis en musique par Nicolas Metru, chantez en l'allégresse de l'heureux retour du Roy* (Paris: Pierre Ballard, 1628).

97. Laugier de Porchères, *Camp de la place Royale,* 136. "En cest endroit ils commencerent à la cadence du mouvement des airs Metriques composez selon les nombres ou mesures anciennes." The music for the first song, "Approchons du tournoy," is the first *air* given in Philidor (for which see ex. 6.1); "Comme à l'assaut" is set to the first section of the second *air,* and "Astre qui luisés" is set to the twelve bars that follow.

tifies the ballet as a classical art recovered, for just as Baïf's project hinged upon the perfect conjunction of music, language, and the body at the level of the single note, syllable, and step, so too the *ballet à cheval* could arguably be said to be built upon the persuasive meters of classical poetry. Viewers probably did not liken the simple rhythms of the dance music to those of classical metrics, but we do know that they were impressed with the synchronicity among the horses and their synchronicity with the music. François de Rosset reported, for example, that "the Knights advanced in the same figure, making their horses go half-way into the air with such equilibrium and proportion that all rose up and came down at the same time."[98] He notes how the horses maintained their placement in a triangle even while moving forward, how they all rose up to the same height, and how they rose and fell together in time to the music. Music banished chaos and reigned in the horses according to a felicitous set of numbers discovered by musicians of antiquity.

At the end of some figures, the riders all "stopped like statues" when the final cadence of the music sounded, reminding us of the magic significance of immobility, of Circe freezing the dancers in their tracks in the *Balet comique de la Royne,* and of other sorcerers and enchantresses such as Armide, Ismen, and Caelide, who interrupted court ballets with waves of their wands. In the *ballet à cheval,* the music that mobilized the spectacle ends, halting the horses and in this way drawing attention to its charm.[99] Even this simultaneous halt showed the command the riders (and the music) exerted over the horses, because horses suitable for dancing had little patience for standing still. In this light it is all the more telling that the engraving of the event captures the horses in midair, their haunches down and their front legs daintily drawn in as they hold themselves aloft. They balance in eternal equilibrium in the plates of Crispan de Pas (see fig. 6.6), performing one of the *courbettes* that Pluvinel used for the knights (for the squires he used *terre à terre*s).

Pluvinel's figures were geometric, beginning with a triangle for a march, then regrouping in a set of two nested circles for most of the ballet, with the knights on the inside and the squires on the outside (fig. 6.6). The circles broke apart and reformed as the horsemen drew close to one another and made *voltes* (tight circles) before returning to their places; they also exchanged places, turned in place, and made *courbettes* in place. At end of the ballet, the knights formed two lines facing one another, rode through each other's lines, turned, and returned to the center, taking one another's hands and turning, after which the squires formed a chain and

98. Rosset, *Romant des chevaliers,* fol. 72v. "Les Chevaliers s'avancerent en la mesme figure, faisants aller leurs chevaux à mez-air, avec tant d'esgalité, & de proportion, que tous s'eslevoient, & baissoient à mesme temps. La mesme cadance perduë, sonnant à la fin de cest air, les fit tous arrester comme immobiles."

99. Ibid.

FIGURE 6.6. "Ballet à cheval." Engraving by Crispan de Pas and Crispan de Pas le Jeune in Antoine de Pluvinel, *L'instruction du roy* (1625). By permission of the British Library, shelfmark 788.g.1

passed around one another other in succession. The triangles, circles, and chains alluded unmistakably to the cosmology and military formations of court ballet (Beaujoyeulx's *bataille rangée* was a similarly interlaced chain), even while spectators with an eye for military horsemanship could have recognized figures based directly on the *passade à la soldate* and caracole. The choreography was at once courtly and military, graceful and powerful, musical and bellicose, just like the courtiers themselves, and at the end of the ballet, the knights formed up in two lines that "charged" though one another, an "assault" resolved as the riders returned to the center, joined hands, turned, and then restored their perfect circle, in this way truly "contrefaisant la guerre au semblant d'une paix."

The innumerable turns to right and left showed off the even handling of the horses, while their crossed paths, the way they turned together while the knights held hands, and their weaving between one another evinced their willingness to work hard in close proximity to one another. The relentless *courbettes* and *terre à terres* were also hard work, showing the horses' strength as they arched their necks, tipped their heads down, and foamed at the mouth with deep breaths. Such moves must have taxed even the strongest horses, so the ballet was rather short (about ten minutes), and Pluvinel designed it in a series of figures that the knights and the squires performed in alternation, which allowed the horses to rest. For while the riders maintained a quiet poise, the horses exerted themselves tremendously during the ballet in a way that denied the aesthetic of effortlessness projected by court dance and suggested instead the tirelessness of military might. The long skirts of their costumes hid the sheen of sweat on the horses' coats and drew attention downward to what was visible of their legs, but there is no question that their labor must have projected, even if only through the heavy and insistent thud of hoofs in time to the music.

THE MUSIC for the ballet was likely by Robert Ballard, who served as *valet de chambre* and *joueur de luth ordinaire* to Henry IV, taught Louis XIII the lute, and probably taught at Pluvinel's academy as well.[100] Ballard published intabulations of the ballet music in his second book of *Diverses pièces mises sur le luth* of 1614—no doubt as a marketable memento of the carrousel, much like the music from *ballets de cour* included in the same print.[101] A five-part version of the dance music, copied much later, survives in the Philidor collection, a series of manuscripts produced by a professional workshop of copyists associated with the music librarian of Louis XIV,

100. On Ballard's life see Ballard, [*Premier livre de luth*], xi–xii.

101. Robert Ballard, *Diverses pièces mises sur le luth* (Paris, 1614), ed. André Souris, Sylvie Spycket, and Jacques Veyrier as *Deuxième livre (1614) et pièces diverses* (Paris: Centre National de la Recherche Scientifique, 1964).

André Danican Philidor (see ex. 6.1).[102] A few volumes are retrospective, organized chronologically and including music from earlier monarchies. The *Ballet à cheval* is one of several ballets from the reign of Louis XIII, works that evince the politics of a music history that found its origin in the backward-trained gaze of Renaissance

Example 6.1. [Robert Ballard], "Ballet à cheval." From the arrangement included in the Philidor Manuscript, Bibliothèque nationale de France, Rés. F. 494.

102. Collection Philidor, Rés. F494, 106–110, Département de la Musique, Bibliothèque nationale de France. Philidor (1647–1730) was the organizer and principal copyist of the collection, which is widely dispersed and has been partly destroyed. The music to 1700 includes some thirty-five volumes in the collection of the Paris Conservatoire, most of it dance music. See Catherine Massip, "La collection musicale Toulouse-Philidor à la Bibliothèque nationale," *Fontes artis musicae* 30 (1983): 184–207; and Denis Herlin, *Catalogue du fonds musical de la Bibliothèque de Versailles* (Paris: Klincksieck, 1995), xxv–xxx.

Example 6.1. (*continued*)

Continued on next page

Example 6.1. (*continued*)

Example 6.1. (*continued*)

4me Air

*MS has *d′*.

**MS has *e*.

Continued on next page

Example 6.1. (*continued*)

Fin du Ballet à cheval

humanists. Like the *Livre de meslanges* of 1560, for which Ronsard wrote the preface, historical anthologies contributed to definitions of Frenchness and, ultimately, to the idea of a French state.[103]

With the laurels of antiquity and political significance wreathing the ballet music, one anticipates something extraordinary, but the score is straightforward in-

103. Kate van Orden, "Imitation and 'la musique des anciens' in Le Roy & Ballard's 1572 *Mellange de chansons*," *Revue de musicologie* 80 (1994): 5–37.

deed. The suite opens with a military march and proceeds with a series of branles, the simplest of all the social dances choreographically and hence the dance with the bluntest musical idioms. The twenty-four branles in *Orchésographie* certainly attest to the dance's popularity, as do those in Ballard's *Diverses pièces.* Branles came in both duple and triple time, and Ballard made good use of both in his suite, providing branles with dotted figures for the knights, who had the most difficult passages to perform, and furnishing a brighter *branle gay* in three for the squires. Rosset calls these tunes the *air des courbettes* and the *air terre à terre* and specifies that the music changed to a "gayer air" for the squires, which suggests the faster tempo that would have been appropriate for their lower and quicker steps.[104]

The only extraordinary feature of the ballet music is a short coda in triple time that ends the second, third, and (in transposition) fourth *airs* as they appear in the Philidor manuscript. This is most likely the "final cadence" referred to in Rosset's description that signaled the end of each figure, an ingenious place-marker for the riders, who may well have had trouble hearing the music amid the clamor of the surrounding procession. The orchestra consisted of instrumentalists who, according to the engraving in Pluvinel's *Instruction du roy,* stood in groups at each corner of the dance space, providing a sort of surround-sound that must have brought its own challenges—another reason the occasion required bold music with clear endings (see fig. 6.6). The engraving shows four trumpets, two drums, four fifes, four shawms, and two cornetti, though sackbuts would certainly have been necessary to play the lower parts of a five-voice arrangement. This larger scoring, we might also note, is unusual compared to the other ballets in the Philidor source, for which a two-voice shorthand is used, though it is a typical sonority for the early seventeenth century. In any case, the five-voice notation distinguishes the ballet as it appears in this later source.

The contrasting *airs* for the knights and the squires built a social distinction into the ballet suggestive of the traditional military hierarchy in which *gens d'armes* rode into battle on the best horses in the kingdom, accompanied by light horsemen and an entourage of grooms, squires, and pages. Although the horse ballet rehearsed the cutting-edge synchronized riding of light cavalry corps, the narrative trappings of the carrousel told the story of military virtue two ways, with the lances of the *gens d'armes* (transported to the joust by other squires) and the precision riding of the *carabins.* The ballet actually brought the new collectivity of the cavalry corps to the *gens d'armes,* enticing them to set aside their cumbersome lances and giving them a new grammar for their heroic acts. What the *reîtres* and *carabins* had proven to be effective in battle was here, in the synthetic space of spectacle, tried on by French

104. Rosset, *Romant des chevaliers,* fols. 72v–73v.

noblemen. The carrousel created an indisputably chivalric context for the display of disciplined ensemble riding, instilling the mental and physical foundations required for collective efforts in the field.

Here we might recall that Zamet was a military man who served as a captain in the guard of Henry IV. Pluvinel, too, began his career in the service of the future Henry III. As for so many Frenchmen, their outward manifestation of military valor eventually garnered them noble status (even though neither seems to have seen battle): Pluvinel and Zamet both became *gentilshommes de la chambre du roy,* and Pluvinel received the title of seigneur de la Baume. Styled as nobles of the sword, these men rode the cusp of an age in which style itself would define the nobleman. The mannerisms of high dressage and the years of training required to perfect them produced facile horses, it is true, but in the greater scheme of life in and around the court, it was the harmonic behavior of the riders that counted, for it emphasized the king as the coordinating power of the aristocracy. The age of jousting, duels, private armies, and aristocratic rebellion would come to an end, superseded by the external coordination of the nobility through, among other things, music and ballet. While the *noblesse d'épée* redefined itself during this transition through stylized martial arts, such definition bore the imprint of royal supremacy and a loss of autonomy.

Just as Pluvinel's academy readied a new breed of nobility to serve a monarchy that would become more absolute as the century progressed, the music in Pluvinel's ballet hinted at its value as an instrument of the state.[105] Music regulated a performance of proportion, coordination, and harmony, and as subjects of collective behavior—at least fleetingly—the Knights of the Lily suppressed individual action in favor of consistent motion. Before they faced off one-on-one in the tourney, the dancers faced the monarch like a cavalry squadron ready to serve his majesty. Although France would soon enough be plunged back into civil war, and Louis would shortly be plagued with worries about the strength of Vendôme's army, the knights' ritual defense of the court was carried out within a musical order that was analogous with the harmonical governance of the king.

The knights' participation in the festive rule of music reharmonized their militaristic traditions to the new demands of life at court. But such art forms were not to remain useful forever. Gradually the synonymy between the military career and the noble profession loosened, and while the martial arts continued to register nobility of character, increasingly it was the form of noble display and not the content

105. For a few of the sources that express the view that absolutism was "always in the making, but never made"—an interpretation that of course implicates public spectacles all the more directly in its fabrication—see Cosandey and Descimon, *Absolutisme en France;* Burke, *Fabrication of Louis XIV;* and David Parker, *Making of French Absolutism.*

itself that came to be so important. Ultimately, it would be the stylishness of ballet—its demonstration of the disciplined behavior defining civility, and not its military discipline per se—that expressed nobility.

SUCH A MAGNIFICENT carrousel would not be mounted again in Paris before 1662, when Louis XIV appeared as the Emperor of the Romans, his shield emblazoned with a sun and the Caesarean motto UT VIDI VICI (as I saw, I conquered). Quadrilles were followed by running at the quintain and tilting for rings, and whereas Louis XIII had been too young to ride in 1612, in 1662 Parisians saw their twenty-three-year-old king parade on horseback at the center of his court and expertly spear sixteen heads of Turks, Negroes, and Medusas at the tourney. This was the first great public spectacle of Louis's reign, and not only was the young king sure to employ the solar device that would serve him so well, but his magnificent imperial garb reaffirmed his political power in the face of a lost bid to acquire the title of emperor in 1658 (it went to Leopold I). The Roman allusion likewise evoked Julius Caesar even as its references traversed the equestrian events of the late Renaissance that had themselves harnessed the symbols of ancient triumph to the knightly culture of the *gendarmerie*.

The timing of the 1662 carrousel coincided with Louis's seizure of power, a fact that implicates the 1612 carrousel as a ceremonial model. Both events were precipitated by the quest for public acclamation at a juncture of political sensitivity. Beneath the thematic and formal parallels, then, the carrousels aimed the tried-and-true strategies of courtly divertissement outward into the city, inviting Paris to fete the king in a genre developed by the last Valois. One notices how little the *marques de noblesse* have changed, how essential to courtiership remains the ability to ride and handle the lance, and the striking political and ceremonial continuity from earlier reigns.

In his memoirs, Louis XIV lingers over the 1662 carrousel as an example of how princely spectacles affirmed the social contracts between the king and his people. There are nations, he says, where majesty consists in hiding oneself, and kings govern by fear and terror. But French subjects, in contrast, have free access to their prince. Spectacles for the court and the people maintain this right of passage, which "is an equality of justice between the king and his subjects that they maintain as in a gentle and honest society, notwithstanding the almost infinite differences of birth, rank, and power."[106] In Louis's telling, spectacle ratifies a social order based not on

106. Louis XIV, *Mémoires de Louis XIV: le métier de roi*, ed. Jean Longnon (Paris: Editions Tallandier, 2001), 133. "C'est une égalité de justice entre lui et eux, qui les tient pour ainsi dire dans une société douce et honnête, nonobstant la différence presque infinie de la naissance, du rang et du pouvoir."

servitude, but on just relations in which all participate willingly, as if the éclat of kingship emanated from the very light of day in which French sovereigns appeared openly before their subjects.

The relevance of spectacle to politics, Louis insists, is immediate, for it was the confrontations of the Fronde that led him to seek order through pleasant means: "The more I was obliged to cut off from this excess, and by the most agreeable remedies, the more it was necessary to conserve and cultivate with care all that—without diminishing my authority and the respect that was due to me—bound my people and especially people of quality with me by affection."[107] Confronted with aristocratic rebellion, in which—notably—Condé and Turenne turned the "royal" techniques of military discipline against the king, Louis responded with the tactics of civility. In addition to public spectacles for the people, he diverted nobles with courtly festivities: "This society of pleasures, which gives courtiers an honest familiarity with us, touches them and charms them more than one can say."[108]

Pleasurable as they may have been, elaborate divertissements such as the musical garden parties held at Versailles in the summer of 1664 were far from innocent bids for renewed affections between a king and his subjects. For the very society of pleasures of which Louis spoke constituted an artistic disarming of the nobility. The centralization of patronage at the court that would move to Versailles—indeed, work on the palace itself—pulled the finest artists from noble service into royal employ. The painter Charles Le Brun, the architect Louis Le Vau, and the garden designer André Le Nôtre were all to collaborate on the palace beginning in the 1660s. But before Versailles, they built the famous chateau at Vaux-le-Vicomte for Louis's finance minister, Nicolas Fouquet, an edifice that became the object of some envy. Never completed, the work on Vaux-le-Vicomte halted after a series of notoriously splendid fetes held there in August 1661. The chateau, gardens, and divertissements so upstaged the pleasures of Louis's court that Fouquet was imprisoned in the Bastille on 5 September 1661 (the king's birthday). In the *société de plaisirs* that followed, nobles no longer participated in spectacles, but watched spectacles paid for by the king. Not only were they in this way disarmed of artistic power, but divertissement itself would become demilitarized, for the 1662 carrousel was the last spectacular expression of the noble culture of arms in this form.

It may well be, as David Parker has it, that "absolutism was always in the making but never made," that it represents a persistent attempt to instate a royal au-

107. Ibid., 134. "Mais plus j'étais obligé à retrancher de cet excès, et par des remèdes plus agréables, plus il fallait conserver et cultiver avec soin tout ce qui, sans diminuer mon autorité et le respect qui m'étais dû, liait d'affection avec moi mes peuples et surtout les gens de qualité."

108. Ibid. "Cette société de plaisirs, qui donne aux personnes de la cour une honnête familiarité avec nous, les touche et les charme plus qu'on ne peut dire."

thority never fully manifest.[109] If so, the performance of power proves all the more foundational to the absolutist project. Ceremonial codified the stock of symbolic actions, objects, emblems, and words that conveyed *royale grandeur,* while the performance of ceremonies made it manifest. Alongside ceremonial, court spectacle was a highly controllable form in which to establish hierarchies and initiate longer-lasting social patterns. This is all to say that without the perpetual reenactment of the founding myths of kingship—through ceremonies that by the time of Louis XIV must have seemed as eternal and God-given as kingship itself—absolutism might have ceased to exist. The establishment by Henry III of a grand master of ceremonies at the moment of most extreme threat to his dynasty evinces this essential dependence of sovereignty on the symbolics of power, while the inheritance of ceremonials from the time of the last Valois marked the reign of Louis XIV with its imprint.

The continuity of musical genres from the early seventeenth century to the golden age of the Sun King was not great—the *air de cour* disappeared after mid-century, and the *ballet de cour* gave way to *tragédie en musique,* though the traditional taste of French kings for Low Mass accompanied by motets blossomed at the Royal Chapel later in the century. What the crucible of the religious wars bequeathed to future kings was systems through which social, religious, military, and political orders could be elaborated and internalized by French subjects. And here music maintained its own particular sort of sovereignty. For if the arts of power aimed to engage subjects in the orders of absolutism—its hierarchies, armies, chivalric confraternities, Gallicanism, academies, and court—few could surpass the effectiveness of music. In social dance, marching, weapons drills, and musical equitation, men submitted physically to the rule of music, in psalm-singing and hymnody they professed communal beliefs, and music always made becoming part of something outside oneself pleasurable, tuneful, and physically memorable, so that should the king's fiddler call the tune again, the right steps came immediately to mind.

The usefulness of dance to the exercise of power was apparently something Louis XIII recognized almost instinctively from a very early age. Héroard describes a scene with the four-year-old dauphin in 1606 in which he came across a dozen soldiers receiving a dance lesson in the Louvre. First Louis ordered them to take up their arms and march about the hall to the *tambour* provided by Boilleau, the dancing master, who played the violin, and then he ordered them to set down their arms and dance a branle with him. The young dauphin had found a way to exercise visible command over a group of people, to have them perform their obedience by marching

109. David Parker, *Making of French Absolutism.* For a history—and historiography—of absolutism that reaffirms Parker's conclusions see Cosandey and Descimon, *Absolutisme en France.*

and dancing upon his order. Unadorned by any of the explanations contemporaries elaborated to account for the effect of music—the Orphic and Apollonian commonplaces representative of musical persuasion, the tales of modal ethos or ancient musical measures, the histories of the Parthian horsemen, Macedonian phalanxes, and imperial triumphs, or philosophies of divine right—the instincts of this child-king captured an essential truth. Music was a magic tool of command, an *instrumentum regni.* Perhaps for this reason, one of his favorite toys in those days was a drum.[110]

110. See Héroard, *Journal,* 1:920. By that time Louis knew all the drum calls, which he beat out when playing with his toy soldiers.

BIBLIOGRAPHY

Manuscript Sources

Cabinet des Estampes, QB1-1588. Bibliothèque nationale de France.

Chansonnier Maurepas. Fonds français 12616–12659. Bibliothèque nationale de France.

Chansonnier Clairambault. Fonds français 12676–12743. Bibliothèque nationale de France.

Collection Michel Henry. Fonds français 24357. Bibliothèque nationale de France.

Collection Philidor. Rés. F494. Département de la Musique. Bibliothèque nationale de France.

Fonds Godefroy 398. Bibliothèque de l'Institut, Paris.

Fonds Godefroy 426. Bibliothèque de l'Institut, Paris.

Manuscrit Deslauriers. Rés Vma ms 571. Département de la Musique. Bibliothèque nationale de France.

Nouvelles acquisitions françaises 7239. Bibliothèque nationale de France.

O^1 915 (32). Archives nationales, Paris.

O^1 915 (181–201). Archives nationales, Paris.

O^1 917 (211). Archives nationales, Paris.

[*Réglements* for Royal Households.] Fonds Dupuy 489. Bibliothèque nationale de France.

Printed Sources

Abdounur, Oscar João. "Ratios and Music in the Late Middle Ages: A Preliminary Survey." In proceedings of the conference "Musique et mathématique à la Renaissance," Tours, France, February 2000, edited by Philippe Vendrix. Tours: Centre d'Études Supérieures de la Renaissance, forthcoming.

Action de graces a Dieu et chants de triomphe au Roy, pour sa victoire. Paris: Pierre Ramier, 1588.

Agrippa, Heinrich Cornelius. *Declamation sur l'incertitude, vanité, et abus des sciences.* N.p.: Jean Durand, 1582.

Airs à quatre de différents auteurs. [troisième livre]. Paris: Pierre Ballard, 1613.

Airs de différents auteurs, mis en tablature de luth, huictiesme livre. Paris: Pierre Ballard, 1618.

Airs de différents auteurs, mis en tablature de luth, quatriesme livre. Paris: Pierre Ballard, 1613.

Airs de différents auteurs, mis en tablature de luth, septiesme livre. Paris: Pierre Ballard, 1617.

Alder, Ken. "A Revolution to Measure: The Political Economy of the Metric System in France." In *The Values of Precision,* edited by M. Norton Wise. Princeton: Princeton University Press, 1995.

Anglo, Sydney. *The Martial Arts of Renaissance Europe.* New Haven: Yale University Press, 2000.

Anthony, James R. *French Baroque Music from Beaujoyeulx to Rameau.* New York: Norton, 1978.

Arbeau, Thoinot. *Orchésographie.* Langres, 1589. Facsimile of the Paris 1888 reprint, edited by Laure Fonta. Bologna: Arnaldo Forni, 1969.

———. *Orchesography.* Translated by Mary Stewart Evans, with notes by Julia Sutton. New York: Dover, 1967.

Aristotle. *Politics.* With translation by H. Rackham. Loeb Classical Library. London: William Heinemann; Cambridge, Mass.: Harvard University Press, 1967.

Aubigné, Agrippa d'. *Histoire universelle.* Maillé, 1616–1620. Edited by André Thierry, 11 vols. Geneva: Droz, 1981–2000.

Auger, Edmond. *Metanoeologie, sur le suget de l'Archicongregation des Penitens de l'Annonciation de nostre Dame, & de toutes telles autres devotieuses assemblées, en l'Eglise sainte.* Paris: Jamet Mettayer, 1584.

———. *Statuts de la Congregation des pénitens.* Paris: Jamet Mettayer, 1583.

Augst, Bertrand. "Descartes's Compendium on Music." *Journal of the History of Ideas* 26 (1965): 119–132.

Babelon, Jean-Pierre. *Paris au XVIe siècle.* Paris: Hachette, 1986.

Bakhtin, Michael. *Rabelais and His World.* Translated by Hélène Iswolsky. Bloomington: Indiana University Press, 1984.

Ballard, Robert. *Diverses pièces mises sur le luth.* Paris, 1614. Edited and transcribed by André Souris, Sylvie Spycket, and Jacques Veyrier as *Deuxième livre (1614) et pièces diverses.* Paris: Centre National de la Recherche Scientifique, 1964.

———. [*Premier livre de luth*]. Paris, 1611. Edited by André Souris and Sylvie Spycket as *Premier livre (1611),* with an introduction and concordances by Monique Rollin. 2nd ed. Paris: Centre National de la Recherche Scientifique, 1976.

Ballet du roy faict dans la salle du petit Bourbon le 14 février. Paris: N. Rousset, 1621.

Le ballet du Roy représentant la furie de Roland. 1618.

Barriffe, William. *Military Discipline: or the Yong Artillery Man.* London: Thomas Harper for Ralph Mab, 1635.

Bascetta, Carlo, ed. *Sport e giuochi: Trattati e scritti dal XV al XVIII secolo.* Milan: Il Polifilo, 1978.

Bassompierre, François de. *Mémoires du Maréschal de Bassompierre.* Edited by Petitot. 3 vols. Paris: Foucault, 1822–1823.

Bataille, Gabriel. *Airs de différents auteurs, mis en tablature de luth.* 6 vols. Paris: Pierre Ballard, 1608–1615.

Bayrou, François. *Henri IV, le roi libre.* Paris: Flammarion, 1994.

Beaujoyeulx, Balthasar de. *Balet comique de la Royne.* Paris: Le Roy & Ballard, 1582. Facsimile ed. edited and introduced by Margaret M. McGowan as *"Le balet comique," 1581: A Facsimile.* Binghamton, N.Y.: Center for Medieval and Early Renaissance Studies, 1982.

Beaunis de Chanterain, Pierre, sieur des Viettes. *La resjouyssance des Compagnies qui auroyent faict monstre devant le Roy & la Royne Regente, estans armez à Paris.* [Paris, 1612].

Beccarie de Pavie, Raymond, Baron of Fourquevaux. *Instructions sur le faict de la guerre.* Paris, 1548. Facsimile ed. edited by G. Dickinson as *The* Instructions sur le Faict de la Guerre *of Raymond de Beccarie de Pavie, sieur de Fourquevaux.* London: University of London, Athlone Press, 1954.

Benedict, Philip. *Rouen during the Wars of Religion.* Cambridge: Cambridge University Press, 1981.

Benoist, René. *L'ordre et les ceremonies du sacre et couronnement du tres-Chrestien Roy de France.* Paris: Nicholas Chesneau, 1575.

Bergier, Nicolas. *La musique speculative.* Edited and translated by Ekkehard Jost. Cologne: Arno Volk Verlag, 1970.

Bertrand, Anthoine de. *Airs spirituels contenant plusieurs hymnes & cantiques.* Paris: Le Roy & Ballard, 1582.

Béthune, Maximilien de, Duke of Sully. *Mémoires.* Edited by Louis-Raymond Lefèvre. 2nd ed. Paris: Gallimard, 1942.

Bie, Jacques de. *Les familles de la France illustrées par les monumens des medailles anciennes et modernes.* Paris: Jacques de Bie, 1634.

———. *La France metallique contenant les actions célèbres tant publiques que privées des Rois et Reynes.* Paris: Jacques de Bie, 1634.

Billacois, François. *Le duel dans la société française des XVIe–XVIIe siècles: Essai de psychosociologie historique.* Paris: École des Hautes Études en Sciences Sociales, 1986.

Billon, Jean de. *Les principes de l'art militaire.* Rouen: Jean Berthelin, 1641.

Bitton, Davis. *The French Nobility in Crisis, 1560–1640.* Stanford: Stanford University Press, 1969.

Blackburn, Bonnie J. "*Te Matrem Dei laudamus:* A Study in the Musical Veneration of Mary." *Musical Quarterly* 53 (1967): 53–76.

Bodin, Jean. *The Six Bookes of a Commonweale: A Facsimile Reprint of the English Translation of 1606.* Translated by Richard Knolles. Edited by Kenneth Douglas McRae. Cambridge, Mass.: Harvard University Press, 1962.

———. *Les six livres de la République.* Paris, 1576. Edited by Christiane Frémont, Marie-Dominique Couzinet, and Henri Rochais, 6 vols. Paris: Fayard, 1986.

Boësset, Antoine. *Airs de cour à 4 et 5 parties.* Paris: Pierre Ballard, 1617.

Bohanan, Donna. *Old and New Nobility in Aix-en-Provence, 1600–1695: Portrait of an Urban Elite.* Baton Rouge: Louisiana State University Press, 1992.

Boni, Guillaume. *Psalmi Davidici novis concentibus sex vocibus modulati cum oratione Regia 12. voc.* Paris: Le Roy & Ballard, 1582.

Bonniffet, Pierre. *Un ballet démasqué: l'union de la musique au verbe dans* Le printans *de Jean-Antoine de Baïf et de Claude Le Jeune.* Paris: Champion-Slatkine, 1988.

Bordeaux, Christophe de. *Beau recueil de plusieurs belles chansons spirituelles.* Paris: For Magdeleine Berthelin, [after 1563].

Boucher, Jacqueline. "Auger." In *Histoire et dictionnaire des guerres de religion,* edited by Arlette Jouanna, Jacqueline Boucher, Dominique Biloghi, and Guy Le Thiec. Paris: Robert Laffont, 1998.

———. *La cour de Henri III.* Rennes: Ouest France, 1986.

———. "Mayenne." In *Histoire et dictionnaire des guerres de religion,* edited by Arlette Jouanna, Jacqueline Boucher, Dominique Biloghi, and Guy Le Thiec. Paris: Robert Laffont, 1998.

Boyssat, P. de. *Recherches sur les duels.* Lyons: Irenee Barlet, 1610.

Brantôme, Pierre Bourdeille de. *Oeuvres complètes.* Edited by Ludovic Lalanne. 11 vols. Paris: Mme Ve Jules Renouard, 1864–1882.

———. *Recueil des dames: poésies et tombeaux.* Edited by Etienne Vaucheret. Paris: Gallimard, 1991.

Breen, Adam van. *Le maniement d'armes de Nassau.* The Hague: Tavernier, 1618.

Brenet, Michel [Marie Bobillier]. "La musique dans les processions." *La tribune de Saint-Gervais* 2 (1896): 65–68, 81–84, 99–103.

———. *Notes sur l'histoire du luth en France.* Geneva: Minkoff Reprint, 1973.

Brioist, Pascal, Hervé Drévillon, Pierre Serna. *Croiser le fer: violence et culture de l'épée dans la France moderne, XVIe–XVIIIe siècle.* Seyssel: Champ Vallon, 2002.

Brobeck, John T. "Musical Patronage in the Royal Chapel of France under Francis I (r. 1515–1547)." *Journal of the American Musicological Society* 48 (1995): 187–239.

Brooks, Jeanice. *Courtly Song in Late Sixteenth-Century France.* Chicago: University of Chicago Press, 2000.

Brooks, Jeanice, Philip Ford, and Gillian Jondorf, eds. *Poetry and Music in the French Renaissance.* Cambridge: Cambridge French Colloquia, 2001.

Brown, Howard Mayer. *Instrumental Music Printed before 1600: A Bibliography.* Cambridge, Mass: Harvard University Press, 1965.

Burgess, Geoffrey. "The Chaconne and the Representation of Sovereign Power in Lully's *Amadis* (1684) and Charpentier's *Medée* (1693)." In *Dance and Music in French Baroque Theatre: Sources and Interpretations,* edited by Sarah McCleave. Institute of Advanced Musical Studies, Study Texts 3. London: King's College London, 1998.

Burke, Peter. *The Fabrication of Louis XIV.* New Haven: Yale University Press, 1992.

Burney, Charles. *A General History of Music from the Earliest Ages to the Present Period.* London, 1789. Edited by Frank Mercer, 2 vols. New York: Dover, 1957.

Cagin, Dom Paul. *Te Deum ou Illatio?* Solesmes: Abbaye de Solesmes, 1906.

Careri, Giovanni, ed. *La Jérusalem délivrée du Tasse: poésie, peinture, musique, ballet; actes du colloque organisée au Musée du Louvre par le Service culturel en collaboration avec l'Istituto italiano di cultura de Paris, les 13 et 14 novembre 1996.* Paris: Klincksieck, 1999.

Castagna, Paulo. "The Use of Music by the Jesuits in the Conversion of the Indigenous Peoples of Brazil." In *The Jesuits: Cultures, Sciences, and the Arts, 1540–1773,* edited by John W. O'Malley, Gauvin Alexander Bailey, Steven J. Harris, and T. Frank Kennedy. Toronto: University of Toronto Press, 1999.

Castiglione, Baltassar. *The Book of the Courtier.* Translated by George Bull. London: Penguin Classics, 1976.

Catalogue de l'histoire de France. 16 vols. Paris, 1855–1895. Reprint, Paris: Bibliothèque nationale, 1968.

Cave, Terence. *The Cornucopian Text: Problems of Writing in the French Renaissance.* Oxford: Clarendon Press, 1979.

Cazaux, Christelle. *La musique à la cour de François Ier.* Paris: École des Chartes and Programme "Ricercar," 2002.

La celebre et magnifique entrée de Charles neufiesme, Tres-chrestien Roy de France, faicte en sa Ville & Cité de Metz. Paris: Jean Dallier, 1569.

Les Ceremonies observees a la solennisation de la Paix, en l'eglise nostre Dame de Paris, le 21 Juin 1598. Paris: Denis Binet, 1598.

Champagne, Jean. *Discours du sacre et coronnement . . . en forme d'Epistre.* Reims: Jean de Foigny, 1575.

Charron, Pierre. *De la sagesse livres trois.* Bordeaux: Simon Millanges, 1601. Translated by Samson Lennard as *Of wisdome three bookes written in French by Peter Charron.* London: Blount & Aspley, [1613].

Chartier, Roger. *The Cultural Uses of Print in Early Modern France.* Translated by Lydia G. Cochrane. Princeton: Princeton University Press, 1987.

Chartier, Roger, Dominique Julia, and Marie-Madeleine Compère. *L'éducation en France du XVIe au XVIIIe siècle.* Paris: Société d'Édition d'Enseignement Supérieur, 1976.

Chastillon, Guillaume de, sieur de La Tour. *Airs.* Caen: Jacques Mangeant, 1593.

Chevalier, Guillaume de. *Discours des querelles et de l'honneur.* Paris: Matthieu Guillemot, 1598.

Chevallier, Pierre. *Louis XIII, roi cornélien.* Paris: Fayard, 1979.

Cholières, Nicolas de. *Oeuvres.* Edited by Edouard Tricotel. Paris: Librairie des Bibliophiles, 1879.

Les choses horribles, contenues en une lettre envoyee à Henry de Valois, par un Enfant de Paris, le vingthuitiesme de Janvier 1589. Paris: Jacques Gregoire, 1589.

Cohen, H. Floris. *Quantifying Music: The Science of Music at the First Stage of the Scientific Revolution, 1580–1650.* Dordrecht and Boston: D. Reidel, 1984.

Cohen, Selma Jeanne, founding ed. *International Encyclopedia of Dance.* New York: Oxford University Press, 1998.

Coignet, Martin. *Instruction aux princes pour garder la foy promise.* Paris: Jacques du Puys, 1584. Translated by Sir Edward Hoby as *Politique Discourses upon Truth and Lying: An Instruction to Princes to Keep Their faith and Promise containing the Summe of Christian and Morall Philosophie, and the deutie of a good man in sundrie politique discourses upon the truth and lying.* London: Ralfe Newberie, 1586.

Colin, Marie-Alexis. "Eustache Du Caurroy: un compositeur français aux confins du XVIe et du XVIIe siècle." *Acta Musicologica* 73 (2001): 189–258.

Collins, James B. *Classes, Estates, and Order in Early Modern Brittany.* Cambridge: Cambridge University Press, 1994.

Compère, Marie-Madeleine, and Dominique Julia. *Les collèges français, 16e–18e siècles.* 2 vols. Paris: Centre National de la Recherche Scientifique, 1984–1988.

Congratulation a la France sur les victoires obtenues par le Roy contre les Estrangers, & son heureux retour en sa bonne ville de Paris le 23. Decembre. Avec les magnificences qui ont esté faictes à son arrivee. Paris: Pierre Chevillot, 1587.

Corte, Claudio. *The Art of Riding.* Translated by Thomas Bedingfield. London: H. Denham, 1584.

———. *Il cavallarizzo.* Venice: G. Ziletti, 1562; 2nd, expanded ed. published as *Il cavallerizzo: nel qual si tratta della natura de' cavalli, delle razze, del modo di governarli, domarli e frenarli.* Lyons: Pierre Roussin, 1573.

Corvisier, André, ed. *Histoire militaire de la France.* 4 vols. Paris: Presses Universitaires de France, 1992–1995.

Cosandey, Fanny, and Robert Descimon. *L'absolutisme en France.* Paris: Seuil, 2002.

Coyssard, Michel. *Paraphrase des Hymnes et Cantiques spirituelz pour chanter avecque la Doctrine chrestienne.* Lyons: Jean Pillehotte, 1592.

———. *Traicté du profit que toute personne tire de chanter en la Doctrine Chrestienne.* Lyons: Jean Pillehotte, 1608.

Crawford, Katherine B. "Love, Sodomy, and Scandal: Controlling the Sexual Reputation of Henry III." *Journal of the History of Sexuality* 12 (2003): 513–542.

Crocker, Richard. *An Introduction to Gregorian Chant.* New Haven: Yale University Press, 2000.

Crosby, Alfred W. *The Measure of Reality: Quantification and Western Society, 1250–1600.* Cambridge: Cambridge University Press, 1997.

Crouzet, Denis. *Les guerriers de Dieu: La violence au temps des troubles de religion, vers 1525–vers 1610.* 2 vols. Seyssel: Champ Vallon, 1990.

Culley, Thomas. "Musical Activity in Some Sixteenth-Century Jesuit Colleges, with Special Reference to the Venerable English College in Rome from 1579 to 1589." *Analecta Musicologica* 19 (1980): 1–29.

[Daneau, Lambert]. *Traité des danses.* [Geneva]: François Estienne, 1579.

Daston, Lorraine, and Katherine Park. *Wonders and the Order of Nature, 1150–1750.* New York: Zone Books, 2001.

Davis, Natalie Zemon. *Society and Culture in Early Modern France: Eight Essays.* Stanford: Stanford University Press, 1975.

Delbrück, Hans. *The Dawn of Modern Warfare.* Translated by Walter J. Renfroe Jr. Vol. 4 of *History of the Art of War.* Lincoln: University of Nebraska Press, 1990.

———. *Warfare in Antiquity.* Translated by Walter J. Renfroe Jr. Vol. 1 of *History of the Art of War.* Lincoln: University of Nebraska Press, 1990.

della Casa, Giovanni. *Le Galatee, premierement composé en Italien par J. de la Case, & depuis mis en François, Latin, & Espagnol par divers auteurs.* Lyons: Jean de Tournes, 1598.

Desan, Philippe. *Les commerces de Montaigne: le discours économique des* Essais. Paris: Nizet, 1992.

———. *Naissance de la méthode: Machiavel, La Ramée, Bodin, Montaigne, Descartes.* Paris: Nizet, 1987.

Descartes, René. *Abrégé de musique; Compendium Musicae.* Edited and translated by Frédéric de Buzon. Paris: Presses Universitaires de France, 1987.

———. *Oeuvres de Descartes.* Edited by Charles Adam and Paul Tannery. Augmented new ed. 11 vols. Paris: Vrin, 1974–1989.

———. *The Philosophical Writings of Descartes.* Translated by John Cottingham, Robert Stoothoff, and Dugald Murdoch. 3 vols. Cambridge: Cambridge University Press, 1985–1991.

———. *Renatus Descartes Excellent Compendium of Musick: with Necessary and Judicious Animadversions Thereupon by a Person of Honour.* Translated by Lord William Brouncker. London: Thomas Harper for Humphrey Moseley, 1653.

Des Caurres, Jean. *Oeuvres morales et diversifiées en histoire.* Paris: Guillaume de la Nouë, 1584.

Descimon, Robert. "The Birth of the Nobility of the Robe: Dignity versus Privilege in the Parlement of Paris, 1500–1700." In *Changing Identities in Early Modern France,* edited by Michael Wolfe. Durham, N.C.: Duke University Press, 1997.

Description du balet dansé a Rome par des Cavalliers François. Rome, 1615.

Description du balet du veritable amour de madame la duchesse de Montmorency. Béziers: Jean Pech, 1618.

Desportes, Philippe. *Cartels et masquarades; épitaphes.* Edited by Victor E. Graham. Geneva: Droz, 1958.

Dewald, Jonathan. *Aristocratic Experience and the Origins of Modern Culture: France, 1570–1715.* Berkeley and Los Angeles: University of California Press, 1993.

———. *The European Nobility, 1400–1800.* Cambridge: Cambridge University Press, 1996.

Diefendorf, Barbara B. *Beneath the Cross: Catholics and Huguenots in Sixteenth-Century Paris.* New York: Oxford University Press, 1991.

Digges, Thomas. *An Arithmeticall Militaire Treatise Named Stratioticos (London, 1571).* Amsterdam: Da Capo Press, 1968.

Dionysius of Halicarnassus. *The Roman Antiquities of Dinoysius of Halicarnassus.* With translation by Earnest Cary. 7 vols. Loeb Classical Library. London: William Heinemann; Cambridge, Mass.: Harvard University Press, 1937–1950.

Discours au vray du ballet [*de la délivrance de Renault*] *dansé par le roy.* Paris: Pierre Ballard, 1617.

Discours de la joyeuse et triomphante entrée de tres-haut, tres-puissant et tres-magnanime Prince Henry IIII de ce nom . . . faicte en sa ville de Rouën, capital de la province & duché de Normandie, le Mercredy saiziéme jour d'Octobre, 1596. Rouen: Jean Crevel, 1599.

Discours de l'entrée faicte par treshaut et trespuissant prince Henry IIII, Roi de France et de Navarre, et Tres illustre princesse Marie de Médicis, La Royne son epouse, en leur ville de Caen, au mois de septembre. 1603. Reprint, Caen: Mancel, 1842.

Discours veritable des ceremonies, feux de joye, solemnitez, et autres resjouissances publiques, faictes en la ville de Tolose, sur sa reduction à l'obeyssance du tres-Chrestien Henry quatriesme Roy de France & de Navarre. Lyons: Thibaud Ancelin, 1596.

Dobbins, Frank. *Music in Renaissance Lyons.* Oxford: Clarendon Press, 1992.

Dorat, Jean. *Magnificentissimi spectaculi, a Regina Regem Matre in hortis suburbanis editi.* Paris: Frederic Morel, 1573.

Downey, Charles T. "Musical-Dramatic Productions Derived from Ariosto and Tasso in the City of Paris, 1600–1800." Ph. D. diss., Catholic University of America, 1998.

Du Caurroy, Eustache. *Preces ecclesiasticae.* Paris, 1609. Edited by Marie-Alexis Colin. Paris: Klincksieck, 2000.

Du Choul, Guillaume. *Discours de la religion des anciens Romains.* Lyons: Guillaume Rouillé, 1556.

———. *Discours sur la castrametation et discipline militaire des Romains . . . Des bains & antiques exercitations Grecques & Romaines; De la Religion des anciens Romains.* Lyons: Guillaume Rouillé, 1555.

Du Fail, Noël. *Oeuvres facétieuses.* Edited by Jules Assézat. 2 vols. Paris: Paul Daffis, 1874.

Duggan, Mary Kay. "Books of Hours: Devotional and Didactic." Paper read at the conference "The Culture of Books." Berkeley, Calif., April 2002.

Dumolin, Maurice. "Les académies parisiennes d'équitation." *Bulletin de la Société Archéologique, Historique et Artistique* 26 (1925): 417–428.

Dunning, Albert. *Die Staatsmotette, 1480–1555.* Utrecht: A. Oosthoek, 1970.

Dürer, Albrecht. *Les quatre livres de la proportion des parties et pourtraicts des corps humains.* Translated by Louis Meigret. Paris: Charles Perier, 1557.

Durosoir, Georgie. *L'air de cour en France, 1571–1655.* Liège: Mardaga, 1991.

Elias, Norbert. *The Civilizing Process: Sociogenetic and Psychogenetic Investigations.* Translated by Edmund Jephcott, with some notes and corrections by the author. Edited by Eric Dunning, Johan Goudsblom, and Stephen Mennell. Oxford: Blackwell, 2000.

Fabbri, Paolo, ed. *Musica in torneo nell'Italia del Seicento.* Lucca: Libreria Musicale Italiana, 1999.

Faret, Nicolas. *L'honneste homme ou l'art de plaire à la court.* Paris, 1630. Edited by Maurice Magendie. Paris: Presses Universitaires de France, 1925.

Farrell, Allan P. *The Jesuit Code of Liberal Education: Development and Scope of the Ratio Studiorum.* Milwaukee: Bruce, 1938.

Les feux de joye faicts a Paris pour l'arrivee du Roy en France, Avec l'Ordre tenu à son entree et reception en la ville de Lyon. Paris: Denis du Pré, 1574.

Les feux de joye faits a Paris sur la resjouyssance de la publication de la Paix, le 12 Juin, 1598. Lyons: Jean Jullieron, 1598.

Les feuz de joye, faicts a Paris, touchant la reddition des Reistres: au Roy Tres-Chrestien de France et de Pologne. Avec un Te Deum solennel chanté en l'eglise nostre D'ame lundy quatoriesme Jour de Decembre. Paris: Hubert Velu, 1587.

Fiaschi, Cesare. *Traicté de la manière de bien emboucher, manier, et ferrer les chevaux.* Paris: Guillaume Auvray, 1578.

———. *Traicté de la manière de bien embrider, manier, et ferrer les chevaux.* Translated by François de Provane. Paris: Charles Perier, 1564.

———. *Trattato del modo dell'imbrigliare, maneggiare et ferrare cavalli.* Bologna: A. Giaccarelli, 1556.

Ficino, Marsilio. *De vita libri tres; De triplici vita; Apologia; Quod Necessaria Sit ad Vitam Securitas.* Florence, 1489. Edited and translated by Carol V. Kaske and John R. Clark as *Three Books on Life.* Binghamton, N.Y.: Medieval and Renaissance Texts and Studies, in conjunction with the Renaissance Society of America, 1989.

Fleurance-Rivault, David de. *Le dessein d'une academie, et de l'introduction d'icelle en la Cour.* Paris: Pierre le Court en l'Imprimerie de Robert Estienne, 1612.

La fleur des chansons nouvelles, traittans partie de l'amour, partie de la guerre, selon les occurrences du temps present. Lyons: Benoist Rigaud, 1580.

Fogel, Michèle. *Les cérémonies de l'information dans la France du XVIe au milieu du XVIIIe siècle.* Paris: Fayard, 1989.

[Foigny, Jean de]. *Le sacre et coronnement du Roy de France, avec toutes les Ceremonies, prieres et oraisons.* Reims: J. de Foigny, 1575.

Folly, Albert. "Les académies d'armes, XVIe et XVIIe siècles." *Bulletin de la Société du VIe arrondissement du Paris* 2 (1899): 163–171.

Fontaine, Marie Madeleine. "Nager, voler." In *Ronsard et les elements: actes du colloque tenu les 14 et 15 avril 1989 à la Faculté des letters de l'Université de Neuchâtel,* edited by André Gendre. Geneva: Droz, 1992.

Foucault, Michel. *Discipline and Punish: The Birth of the Prison.* Translated by Alan Sheridan. 2nd ed. New York: Vintage Books, 1995.

Fouqueray, Henri. *Histoire de la Compagnie de Jésus en France des origines à la suppression, 1528–1762.* 5 vols. Paris: Bureaux des Études, 1910–1925.

Franklin, Julian H. *Jean Bodin and the Rise of Absolutist Theory.* Cambridge: Cambridge University Press, 1973.

Franko, Mark. *Dance as Text: Ideologies of the Baroque Body.* Cambridge: Cambridge University Press, 1993.

———. *The Dancing Body in Renaissance Choreography, c. 1416–1589.* Birmingham, Ala.: Summa, 1986.

———. "Double Bodies: Androgyny and Power in the Performances of Louis XIV." *Drama Review* 38 (1994): 71–82.

———. "Jouer avec le feu: la subjectivité du roi dans *La délivrance de Renault.*" In *La Jérusalem délivrée du Tasse: poésie, peinture, musique, ballet; actes du colloque organisé au Musée du Louvre par le Service culturel en collaboration avec l'Istituto italiano di cultura de Paris, les 13 et 14 novembre 1996,* edited by Giovanni Careri. Paris: Klincksieck, 1999.

Friedmann, Adrien. *Paris: ses rues, ses paroisses du Moyen Âge à la Révolution.* Paris: Plon, 1959.

Fritsch-Pinaud, Laurence. "La vie paroissiale à Saint-Jacques-de-la-Boucherie au XVe siècle." *Paris et Île-de-France* 33 (1982): 7–97.

Gaffurius, Franchinus. *Practica music*e. Milan, 1496. Translated and transcribed by Clement A. Miller as *Practica musicae,* Musicological Studies and Documents 20. Dallas, Tex.: American Institute of Musicology, 1968.

Galilei, Galileo. *Il saggiatore.* Rome, 1623. Edited by Libero Sosio. Milan: Feltrinelli Editore, 1965.

Garrisson, Janine. *Henri IV, le roi de la paix.* Paris: Tallandier, 2000.

Gaukroger, Stephen. *Descartes: An Intellectual Biography.* Oxford: Clarendon Press; New York: Oxford University Press, 1995.

Geertz, Clifford. *Local Knowledge: Further Essays in Interpretive Anthropology,* 3rd ed. New York: Basic Books, 2000.

Gérold, Théodore. *L'art du chant en France au XVIIe siècle.* Strasbourg: Université de Strasbourg, 1921.

Gheyn, Jacob de. *Maniement d'armes, d'arquebuses, mousquetz, et piques en conformité de l'ordre de Monseigneur le Prince Maurice.* 2nd ed. The Hague, 1608.

———. *Wapenhandelinghe.* The Hague, 1607. Published in facsimile ed. with a commentary by J. B. Kist. Lochem: De Tijdstroom, 1971.

Giacone, Franco. "Le catholicisme dans les *Hymnes Ecclesiastiques.*" In *Poésie encyclopédique et kabbale chrétienne: onze études sur Guy Le Fèvre de La Boderie,* edited by François Roudaut. Paris: Honoré Champion, 1999.

Giesey, Ralph E. "The King Imagined." In *The French Revolution and the Creation of Modern Political Culture,* edited by Keith Michael Baker, 4 vols. Oxford: Pergamon Press, 1987–1994.

———. "Models of Rulership in French Royal Ceremonial." In *Rites of Power: Symbolism, Ritual, and Power since the Middle Ages,* edited by Sean Wilentz. Philadelphia: University of Pennsylvania Press, 1985.

———. *The Royal Funeral Ceremony in Renaissance France.* Geneva: Droz, 1960.

Giraldi, Giuliano. *Esequie d'Arrigo IV, Cristianissimo re di Francia e di Navarra, celebrate in Firenze dal Serenissimo don Cosimo II granduca di Toscana.* Florence: Sermartelli, 1610.

[Giron, Matthieu]. *Au roy sur le sacre de sa Majesté.* Paris: Pierre Chevillot, 1594.

Glareanus, Heinrich. *Dodecachordon.* Basel: Henrichum Petri, 1547.

Godefroy, Théodore, and Denis Godefroy le Jeune. *Le cérémonial françois, recueilly par Théodore Godefroy et mis en lumière par Denis Godefroy.* 2 vols. Paris: Cramoisy, 1649.

Godt, Irving. "Guillaume Costeley: Life and Works." Ph.D. diss., New York University, 1969.

Gordon-Seifert, Catherine E. "'La Réplique Galante': Sébastien de Brossard's Airs as Conversation." In *Sébastien de Brossard, musicien,* edited by Jean Duron. [Versailles]: Centre de Musique Baroque de Versailles; [Paris]: Klincksieck, 1998.

Gouk, Penelope. *Music, Science, and Natural Magic in Seventeenth-Century England.* New Haven: Yale University Press, 1999.

Goulart, Simon. *Mémoires de l'estat de France sous Charles Neuviesme.* 2nd ed. Meidelbourg: H. Wolf, 1578.

Goyard-Fabre, Simone. *Jean Bodin et le droit de la république.* Paris: Presses Universitaires de France, 1989.

Grafton, Anthony, and Lisa Jardine. *From Humanism to the Humanities: Education and the Liberal Arts in Fifteenth- and Sixteenth-Century Europe.* Cambridge, Mass.: Harvard University Press, 1986.

Graham, Victor E., and W. McAllister Johnson. *The Paris Entries of Charles IX and Elisabeth of Austria, 1571: With an Analysis of Simon Bouquet's* Bref et sommaire recueil. Toronto: University of Toronto Press, 1974.

———. *The Royal Tour of France by Charles IX and Catherine de' Medici: Festivals and Entries, 1564–6.* Toronto: University of Toronto Press, 1979.

Greenblatt, Stephen. *Renaissance Self-Fashioning: From More to Shakespeare.* Chicago: University of Chicago Press, 1980.

Greene, Thomas. "The King's One Body in the *Balet comique de la Royne.*" *Yale French Studies* 86 (1994): 75–93.

Greengrass, Mark. *The French Reformation.* Oxford: Basil Blackwell, 1987.

Griselle, Eugène. *État de la Maison du roi Louis XIII.* Paris: Éditions de Documents d'Histoire, 1912.

———. *Supplément à la Maison du roi Louis XIII.* Paris: Éditions de Documents d'Histoire, 1912.

Grisone, Federico. *Gli ordini di cavalcare.* Naples: G. P. Suganappo, 1550.

Guédron, Pierre. *Quatriesme livre d'airs de cour à 4 et 5 parties.* Paris: Pierre Ballard, 1618.

———. *Second livre d'airs de cour à 4 et 5 parties.* Paris: Pierre Ballard, 1613.

Guéraud, Jean. *La chronique lyonnaise de Jean Guéraud, 1536–1562.* Edited by Jean Tricou. Lyons: Imprimerie Audinienne, 1929.

Guéry, Alain. "Le Roi est Dieu, Le Roi et Dieu." In *L'état ou le roi: Les fondations de la modernité monarchique en France, XIVe–XVIIe siècles,* edited by Neithard Bulst, Robert Descimon, and Alain Guerreau. Paris: Maison des Sciences de l'Homme, 1996.

Guichard, Claude. *Funerailles, & diverses manieres d'ensevelir des Rommains, Grecs, & autres nations, tant anciennes que modernes.* Lyons: Jean de Tournes, 1581.

Guiffrey, Georges, ed. *Chronique du Roy François.* Paris: Renouard, 1860.

Guillo, Laurent. *Pierre I Ballard et Robert III Ballard: imprimeurs du roy pour la musique, 1599–1673.* 2 vols. Liège: Mardaga, 2003.

Guillot, Pierre. *Les Jésuites et la musique: le Collège de la Trinité à Lyon, 1565–1762.* Liège: Mardaga, 1991.

Haar, James. "Orlando di Lasso: Composer and Print Entrepreneur." In *Music and the Cultures of Print,* edited by Kate van Orden. New York: Garland, 2000.

Haggh, Barbara. "The Archives of the Order of the Golden Fleece and Music." *Journal of the Royal Musical Association* 120 (1995): 1–43.

Hale, J. R. "The Military Education of the Officer Class in Early Modern Europe." In *Cultural Aspects of the Italian Renaissance: Essays in Honour of Paul Oskar Kristeller,* edited by Cecil H. Clough. New York: A. F. Zambelli, 1976.

Heartz, Daniel. "Un divertissement de palais pour Charles Quint à Binche." In *Les fêtes de la Renaissance,* ed. Jean Jacquot, 2nd ed. 3 vols. Paris: Centre National de la Recherche Scientifique, 1973–1975.

———. "*Voix de ville:* Between Humanist Ideals and Musical Realities." In *Words and Music: The Scholar's View; A Medley of Problems and Solutions Compiled in Honor of A. Tillman Merritt by Sundry Hands,* edited by Laurence Berman. Cambridge, Mass.: Harvard University Press, 1972.

Herlin, Denis. *Catalogue du fonds musical de la Bibliothèque de Versailles.* Paris: Klincksieck, 1995.

Héroard, Jean. *Journal.* Edited by Madeleine Foisil. 2 vols. Paris: Fayard, 1989.

Heures de Nostre Dame a l'usage de Rome, selon la réformation de nostre S. Pere Pape Pie V. Paris: Jamet Mettayer, 1583, with musical supplement by Le Roy & Ballard.

Hilton, Wendy. *Dance of Court and Theater: The French Noble Style, 1690–1725.* Princeton: Princeton Book Co., 1981.

His, Isabelle. *Claude Le Jeune, v. 1530–1600: un compositeur entre Renaissance et baroque.* Arles: Actes Sud, 2000.

Histoire generale de tout ce qui s'est passé au Parc Royal sur la resjouïssance du Mariage du Roy avec l'Infante d'Espagne. Paris: Anthoine du Brueil, 1612.

Hoffmann, George. *Montaigne's Career.* Oxford: Clarendon; New York: Oxford University Press, 1998.

Hoffmann, Wilhelm. *Bericht von den Soldaten in drei Theilen.* Frankfurt, 1609.

Holt, Mack P. *The French Wars of Religion, 1562–1629.* Cambridge: Cambridge University Press, 1995.

———. "Wine, Community, and Reformation in Sixteenth-Century Burgundy." *Past and Present* 138 (1993): 58–93.

———, ed. *Society and Institutions in Early Modern France.* Athens: University of Georgia Press, 1991.

Hotman, François. *Francogallia.* Geneva, 1573. Latin text edited by Ralph E. Giesey. Translated by J. H. M. Salmon. Cambridge: Cambridge University Press, 1972.

Howard, Skiles. *The Politics of Courtly Dancing in Early Modern England.* Amherst: University of Massachusetts Press, 1998.

Huizinga, Johan. *The Autumn of the Middle Ages.* Translated by Rodney J. Payton and Ulrich Mammitzsch. Chicago: University of Chicago Press, 1996.

Huppert, George. *Les Bourgeois Gentilshommes: An Essay on the Definition of Elites in Renaissance France.* Chicago: University of Chicago Press, 1977.

———. *Public Schools in Renaissance France.* Urbana: University of Illinois Press, 1984.

Jackson, Richard A. *Vive le Roi! A History of the French Coronation from Charles V to Charles X.* Chapel Hill: University of North Carolina Press, 1984.

Jacquot, Jean, ed. *Les fêtes de la Renaissance.* 2nd ed. 3 vols. Paris: Centre National de la Recherche Scientifique, 1973–1975.

Janequin, Clément. *Chansons de maistre Clement Janequin.* Paris: Pierre Attaingnant, [1528].

———. *Chansons polyphoniques.* Edited by A. Tillman Merritt and François Lesure. 6 vols. Monaco: Éditions de l'Oiseau-Lyre, 1965–1971.

———. *Missa La bataille.* Lyons, 1532. Edited by Frank Dobbins. Espoo, Finland: Fazer Music, 1995.

———. *Verger de musique.* Paris: Le Roy & Ballard, 1559.

Jeanneret, Michel. *Poésie et tradition biblique au XVIe siècle: Recherches stylistiques et paraphrases des "Psaumes," de Marot à Malherbe.* Paris: Librairie José Corti, 1969.

Jouanna, Arlette, Jacqueline Boucher, Dominique Biloghi, and Guy Le Thiec. *Histoire et dictionnaire des guerres de religion.* Paris: Robert Laffont, 1998.

Jouhaud, Christian. "Printing the Event: From La Rochelle to Paris." In *The Culture of Print: Power and the Uses of Print in Early Modern Europe,* edited by Roger Chartier and translated by Lydia G. Cochrane. Princeton: Princeton University Press, 1989.

Kantorowicz, Ernst H. *The King's Two Bodies: A Study in Mediaeval Political Theology.* New preface by William Chester Jordan. Princeton, N.J.: Princeton University Press, 1997.

Kassler, Jamie Croy. "Music as a Model in Early Science." *History of Science* 20 (1982): 103–139.

Keegan, John. "Keeping in Time: The Rise of Foot Drill and the Decline of the Minuet." *Times Literary Supplement.* 12 July 1996, 3–4.

Kennedy, T. Frank. "Jesuits and Music: Reconsidering the Early Years." *Studi Musicali* 17 (1988): 71–100.

Kirsch, Winfried. *Die Quellen der mehrstimmigen Magnificat- und Te Deum-Vertonungen bis zur Mitte des 16. Jahrhunderts.* Tutzing: Hans Schneider, 1966.

La Broue, Salomon de. *Le cavalerice françois.* 3 vols. Paris: Abel l'Angelier, 1602.

Lacroix, Paul. *Ballets et mascarades de cour de Henri III à Louis XIV, 1581–1652.* 6 vols. Geneva: Gay et Fils, 1868–1870.

La Grotte, Nicolas de. *Chansons de P. de Ronsard, Ph. Desportes, et autres.* Paris: Le Roy & Ballard, 1569.

La Motte-Messemé, François Le Poulchre de. *Le passe-temps.* 2nd ed. Paris: Jean le Blanc, 1597.

La Noue, François de. *Discours politiques et militaires.* Geneva, 1587. Edited by F. E. Sutcliffe. Geneva: Droz, 1967.

La Noue, Pierre de. *La cavalerie françoise et italienne.* Lyons: Cl. Morillon, 1620.

La Primaudaye, Pierre de. *Academie françoise.* 3rd ed. Paris: Guillaume Chaudière, 1580.

La Roche, Commandant de. *Les académies militaires sous l'Ancien Régime.* Paris: Auguste Picard, 1929.

Laugier de Porchères, Honoré. *Le camp de la place Royale.* Paris: Jean Micard and Toussaint Du Bray, 1612.

Launay, Denise. *La musique religieuse en France du Concile de Trente à 1804.* Paris: Société Française de Musicologie, 1993.

Lauze, François de. *Apologie de la Danse et de la parfaite methode de l'enseigner tant aux cavaliers qu'aux dames.* 1623. Facsimile ed. Geneva: Minkoff Reprint, 1977.

Laval, Antoine Mathé de. *Desseins de professions nobles et publiques.* Paris: Abel l'Angelier, 1605.

La Via, Stefano. "Le *combat* retrouvé: Les 'passions contraires' du 'divin Tasse' dans la représentation musicale de Monteverdi." In *La Jérusalem délivrée du Tasse: poésie, peinture, musique, ballet; actes du colloque organisé au Musée du Louvre par le Service culturel en collaboration avec l'Istituto italiano di cultura de Paris, les 13 et 14 novembre 1996,* edited by Giovanni Careri. Paris: Klincksieck, 1999.

Le Fèvre de La Boderie, Guy. *Hymnes Ecclesiastiques.* 2nd ed. Paris, 1582.

Le Jeune, Claude. *Airs.* Paris, 1608. Edited by D. P. Walker. 4 vols. Rome: American Institute of Musicology, 1951–1959.

———. *Le printemps.* Paris, 1603. Edited by Henry Expert. Maîtres musiciens de la Renaissance française 12–14. Paris: Leduc, 1900–1901.

Leroquais, Abbé Victor. *Les livres d'heures manuscrits de la Bibliothèque nationale.* 2 vols. Paris: Bibliothèque nationale, 1927.

Le Roy Ladurie, Emmanuel. *Carnival in Romans.* Translated by Mary Feeney. New York: G. Braziller, 1979.

L'Estoile, Pierre de. *Mémoires-journaux de Pierre de L'Estoile.* Edited by Gustave Brunet,

Aimé Louis Champollion-Figeac, Eugène Halphen, Paul Lacroix, Charles Read, Tamizey de Larroque, and Edouard Tricotel. 12 vols. Paris: Alphonse Lemerre, 1888–1896.

———. *Registre-Journal du règne de Henri III.* Edited by Madeleine Lazard and Gilbert Schrenck. 6 vols. Geneva: Droz, 1992–2003.

Lesure, François. "France (I)." In *The New Grove Dictionary of Music and Musicians,* edited by Stanley Sadie. 2nd ed. 29 vols. London: Macmillan, 2002.

———. "Le recueil de ballets de Michel Henry." In *Les fêtes de la Renaissance.* Edited by Jean Jacquot. 2nd ed. 3 vols. Paris: Centre Nationale de la Recherche Scientifique, 1973–1975.

Lesure, François, and Geneviève Thibault. *Bibliographie des éditions d'Adrian Le Roy et Robert Ballard, 1551–1598.* Paris: Société Française de Musicologie, 1955.

Le Vavasseur, Nicolas. *Airs à III. IIII. et V parties par Nicol. Le Vavasseur.* Paris: Pierre Ballard, 1626.

Lipsius, *De militia Romana libri quinque, commentarius ad Polybium.* Antwerp: Plantin-Moretus, 1596.

———. *Politicorum sive Civilis Doctrinae libri sex.* Leiden: ex officina Plantiniana apud Franciscum Raphelengium, 1589.

Lorenzetti, Stefano. *Musica e identità nobiliare nell'Italia del Rinascimento: educazione, mentalità, immaginario.* Florence: Olschki, 2003.

Louis XIV, King of France. *Mémoires de Louis XIV.* Edited by Charles Dreyss. 2 vols. Paris: Didier, 1860.

———. *Mémoires de Louis XIV: le métier de roi.* Edited by Jean Longnon. Paris: Éditions Tallandier, 2001.

Love, Ronald S. "All the King's Horsemen: The Equestrian Army of Henri IV, 1585–1598." *Sixteenth Century Journal* 22 (1991): 510–533.

Lowinsky, Edward E. *The Medici Codex of 1518: A Choirbook of Motets Dedicated to Lorenzo de' Medici, Duke of Urbino.* Monuments of Renaissance Music 3–5. Chicago: University of Chicago Press, [1968].

Lucian of Samosata. *Les oeuvres de Lucian de Samosate.* Translated by Filbert Bretin. Paris: Abel l'Angelier, 1583.

Lynn, John A. *Giant of the* Grand Siècle: *The French Army, 1610–1715.* Cambridge: Cambridge University Press, 1997.

Machiavelli, Niccolò. *Arte della guerra.* Florence, 1521. Translated by Ellis Farneworth as *The Art of War,* revised and with an introduction by Neal Wood. New York: Da Capo Press, 1990.

———. *Selected Political Writings.* Edited and translated by David Wootton. Indianapolis: Hackett, 1994.

Magendie, Maurice. *La politesse mondaine et les théories de l'honnêteté en France au XVIIe siècle, de 1600 à 1660.* 2 vols. Paris: Librairie Félix Alcan, 1925.

La magnificence de la superbe et triumphante entrée de la noble & antique cité de Lyon. Lyons: Guillaume Rouillé, 1549. Facsimile edition published as Maurice Scève, *The Entry of Henri II into Lyon, September 1548,* introduction by Richard Cooper. Tempe, Ariz.: Medieval and Renaissance Texts and Studies, 1997.

Malingre, Claude. *Les annales generales de la ville de Paris.* Paris: Pierre Rocolet, Cardin Besongne, Henry Le Gras, vefve Nicolas Trabouilliet, 1640.

Mamone, Sara. *Firenze e Parigi: due capitali dello spettacolo per una regina, Maria de' Medici.* Milan: Amilcare Pizzi, 1987.

Marin, Louis. *Le portrait du roi.* Paris: Éditions de Minuit, 1981.

Mars, his Field, or, The Exercise of Armes. London: Roger Daniell, [1611].

Martin, Michel. *Les monuments équestres de Louis XIV: une grande entreprise de propagande monarchique.* Paris: Picard, 1986.

Massip, Catherine. "La collection musicale Toulouse-Philidor à la Bibliothèque nationale." *Fontes artis musicae* 30 (1983): 184–207.

[Matthieu, Pierre]. *Les deux plus grandes, plus celebres et memorables resjouissances de la ville de Lyon. La premiere, Pour l'entrée de . . . Henry IIII . . . La seconde, Pour l'heureuse publication de la paix.* Lyons: Thibaud Ancelin, 1598.

———. *L'entrée de . . . Marie de Médicis . . . en la ville de Lyon.* Rouen: Jean Osmont, 1601.

———. *L'entrée de tres-grand, tres-chrestien, tres-magnanime, et victorieux Prince Henry IIII. Roy de France & de Navarre, en sa bonne ville de Lyon, Le IIII. Septembre l'an M.D.XCV.* Lyons: Pierre Michel, n.d.

———. *Pompe funèbre des pénitens de Lyon.* Lyons: J. Roussin, 1589.

McClelland, John. "Measuring Poetry, Measuring Music: From the Rhétoriqueurs to the Pléiade." In *Poetry and Music in the French Renaissance,* edited by Jeanice Brooks, Philip Ford, and Gillian Jondorf. Cambridge: Cambridge French Colloquia, 2001.

McGowan, Margaret M. *L'art du ballet de cour en France, 1581–1643.* Paris: Centre National de la Recherche Scientifique, 1978.

———. "The Arts Conjoined: A Context for the Study of Music." *Early Music History* 13 (1994): 171–198.

———. *Ideal Forms in the Age of Ronsard.* Berkeley and Los Angeles: University of California Press, 1985.

———. "A Renaissance War Dance: The Pyrrhic." *Dance Research* 3 (1984): 29–38.

McNeill, William H. *Keeping Together in Time: Dance and Drill in Human History.* Cambridge, Mass.: Harvard University Press, 1995.

Melzo, Lodovico. *Regole militari sopra il governo e servitio particolare della cavalleria.* Antwerp: G. Trognoesius, 1611.

———. *Reigles militaires du chevalier frere Louys Melzo/Kriegs Regeln dess Ritters L. Meltzo.* Frankfurt: Caspar Rodtel, 1625.

Ménestrier, Claude François. *Traité des tournois, joustes, carrousels, et autres spectacles publics.* Lyons, 1669. Facsimile reprint. Lyons: Editions Horvath, 1978.

Menou, René de, seigneur de Charnizay. *La practique du cavalier.* 4th ed. Paris: Jean Corrozet, 1629.

Mercure François. Vol. 14, 1629.

Mercurialis, Hieronymus. *Artis gymnasticae . . . libri sex.* Venice: Giunti, 1569.

Mersenne, Marin. *Correspondance du P. Marin Mersenne.* Edited by Cornelis de Waard. 17 vols. Paris: G. Beauchesne, 1932–1988.

———. *Harmonie universelle.* Paris, 1636. Facsimile ed. edited by François Lesure. 3 vols. Paris: Centre National de la Recherche Scientifique, 1965.

Métru, Nicolas. *Recueil des vers du sieur Guil. de Baïf, mis en musique par Nicolas Metru, chantez en l'allégresse de l'heureux retour du Roy.* Paris: Pierre Ballard, 1628.

Meurier, Hubert. *Traicté de l'institution et vray usage des processions tant ordinaires, qu'extraordinaires, qui se font en l'Eglise Catholique, contenant un ample discours de ce qui s'est passé pour ce regard en la Province de Champaigne, depuis le 22. de Juillet jusques au 25. d'Octobre, 1583.* Rheims: Jean de Foigny, 1584.

Meursius, Johannes. *De ludis graecorum liber singularis.* Leiden: Elzevier, 1625.

The Military Discipline wherein is most Martially showne the order of Drilling for y[e] Musket and Pike. London: Roger Daniell, 1623.

Monluc, Blaise de. *Commentaires, 1521–1576.* Edited by Paul Courteault. Paris: Gallimard, 1964.

Monson, Craig. "The Council of Trent Revisited." *Journal of the American Musicological Society* 55 (2002): 1–37.

Montaigne, Michel de. *Essais.* Bordeaux, 1580, 1588. Edited by Pierre Villey and V.-L. Saulnier. 3 vols. Paris: Presses Universitaires de France, 1988.

Monteverdi, Claudio. *Madrigali guerrieri, et amorosi con alcuni opuscoli in genere rappresentativo . . . Libro ottavo.* Venice: Alessandro Vicenti, 1638.

Montgommery, Louis de. *La milice françoise reduite à l'ancien ordre et discipline militaire des légions.* 2nd ed. Paris: François Rousselet, 1610.

Morel, Horace. *Subject du feu d'artifice fait a l'entree du Roy dans sa ville de Paris Ensemble le Ballet representé sur la riviere de Seine, devant le Louvre.* Nantes: Hilaire Mauclerc, 1629.

Motley, Mark. *Becoming a French Aristocrat: The Education of the Court Nobility, 1580–1715.* Princeton: Princeton University Press, 1990.

Moyer, Ann E. *Musica Scientia: Musical Scholarship in the Italian Renaissance.* Ithaca: Cornell University Press, 1992.

Mukerji, Chandra. *Territorial Ambitions and the Gardens of Versailles.* Cambridge: Cambridge University Press, 1997.

Naïs, Hélène. *Les animaux dans la poésie française de la Renaissance: Science, symbolique, poésie.* Paris: Marcel Didier, 1961.

Nouveau recueil des chansons qu'on chante a present, tant de la guerre & voyage de la Fère, de la Mure: & des Chansons amoureuses. Lyons: Benoist Rigaud, 1581.

Nouveau recueil de toutes les chansons nouvelles, tant de l'amour que de la guerre. Lyons: Benoist Rigaud, [1580].

Nugent, George. "Anti-Protestant Music for Sixteenth-Century Ferrara." *Journal of the American Musicological Society* 43 (1990): 228–291.

Octo cantica divae Mariae virginis . . . quaternis vocibus. Paris: Veuve Robert Ballard and Pierre Ballard, 1599.

Oestreich, Gerhard. *Neostoicism and the Early Modern State.* Edited by Brigitta Oestreich and H. G. Koenigsberger. Translated by David McLintock. Cambridge: Cambridge University Press, 1982.

O'Malley, John W., Gauvin Alexander Bailey, Steven J. Harris, and T. Frank Kennedy, eds. *The Jesuits: Cultures, Sciences, and the Arts, 1540–1773.* Toronto: University of Toronto

Press, 1999.
Oman, Charles. *A History of the Art of War in the Sixteenth Century.* 2nd ed. London: Greenhill Books; Novato, Calif.: Presidio Press, 1991.
Orgel, Stephen. *The Illusion of Power: Political Theater in the English Renaissance.* Berkeley and Los Angeles: University of California Press, 1975.
Ortiz, Diego. *El primo libro nel quale si tratta delle glose sopra le cadenze et altre sorte de punti in la musica del violone.* Rome: Dorico, 1553.
———. *Libro secondo.* Rome: Dorico, 1553.
Ouvrard, Jean-Pierre. "Modality and Text Expression in Sixteenth-Century French Chansons: Remarks Concerning the E Mode." *Basler Jahrbuch für historische Musikpraxis* 16 (1992): 89–116.
Palfy, Barbara. "Pyrrhic." In *International Encyclopedia of Dance,* founding editor Selma Jeanne Cohen. New York: Oxford University Press, 1998.
Paradin, Guillaume. *Le blason des danses.* Beaujeu, 1556. Paris: Didot, 1830.
Parker, David. *The Making of French Absolutism.* London: Edward Arnold, 1983.
Parker, Geoffrey. *The Military Revolution: Military Innovation and the Rise of the West, 1500–1800.* 2nd ed. Cambridge: Cambridge University Press, 1996.
Parrott, David. *Richelieu's Army: War, Government, and Society in France, 1624–1642.* Cambridge: Cambridge University Press, 2001.
Parthenay, Catherine de, Duchess of Rohan. *Ballets allégoriques en vers, 1592–1593.* Tours, [1593]. Edited by Raymond Ritter. Paris: Champion, 1927.
Pau, Gérald. "De l'usage de la chanson spirituelle par les Jésuites au temps de la contre-réforme." In *La chanson à la Renaissance: Actes du XXe colloque d'études humanistes du Centre d'Études Supérieures de la Renaissance de l'Université de Tours, juillet 1977,* edited by Jean-Michel Vaccaro. Tours: Éditions Van de Velde, 1981.
Pelletier, Thomas. *La nourriture de la noblesse.* Paris: veufve Mamert Patisson, 1604.
Perella, Lisa. "Bénigne de Bacilly and the *Recueil des plus beaux vers, qui ont esté mis en chant* of 1661." In *Music and the Cultures of Print,* edited by Kate van Orden. New York: Garland, 2000.
———. "Mythologies of Musical Persuasion: The Power and Politics of Song in France, 1653–1673." Ph.D. diss., University of Pennsylvania, 2003.
Perrenet, Pierre. *Estienne Tabourot, sa famille et son temps.* Dijon: Éditions du Raisin, 1926.
Pillorget, René, and Suzanne Pillorget. *France baroque, France classique, 1589–1715.* 2 vols. Paris: Robert Laffont, 1995.
Planchart, Alejandro Enrique. "Guillaume Du Fay's Benefices and his Relationship to the Court of Burgundy." *Early Music History* 8 (1988): 117–171.
Plato. *Laws.* With translation by R. G. Bury. 2 vols. Loeb Classical Library. London: William Heinemann; Cambridge, Mass.: Harvard University Press, 1967–1968.
———. *The Republic.* With translation by Paul Shorey. 2 vols. Loeb Classical Library. London: William Heinemann; Cambridge, Mass.: Harvard University Press, 1982-1987.
———. *Timaeus.* With translation by R. G. Bury. Loeb Classical Library. London: William Heinemann; Cambridge, Mass.: Harvard University Press, 1981.

Pluvinel de la Baume, Antoine de. *L'instruction du roy en l'exercice de monter à cheval.* Edited by René de Menou. Paris: Michel Nivelle, 1625.

———. *Le maneige royal où lon peut remarquer le défaut et la perfection du chevalier en tous les exercices de cet art.* Paris: G. Le Noir, 1623.

Poirier, Guy. *L'homosexualité dans l'imaginaire de la Renaissance.* Paris: Honoré Champion, 1996.

Poisson, Pierre, sieur de La Bodinière. *Harangue au peuple de France sur les louanges des anciens François, & de nostre Treschrestien Roy de France, & de Poloigne Henry 3. n'agueres empesché à repousser les estrangers, pour la manutention de l'Estat, & resjouissance de son heureux retour en sa ville de Paris.* Paris: Denis Cotinet, 1588.

Pontaymery, Alexandre de. *L'academie ou Institution de la Noblesse Françoise.* In *Les oeuvres.* Paris: Jean Richer, 1599.

Poquet, Alexandre-Eusèbe. *Histoire de Château-Thierry.* 2 vols. Château-Thierry: A. Laurent, 1839–1840.

Powers, Harold. "Tonal Types and Modal Categories in Renaissance Polyphony." *Journal of the American Musicological Society* 34 (1981): 428–470.

Prizer, William F. "Music and Ceremonial in the Low Countries." *Early Music History* 5 (1985): 113–153.

Processionale usibus ac ritibus sanctae romanae ecclesiae accommodatum. Paris: Jean Henault, 1667.

Prunières, Henri. *Le ballet de cour en France avant Benserade et Lully.* Paris, 1914. Reprint, Paris: Editions d'Aujourd'hui, 1982.

Pussot, Jean. *Journalier ou mémoires de Jean Pussot, Maître-Charpentier en la Couture de Reims.* Edited by Édouard Henry. Reims: P. Regnier, 1856.

Rabelais, François. *Oeuvres complètes.* Edited by Mireille Huchon, with the collaboration of François Moreau. Paris: Gallimard, 1994.

Rabutin, François de. *Commentaires des guerres en la Gaule belgique, 1551–1559.* Paris, 1555. Edited by Charles Gailly de Taurines. 2 vols. Paris: Henri Didier, 1932–1942.

Ramis de Pareia, Bartolomeo. *Musica practica.* Bologna, 1482. Translated and with commentary by Clement A. Miller. Neuhausen-Stuttgart: American Institute of Musicology, 1993.

Ramsey, Ann W. *Liturgy, Politics, and Salvation: The Catholic League in Paris and the Nature of Catholic Reform, 1540–1630.* Rochester, N.Y.: University of Rochester Press, 1999.

Ranum, Orest. "Courtesy, Absolutism, and the Rise of the French State, 1630–1660." *Journal of Modern History* 52 (1980): 426–451.

———. "Strengthening the Noble Male Body: Guillaume Du Choul on Ancient Bathing and Physical Exercise." In *Politics, Ideology, and the Law in Early Modern Europe: Essays in Honor of J. H. M. Salmon,* edited by Adrianna E. Bakos. Rochester, N.Y.: University of Rochester Press, 1994.

Réception de très-chrestien, très-juste et très-victorieux monarque Louis XIII, roi de France et de Navarre, premier comte et chanoine de l'église de Lyon . . . par Messieurs les Doyens, Chanoines & comtes de Lyon en leur Cloistre & Eglise, le xi Decembre, 1622. Lyons: Jacques Roussin, 1623.

Recueil de plusieurs chansons divisé en trois parties: en la premiere sont les chansons musicales:

en la seconde les chansons amoureuses & rustiques: & en la tierce les chansons de la guerre. Lyons: Benoist Rigaud & Jan Saugrin, 1557.

Rees, Owen. "Guerrero's *L'homme armé* Masses and Their Models." *Early Music History* 12 (1993): 19–54.

Régnier de La Planche, Louis. *La Legende de Charles, Cardinal de Lorraine.* Reims: Jacques Martin, 1576.

Reiss, Timothy J. *Knowledge, Discovery, and Imagination in Early Modern Europe: The Rise of Aesthetic Rationalism.* Cambridge: Cambridge University Press, 1997.

Relation du grand ballet du Roy . . . sur l'adventure de Tancrede. Paris: Jean Sara, 1619.

Resjouissance Chrestienne des vrais et naturels François, pour la reddition generale des Reistres à l'obeissance du Roy. Avec le Te Deum chanté en l'Eglise nostre Dame, & Feux de joyes faits, ce 14 Decembre, 1587. Paris: Pierre Chevillot, 1587.

Revel, Jacques. "The Uses of Civility." In *Passions of the Renaissance,* edited by Roger Chartier. Translated by Arthur Goldhammer. Vol. 3 of *A History of Private Life,* general editors Philippe Ariès and Georges Duby. Cambridge, Mass.: Belknap Press of Harvard University Press, 1989.

Ricquier, Nicolas. *Recueil mémorable d'aulcuns cas advenus depuis l'an du salut 1573 tant à Beauvais qu'ailleurs.* Edited by Victor Leblond. Vol. 2 of *Documents pour servir à l'histoire de Beauvais et du Beauvaisis au XVIe siècle.* Paris: Honoré Champion; Beauvais: Société Académique de l'Oise, 1909.

Roche, Daniel, direction scientifique, and Daniel Reytier, ed. *Le cheval et la guerre du XVe au XXe siècle.* Paris: Association pour l'académie d'art equestre de Versailles, 2002.

Rochemonteix, Camille de. *Un collège de jésuites aux XVIIe et XVIIIe siècles: le collège Henri IV de La Flèche.* 2 vols. Le Mans: Leguicheux, 1889.

Rodis-Lewis, Geneviève. *Descartes: His Life and Thought.* Translated by Jane Marie Todd. Ithaca: Cornell University Press, 1998.

Roelker, Nancy Lyman. *One King, One Faith: The Parlement of Paris and the Religious Reformations of the Sixteenth Century.* Berkeley and Los Angeles: University of California Press, 1996.

Ronsard, Pierre de. *Oeuvres complètes.* Edited by Jean Céard, Daniel Ménager, and Michel Simonin. 2 vols. Paris: Gallimard, 1993–1994.

Rosset, François de. *Le romant des chevaliers de la gloire.* Paris: Veuve Bertault, 1612.

Sadie, Stanley, ed. *New Grove Dictionary of Music and Musicians.* 2nd ed. 29 vols. New York: Grove, 2001; London: Macmillan, 2002.

Sainctes, Claude de. *Discours sur le saccagement des Eglises Catholiques.* Paris: Claude Fremy, 1563. Reprinted in vol. 4 of *Archives curieuses de l'histoire de France,* edited by L. Cimber and F. Danjou, ser. 1. 15 vols. Paris: Institute Historique, 1834–1840.

Saint-Gelais, Mellin de. *Oeuvres poétiques françaises.* Edited by Donald Stone Jr. 2 vols. Paris: Société des Textes Français Modernes, 1993–1995.

Saint-Germain, Samson de, sieur de Juvigny. *Advis de l'établissement de quatre académies en la France.* Rouen: Raphaël du Petit Val, 1596.

Salinas, Francisco. *De Musica libri septem.* Salamanca, 1577. Facsimile ed., Kassel: Bärenreiter, 1958.

Salmon, J. H. M. *Renaissance and Revolt: Essays in the Intellectual and Social History of Early Modern France.* Cambridge: Cambridge University Press, 1987.

———. *Society in Crisis: France in the Sixteenth Century.* London: E. Benn, 1975.

Samard, Evelyne. "Les danses guerrières dans le ballet de cour en Savoie au XVIIe siècle." In *Le ballet aux XVIe et XVIIe siècles en France et à la cour de Savoie,* edited by Marie-Thérèse Bouquet-Boyer. Geneva: Slatkine, 1992.

Sandoval, Prudencio de. *Historia de la vida y hechos del Emperador Carlos V.* 2 vols. Pamplona: Bartholome Paris, 1634.

Saulx-Tavannes, Gaspard de. *Mémoires.* Paris, 1574. Reprinted in *Choix de chroniques et mémoires sur l'histoire de France, Mémoires de Gaspard de Saulx-Tavannes, Mémoires de Boyvin du Villars,* edited by Jean-Alexandre C. Buchon. Paris: A. Desrez, 1836.

Scaglione, Aldo. *The Liberal Arts and the Jesuit College System.* Amsterdam and Philadelphia: John Benjamins, 1986.

Scève, Maurice. *The Entry of Henri II into Lyon, September 1548.* Facsimile with an introduction by Richard Cooper. Medieval and Renaissance Texts and Studies 160. Tempe, Ariz.: Medieval and Renaissance Texts and Studies, 1997.

Schalk, Ellery. *From Valor to Pedigree: Ideals of Nobility in France in the Sixteenth and Seventeenth Centuries.* Princeton: Princeton University Press, 1986.

Sealy, Robert J. *The Palace Academy of Henry III.* Geneva: Droz, 1981.

Second livre d'airs de cour. Paris: Pierre Ballard, 1617.

Simonin, Michel. *Pierre de Ronsard.* Paris: Fayard, 1990.

Simpson, Christopher. *The Division Viol, or, The Art of Playing Ex tempore upon a Ground.* London: W. Godbid, 1665.

La società in costume: Giostre e tornei nell'Italia di Antico Regime. Foligno: Edizioni dell' Arquata, 1986.

Spitzer, John, and Neal Zaslaw. *The Birth of the Orchestra: History of an Institution, 1600–1815.* Oxford: Oxford University Press, 2003.

Strong, Roy. *Art and Power: Renaissance Festivals, 1450–1650.* Suffolk: Boydell Press, 1973.

———. "Festivals for the Garter Embassy at the Court of Henry III." *Dance Research* 1 (1983): 45–58.

———. *Henry, Prince of Wales, and England's Lost Renaissance.* London: Thames & Hudson, 1986.

Supple, James J. *Arms versus Letters: The Military and Literary Ideals in the* Essais *of Montaigne.* Oxford: Clarendon Press, 1984.

Sylla, Edith. "Compounding Ratios: Bradwardine, Oresme, and the First Edition of Newton's *Principia.*" In *Transformation and Tradition in the Sciences: Essays in Honor of I. Bernard Cohen,* edited by Everett Mendelsohn. Cambridge: Cambridge University Press, 1984.

Tabourot, Estienne. *Les bigarrures du Seigneur des Accords (premier livre).* Paris, 1588. 2nd ed. Facsimile ed. edited by Francis Goyet. Geneva: Droz, 1986.

———. *Les bigarrures du Seigneur des Accords, quatriesme livre.* Rouen: Jean Bauchu, 1591.

Tallemant, Gédéon, sieur des Réaux. *Les historiettes de Tallemant des Réaux.* 3rd ed. Edited by Louis Jean Nicolas Monmerqué and Alexis Paulin Paris. 9 vols. Paris: Techener, 1854–1860.

Taruskin, Richard. "Antoine Busnoys and the *L'homme armé* Tradition." *Journal of the American Musicological Society* 39 (1986): 255–293.

Tassone, Ercole Estense. *Il monte di Feronia.* Ferrara: Valente Panizza, 1562.

———. *L'isola beata.* [Ferrara: Francesco Rossi, 1570.]

Taylor, Larissa. *Soldiers of Christ: Preaching in Late Medieval and Reformation France.* New York: Oxford University Press, 1992.

Teasley, David. "The Charge of Sodomy as a Political Weapon in Early Modern France: The Case of Henry III in Catholic League Polemic, 1585–1589." *Maryland Historian* 18 (1987): 17–30.

Thibault, Girard. *Academie de l'espée . . . ou se demonstrent par reigles mathematiques sur le fondement d'un Cercle mysterieux la Theorie et pratique des vrais et jusqu'a present incognus secrets du maniement des armes a pied et a cheval.* 2nd ed. Leiden: B & A Elzevier, 1628.

Thou, Jacques-Auguste de. *Histoire universelle.* 16 vols. London, 1734.

[Thou, Nicolas de.] *Ceremonies observees au sacre & coronement du Tres-Chrestien et Tres-valeureux Henry IIII. Roy de France & de Navarre.* Paris: Jamet Mettayer & Pierre L'Huillier, 1594.

Tomlinson, Gary. *Monteverdi and the End of the Renaissance.* Berkeley and Los Angeles: University of California Press, 1987.

———. *Music in Renaissance Magic: Toward a Historiography of Others.* Chicago: University of Chicago Press, 1993.

Tricou, Jean. "Les Enfants de la Ville." *Bulletin de la Société littéraire, historique et archéologique de Lyon* 14 (1937): 107–138.

La triomphante entrée de tresillustre dame Madame Magdeleine de la Rochefoucaud, Espouse de hault & puissant Seigneur Messire Just-Loys de Tournon, Seigneur & Baron dudict lieu, Comte de Roussillon, &c., Faicte en la Ville & Université de Tournon le dimenche vingtquatriesme du moys d'Avril 1583. Lyons, 1583. Facsimile ed. edited by Maxime Gaume, Images et témoins de l'âge classique 4. Saint-Étienne: Presses de l'Université de Saint-Étienne, 1976.

Le triomphe des Saints Ignace de Loyola fondateur de la Compagnie de Jésus, et François Xavier apôtre des Indes, au collège royal de la même compagnie, à La Flèche. La Flèche, 1622. Reprinted in Camille de Rochemonteix, *Un collège de jésuites aux XVIIe et XVIIIe siècles: le collège Henri IV de La Flèche,* 2 vols. Le Mans: Leguicheux, 1889.

Tuccaro, Arcangelo. *Trois dialogues de l'exercice de sauter, et voltiger en l'air.* Paris: Claude de Monstr'oeil, 1599.

Tyard, Pontus de. *Solitaire premier.* Lyons, 1552. Edited by Silvo F. Baridon. Geneva: Droz, 1950.

———. *Solitaire second.* Lyons, 1555. Edited by Cathy M. Yandell. Geneva: Droz, 1980.

Vaccaro, Jean-Michel. "Le livre d'airs spirituels d'Anthoine de Bertrand." *Revue de musicologie* 56 (1970): 35–53.

Vale, Malcolm. *War and Chivalry: Warfare and Aristocratic Culture in England, France, and Burgundy at the End of the Middle Ages.* Athens: University of Georgia Press, 1981.

Valeri, Valerio. "Regalità." In *Enciclopedia Einaudi.* Vol 11, *Prodotti-Ricchezza,* 742–771. Turin: Einaudi, 1980.

Vallet, Nicolas. *Le secret des Muses.* Amsterdam, 1615–1616. Edited by André Souris and Monique Rollin as *Oeuvres de Nicolas Vallet pour luth seul.* Paris: Centre National de la Recherche Scientifique, 1970.

van Orden, Kate. "*La chanson vulgaire* and Ronsard's Poetry for Music." In *Poetry and Music in the French Renaissance,* edited by Jeanice Brooks, Philip Ford, and Gillian Jondorf. Cambridge: Cambridge French Colloquia, 2001.

———. "Cheap Print and Street Song Following the Saint Bartholomew's Massacres of 1572." In *Music and the Cultures of Print,* edited by Kate van Orden, with an afterword by Roger Chartier. New York: Garland, 2000.

———. "Imitation and 'la musique des anciens' in Le Roy & Ballard's 1572 *Mellange de chansons,*" *Revue de musicologie* 80 (1994): 5–37.

———. "Sexual Discourse in the Parisian Chanson: A Libidinous Aviary." *Journal of the American Musicological Society* 48 (1995): 1–42.

———. "Vernacular Culture and the Chanson in Paris, 1570–1580." Ph.D. diss., University of Chicago, 1996.

———. "*Les vers lascifs d'Horace:* Arcadelt's Latin Chansons." *Journal of Musicology* 14 (1996): 338–369.

———, ed. *Music and the Cultures of Print.* Afterword by Roger Chartier. New York: Garland, 2000.

Vanuxem, Jacques. "Le Carrousel de 1612 sur la Place royale et ses devises." In *Les fêtes de la Renaissance,* ed. Jean Jacquot, 2nd ed., 3 vols. Paris: Centre National de la Recherche Scientifique, 1973–1975.

Vendrix, Philippe. "Nicolas Bergier: Le dernier théoricien de la Renaissance en France." In *"La musique, de tous les passetemps le plus beau": Hommage à Jean-Michel Vaccaro,* edited by François Lesure and Henri Vanhulst. Paris: Klincksieck, 1998.

La victoire entiere et parfaicte obtenue nouvellement par le Roy à l'encontre des Reistres. Plus les feuz de joye: faicts à Paris, & un Te Deum solennel chanté en l'Eglise nostre D'ame, le lundy quatoriesme jour de Decembre. Paris: Hubert Velu, 1587.

Victoire obtenue par monsieur le Mareschal de Biron contre les perturbateurs de le Guienne, dont le Te Deum fut chanté en la grand Eglise nostre Dame de Paris, le vingt troisiesme Septembre, 1580. Paris: Pierre Chevillot, 1580.

Vigarello, Georges. "The Upward Training of the Body from the Age of Chivalry to Courtly Civility." In *Fragments for a History of the Human Body,* part 2, edited by Michel Feher with Ramona Naddaff and Nadia Tazi. New York: Zone Books, 1989.

Le voyage de Reims. Paris: E. Foucault, 1610.

La voye de laict ou le chemin des heros au palais de la gloire ouvert à l'entrée triomphante de Louis XIII . . . en la cité d'Avignon le 16. Nov. 1622. Avignon: J. Bramereau, 1623.

Le vray discours des grandes processions qui se font depuis les frontires [sic] *d'Allemagne jusques à la France, dont jamais n'en fut faicte de semblable, & comme plus amplement vous sera monstré dans le discours.* Paris, 1583.

Vray, Nicole. *Catherine de Parthenay, duchesse de Rohan: protestante insoumise, 1554–1631.* Paris: Librairie Académique Perrin, 1998.

Walker, D. P. "Joan Albert Ban and Mersenne's Musical Competition of 1640." *Music and Letters* 57 (1976): 233–255.

———. *Music, Spirit, and Language in the Renaissance.* Edited by Penelope Gouk. London: Variorum Reprints, 1985.

Wallhausen, Johann Jacobi von. *L'art militaire à cheval.* Frankfurt: Paul Jacques, 1616.

———. *L'art militaire pour l'infanterie.* Translated Theodore de Bry. Oppenheim: Uldrick Balck, 1615.

Wise, M. Norton, ed. *The Values of Precision.* Princeton: Princeton University Press, 1995.

Wistreich, Richard. "'Giulio Cesare Brancaccio and Solo Bass Singing in Sixteenth-Century Italy." Ph.D. diss., Royal Holloway College, University of London, 2002.

Woitas, Monika. "*Combattimento* & Co.: Choreographie und *imitatio* in den Torneo-Tänzen um 1600 und in Monteverdis *Combattimento.*" In *Claudio Monteverdi und die Folgen: Bericht über das Internationale Symposium, Detmold 1993,* edited by Silke Leopold and Joachim Steinheuer. Kassel: Bärenreiter, 1998.

Wolfe, Michael, ed. *Changing Identities in Early Modern France.* Durham, N.C.: Duke University Press, 1997.

Wood, James B. *The King's Army: Warfare, Soldiers, and Society during the Wars of Religion in France, 1562–1576.* Cambridge: Cambridge University Press, 1996.

———. *The Nobility of the Election of Bayeux, 1463–1666: Continuity through Change.* Princeton: Princeton University Press, 1980.

Wright, Craig. *The Maze and the Warrior: Symbols in Architecture, Theology, and Music.* Cambridge, Mass.: Harvard University Press, 2001.

Xenophon. *Anabasis.* With translation by Carleton L. Brownson. Revised by John Dillery. Loeb Classical Library. Cambridge, Mass.: Harvard University Press, 1998.

Yates, Frances A. *Astraea: The Imperial Theme in the Sixteenth Century.* London and Boston: Routledge & Kegan Paul, 1975.

———. "Dramatic Religious Processions in Paris in the Late Sixteenth Century." *Annales Musicologiques* 2 (1954): 215–270.

———. *The French Academies of the Sixteenth Century.* 2nd ed. London: Routledge, 1988.

Žak, Sabine. "Das Tedeum als Huldigungsgesang." *Historisches Jahrbuch* 102 (1982): 1–32.

INDEX

Locators in italics refer to figures, examples, or plates.

Entreé, A. lion. Du. ROY. Henry. IV. Le. 4. de sep